AKADEMIE DER WISSENSCHAFTEN DER DDR

Veröffentlichungen des Forschungsbereichs Geo- und Kosmoswissenschaften

Herausgegeben von Heinz Kautzleben

HEFT 17

High Pressure Geosciences and Material Synthesis

Proceedings of the
XXV. Annual Meeting of the European High Pressure Research Group

Potsdam, GDR August 25—27, 1987

Editor: H. Vollstädt, Potsdam

With 149 Figures and 26 Tables

AKADEMIE-VERLAG BERLIN
1988

ISBN 3-05-500555-4
ISSN 0138-4600

Erschienen im Akademie-Verlag Berlin, DDR-1086 Berlin, Leipziger Straße 3—4

Lizenznummer: 202 · 100/522/88
Printed in the German Democratic Republic
Gesamtherstellung: VEB Druckerei „Thomas Müntzer", 5820 Bad Langensalza
Lektor: Dipl.-Met. Heide Deutscher
LSV 1495
Bestellnummer: 763 872 2 (2156/17)
04500

PREFACE

This volume is based on the International Conference "High Pressure Geosciences and Material Synthesis" held at Potsdam August, 25-27, 1987 as the 25th Annual Meeting of the European High Pressure Research Group (EHPRG). 160 scientists from Europe, the U.S.A., Japan, PR China, and Vietman attended this annivesary meeting, which had been organized by the Central Institute for Physics of the Earth of the Academy of Sciences of the GDR. The papers in the volume concentrate mainly on the different properties of rocks and minerals under high pressures and temperatures, the phase behaviour and the synthesis under these extreme conditions. Because of the character as an annual meeting there are also papers on topics like planetary physics and phase transitions in other materials at high pressure and high temperature conditions. Papers concerning high pressure methods and equipment are also included in this book.

The results of the conference show that in the last few years very good progress has been made,, both in static and dynamic methods of reaching exceedingly high pressures, in generating experimental techniques, collecting new interesting data and developing new models to interprete the structure and processes of the Earth's and planets' interior. This underlines the important role of high pressure in geology, petrology, mineralogy and geophysics.

I should like to thank Mr. Bernd Schröder for his efficient help to prepare the manuscript and to edit the proceedings.

H. Vollstädt

INTERNATIONAL CONFERENCE

"High Pressure Geosciences and Material Synthesis"

Convenors: H. Stiller, H. Vollstädt (GDR)

Organizer: European High Pressure Research Group
(President: H. D. HOCHHEIMER)

International Scientific Committee:

S. Akimoto (Japan)
H. Fukunaga (Japan)
H. D. Hochheimer (FRG)
G. Jenner (France)
R. N. Keeler (USA)
V. Kropacek (Czechoslovakia)
N. V. Novikov (USSR)
S. Porowski (Poland)
R. G. Ross (UK)
A. Sawaoka (Japan)
H. Spetzler (USA)
H. Stiller (GDR)
H. Vollstädt (GDR)
E. N. Yakovlev (USSR)

Sponsored by:
- Academy of Sciences of the GDR
  Central Institute for Physics of the Earth
  (Director: H. KAUTZLEBEN)
- AIRAPT - International Association for the Advancement of High Pressure Science and Technology
  (President: R. N. KEELER)
- EPS - European Physical Society

Organizing Commitee:
H. Vollstädt - Chairman
S. Franck - Vice-Chairman
K. Weinreich - Secretary
B. Schröder - Secretary

Members and Officers of the Organizing Committee of the EUROPEAN HIGH PRESSURE RESEARCH GROUP

H.D. Hochheimer Chairman since 1985	Los Alamos National Laboratory, P-10, MS-K 764 Los Alamos, N.M. 87545, USA
G. Jenner Secretary since 1985	Laboratoire de Piézochimie Organique, Université L. Pasteur, 1 rue Blaise Pascal, 67008 Strasbourg France
B.J. Gunney Treasurer since 1987	University of Surrey, Physics Department, Guildford, Surrey GU2 5XH, U.K.
A.R. Adams Member since 1985	Physics Department, University of Surrey, Guildford, Surrey GU2 5XH, U.K.
K.R. Allakhverdiev Member since 1986	Institute of Physics, Academy of Sciences of Azerbaidjan SSR, 33 Nazimarov Ulitsa, Baku, USSR
B.A. Bilal Member since 1987	Hahn-Meitner-Institut für Kernforschung, Bereich Kernchemie und Reaktor, Postfach 39 01 28, 1000 Berlin 39, W.Berlin
A. Birks Member since 1986	The Queen's University, Department of Mechanical and Industrial Engineering, Ashby Institute, Belfast BT9 5AH, North. Ireland, U.K.
S.N. Biswas Member since 1987	Van der Waals Laboratorium, Universiteit van Amsterdam, Postbus 20216, 1000 HE Amsterdam, The Netherlands
H. Fjellvåg Member since 1986	Department of Chemistry, Universitet i Oslo, P.O. Box 1033 Blindern, 0315 Oslo 3, Norway
I. Gorczyca Member since 1987	Polish Academy of Sciences, High Pressure Re-Research Centre "Unipress", ul. Sokolowska 29/37, 01-142 Warsaw, Poland

N.S. Isaacs Member since 1987	University of Reading, Department of Chemistry, High Pressure Chemistry Unit, Reading, P.O. Box 224, Reading RG6 2AD, U.K.
J.P. Itié Member since 1987	Université Pierre et Marie Curie (Paris VI), Physique des Milieux Très Condensés, 4 place Jussieu, 75230 Paris Cédex, France
G.A. Kourouklis Member since 1985	Department of Physics, National Technical University, Zografou Campus, Athens 15773, Greece
A. Laisaar Member since 1985	Institute of Physics, Academy of Sciences of the Estonian SSR, Riga ul. 142, 202400 Tartu, USSR
A. Merbach Member since 1987	Institut de Chimie Minérale et Analytique, Université de Lausanne, 3 place du Château, 1005 Lausanne, Switzerland
S. Porowski Member since 1985	Polish Academy of Sciences, High Pressure Research Center "Unipress", Sokolowska 29/37, 01-142 Warsaw, Poland
A. Rahm Member since 1985	Laboratoire de Chimie de l'Etain, Université de Bordeaux I, 351 Cours de la Libération, 33405 Talence, France
M. Rosenhauer Member since 1985	Mineralogisch-Petrologisches Institut der Universität, Goldschmidtstrasse 1, 3400 Göttingen, FRG
J.W. Scheeren Member since 1985	Department of Organic Chemistry, Katholieke Universiteit, 6525 Nijmegen, The Netherlands
F. Seifert Member since 1986	Bayrisches Geoinstitut der Universität Bayreuth, Postfach 10 12 51, 8580 Bayreuth, FRG
H. Stiller Member since 1985	Akademie der Wissenschaften der DDR, Zentralinstitut für Physik der Erde, Telegrafenberg, 1561 Potsdam, GDR
B. Sundquist Member since 1986	Department of Physics, Universtity of Umeå, 90187 Umeå, Sweden

K. Tödheide Member since 1985	Institut für Physikalische Chemie und Elektrochemie der Universität Karlsruhe, Kaiserstrasse 12, 7500 Karlsruhe, FRG
R. Van Eldik Member since 1987	Institut für Anorganische Chemie der Universität Witten/Herdecke, Stockumer Strasse 10, 5810 Witten-Annen, FRG

Ex officio members:

R.N. Keeler President of the AIRAPT Executive Committee	Los Alamos National Laboratory, M-6, MS-J 970, Los Alamos, New Mexico 87545, USA
F. Bassani Chairman of the CMD European Physical Society	Scuola Normale Superiore, 56100 Pisa, Italy

September 1987

C O N T E N T S

SECTION E : INSTRUMENTATION AND TECHNIQUE
Chairman: S. FRANCK

OPENING REMARKS

H. Stiller, Vice-President of the Academy of Sciences of the GDR

Ladies and Gentlemen!
Dear colleagues!

On behalf of the President and of the Presidium of the Academy of Sciences I extend a hearty welcome to the participants of the 25th Annual Meeting of the European High Pressure Research Group in Potsdam. I am particularly glad because this event is the 25th anniversary as well as the first annual meeting to take place in a socialist country.
Scientists from all over the world responded to the call to join this event. It is my privilege and I have the great honour to welcome

Prof. Keeler, President of AIRAPT
and Prof. Hochheimer, President of EHPRG

and representatives of the local authorities of Potsdam.
In the Central Institute for Physics of the Earth high pressure research has been developed over the last two decades. It mainly concentrates on application in geosciences. Following international trends, besides mineralogical, geological and planetologic aspects, problems of the synthesis of different systems of materials under high pressures and temperatures are part of this research.
International experience shows that a modern development in different branches of natural sciences and technology is impossible without the application of high pressures and temperatures. EHPRG has established a good tradition to concentrate on current problems of high pressure research and define major trends. There is good co-operation with AIRAPT which especially at its 11th International Conference in Kiev this year gave evidence of its task to combine people in the development of a strategy of sciences and information.
The Academy of Sciences of the GDR with its scientific potential is willing and interested in actively supporting close scientific peaceful cooperation between all scientists of the world.
In this sense I wish the 25th Annual Meeting every success, many new scientific results and interesting discussions.
I wish our guests a pleasant stay in Potsdam and many interesting hours at this place which is rich in traditions. I should like to thank the host, the Potsdam Teachers Training College, for giving us the opportunity to hold our meeting here.

SEISMIC PROPERTIES OF THE CALABRIAN CRUST BASED ON ULTRASONIC MEASUREMENTS ON ROCKS FROM A CRUSTAL SECTION OF THE SERRE, S. ITALY

H. Kern[1)]

Extended Abstract[2)]

1. Introduction

Knowledge of the distribution of seismic velocities in relationship to the petrochemical and structural variation from a well known crustal section can provide clues that aid in the interpretation of seismic horizons in the crust.

An appropriate "crustal ship" for a combined petrophysical and lithologic study is exposed in the Serre which belongs to the southern part of the Calabrian massive.

Detailed field-mapping and petrological studies carried out by Schenk (1981, 1984) suggest that the lithological sequences exposed in the Serre of South-Calabria is a tilted block that is a nearly complete crustal profile. The section is composed of interlayered acidic, mafic and pelitic rocks metamorphosed to granulite- and amphibolite-facies conditions. In addition, a suite of post-tectonic granitoids intrudes the section.

The vertical profile which has been obtained by rotating the tilted lithological sequences consits of the following top-to-bottom sequences:

a) Paleozoic phyllites and schists overlain by Mesozoic and Cenozoic sedimentary rocks,
b) a layer of amphibolite-facies paragneisses containing intercalations of amphibolites and pre-Hercynian orthogneises,
c) post-tectonic Hercynian granitoids, with granodioritic to granitic compositions,

----

1) Mineralogisch-Petrographisches Institut, Olshausenstr. 40, 2300 Kiel (F. R. Germany)

2) A comprehensive paper on this topic by H. Kern and V. Schenk (The model crustal-velocity structure beneath Calabria, southern Italy: Constraints from laboratory determinations) has been submitted to EPSL.

d) a 2 - 3 km thick Hercynian tonalite,
e) a 2 - 3 km thick layer of dioritic gneisses,
f) an approximately 5 km thick series of essentially metapelitic rocks (garnet-rich sillimanite-biotite-cordierite gneisses),
g) a 2 km thick layer of high-grade metamorphic mafic rocks (pyroxene-plagioclase felses) with intercalations of felsic granulites.

The total thickness has been estimated by assuming that the base of the tilted crustal section can be correlated with the high-velocity layer that lies at a depth of 20 km, according the seismic refraction studies of Schütte (1978).

2. Experimental approach

Detailed experimental and petrophysical studies on representative rock samples for each of the major rock units of the crustal cross section provide the necessary data base for the discussion.

The experimental data include compressional (V_P) and shear wave (V_S) velocity measurements at confining pressure ($\leq$ 600 MPa) and temperature ($\leq$ 700 °C), the determination of the pressure and temperature derivatives of V_P and V_S, and the determination of the directional dependence of velocity (velocity anisotropy). The petrophysical data are correlated with chemical and mineralogical characteristics of rocks.

The seismic velocities have been measured with the ultrasonic transmission method in a cubic anvil apparatus on cube-shaped samples (43 mm on edges), using transducers with natural frequencies of 2 MHz (for details see Kern, 1982). The use of the cubic press allows simultaneous measurements of velocities in the three multually orthogonal directions of the sample cubes and direct determination of length changes of the sample with increasing pressure and temperature.

3. Experimental results and discussion

In conjunction with the structural and stratigraphic field data the experimentally determined data have been used to construct a seismic velocity model for the crust beneath Calabria. We assumed the pressure at depth to be equal to the overburden pressure. For the determination of the geothermal gradient we used the data reported by El-Ali and Giese (1978).

In the upper crustal levels, the compressional and shear wave velocities vary in different manners. Below the uppermost amphibolite-facies gneisses (6.1 km/s), a weak velocity inversion of V_P is associated with the post-tectonic granites (6.0 km/s). At the granite/tonalite and the tonalite/dioritic gneiss boundary the compressional wave velocity increase by steps of about 200 and 300 m/s, respectively. In cantrast, the shear wave velocity (3.6 km/s) remains constant or even decreases slightly through this same sequence of rocks. This contrasting behaviour is primarily related to the increase of the Poisson's ratio with depth, which is itself correlated with the quartz/feldspar ratio in the rocks.

The most pronounced velocity jumb occurs at the top of the granulite-facies metapelite unit and probably represents the Conrad-discontinuity. At this depth, both V_P and V_S increase markedly (400 m/s respectively 500 m/s), due to mineralogic changes that accompany increasing metamorphic grade and with greater abundances of mafic and garnet-bearing restites and partially melted pelite. The mafic rocks of the lowermost crustal layer have P- and S-wave velocities that are significantly lower than in the overlying metapelites.

The variation of the Poisson's ratio with depth suggests that this parameter, which is a combination of V_P and V_S, is more sensitive to lithology than the P- and S-wave velocities. In rock units of upper and mid-crustal levels with average P-wave velocities < 6.5 km/s, the Poisson's ratio can be used to discriminate quartz-rich from feldspar-rich rocks.

The velocity structure for P-waves estimated for the Serre of South Calabria is in agreement with the velocity model for the Mid-European continental crust proposed by Müller (1977). It is, therefore, suggested that the results of the present study may even be of more general importance.

References

El-Ali, H. and Giese, P.: A geothermal profile between the Adriatic and the Tyrrhenian Sea. In: H. Closs, D. Roeder and K. Schmidt (Editors), Alps, Apennines, Hellenides. Schweizerbart, Stuttgart (1978) 324 - 327.

Kern, H.: P- and S-wave velocities in crustal and mantle rocks under the simultaneous action of high confining pressure and high temperature and the effect of the rock microstructure. In W. Schreyer (Editor), High Pressure Researches in Geosciences. Schweizerbart, Stuttgart (1982) 15 - 45.

Kern, H. and Schenk, V.: The model crustal-velocity structure beneath Calabria, southern Italy: Constraints from laboratory determinations. Earth and Planetary Science Letters (submitted) (1988).

Müller, St.: A new model of the continental crust. AGU Monograph 20, J. G. Heacock (Editor), The Earth's Crust (1977) 289 - 317.

Schenk, V.: Synchronous uplift of the lower crust of the Ivrea Zone and of southern Calabria and its possible consequences for the Hercynian orogeny in southern Europe. Earth Planet. Sci. Lett. (1981) 56, 305 - 320.

Schenk, V.: Petrology of felsic granulites, metapelites, metabasics, ultramafics, and metacarbonates from southern Calabria (Italy): prograde metamorphism, uplift and cooling of a former lower crust. J. Petrol. (1984) 25, 255 - 298.

Schütte, K.-G.: Crustal structure of southern Italy. In: H. Closs, D. Roeder and K. Schmidt (Editors), Alps, Apennines, Hellenides. Schweizerbart, Stuttgart (1978) 315 - 320.

THERMOBARIC PETROPHYSICAL STUDIES AND SOME PROBLEMS OF THEIR DEVELOPMENT

T. S. Lebedev 1)

Intensive studies of high thermobaric petrophysical parameters are made, which enables us, in particular, to reveal pecularities of possible deep mineral matter physical properties distribution. The programmed modeling method proposed not only considers the general regimes of the tectonosphere but also those of concrete geological bodies. This enables us to bring the experimental PT conditions closer to the real ones in the interior [I-4 et al.].

E. g. the generalized laboratory experimental data obtained by the above methode have shown possibility of significantly different elastic feaures of compositionally similar rocks in different deep thermobaric situations, i.e. the P-T conditions effect the changes of their elastic properties no less than the general composition and numerical mineral proportion do. Zones of low and high elastic wave velocity have been revealed whose nature is due to thermodynamic regimes of respective depth and thermobaric zonality of mineral matter elastic properties distribution for large and compositionally homogene crustal blocks has been established [I-4 et al.].

The elastic properties of ocean type crustal rocks systematically under study in our laboratory. These experiments have made it possible to prognosticate a number of their distributional pecularities with depth in corresponding PT conditions [I, 4 et al].

Complex investigation of the electric peoperties of the mineral objects involved different kind approach. Continental and oceanic rock samples are investigated at various PT conditions. According to corresponding thermobaric program the dry and water-gas-saturated rocks are examinated at the laboratory conditions. The main features of the fluid affect and some general regularities of the electric parameters changes at the large depth have been revealed [I-4 et al.].

Thermobaric variations of the rock magnetic properties have been studied in several aspects. First of all it was made the refinement of the factors which determine the stability of remanence saturation to the affect of P and T. Among them: the external pressure transfer methods, the initial magnetic state the region under study, the type of dominated in it ferromagnetic mineral, the affect of preliminary heating. Besides, account was taken of grain size and structural pecularities of the ferromagnetic minerals, internal pressures and their sourses, unmagnetic matrix and so on. Experiments, analysis and general conclusions have been fulfilled

---

1) Institute of Geophysics Ac. Sci. Ukr. SSR, Kiev, USSR

also in a regional plan [I-4 et al.].
The study of dependence of the rock thermal features have been fulfilled by running baric, thermal and program tests. For example, the investigation of thermoconductivity coefficient changes have been carried out at the samples of granitoids of the Ukrainian Shield. The experiments have shown that up to 350°C the their thermoconductivity falls in a whole due to mineral substance density decreasing and increasing of background scattering at the atmospheric pressure. Within the range of 140-200° C a slight increasing of λ is observed probably due to the extra convectional heat transfer appeared as a result of removing of porous water. The experiments conducted under hydrostatic pressure action and the model program PT tests provide a similar intense increasing of the thermoconductivity [I-4 et al.].
The construction of the regional and general petrophysical models of the lithosphere is based on the data of mining rocks at high PT as well as geophysical, geological, petrological, geochemical and other materials [I-4 et al.].
The petrophysical models are in fact one of the best contemporary form of transition from information about PT changes of physical parameters of mineral substance and geology-geophysical data to prediction of the composition and state of deep seated zones of the Earth. The construction of such a models represents a new stage of investigation the dynamics of development of the substance of the interiors and its distributional regularities. The aim of these models is to synthesize all available geophysical data, the results of thermobaric petrophysical tests, the petrology-geochemical data and other materials obtained for different scale geological regions.
The reliability and information ability of such petrological models depends on the amount of original field and laboratory data. However, only complex analysis of this data will make it possible to construct the general petrophysical cross-section. The variants of petrovelocity and velocity models along the DSS profiles of the Ukrainian Shield are constructed on the base of new developed techniques and approaches [3,4].
Progress of petrophysical laboratory studies made in different modelled thermobaric regimes of the Earth's crust and upper mantle is first of all provided by expanding the experimental high P and T range in new or previous improved apparatus enabling imitation of their simultaneous effect at different depths. Important is the elaboration of more universalautomatic systems registering even very small changes in experimental PT parameters and physical properties. The enlargement of the samples considered enables one to exclude some possible data inhomogeneities in the experiments immediately. Particular requirements are set as to the modernization of specialized PT-chambers for studies of physical and mechanical characteristics of mineral bodies in thermodynamic situations

of their polymorphic and phase transitions including solid phase changes and melting.

As to the general task of high thermobaric parameter petrophysics, as one of the most important fundamental Earth's physics field, the next decade should greatly intensify the following activities [I, 4 et al.]:

1) theoretic and experimental studies of the nature and laws of the general and regional blockwise deep changes and interrelations of physico-mechanical and other lithospheric and deeper Earth's mineral matter properties considering its geostructure, genesis and evolution;

2) modelling of the composition and state of geologic media at different depths for different regions based on the analysis of modern theoretical data, petrophysical PT experiments, the whole set of geologic-geophysical and petrologic-geochemical information, time and scale factors, water and gas saturation, stressed state, rheologic features and anisotropy of mineral associations etc.;

3) finding and detailed studies of factors causing various changes of physical characteristics of rocks in extreme PT-conditions with taking their high and low level transformations strongly into account;

4) search of essentially new methods (including non-linear ones) in planing combined PT petrophysical and other studies;

5) improvement and creation of new specialized equipments, apparatus and methods wide automatization of the experiments using control processors;

6) elaboration of universal methods of the application of laboratory experimental thermobaric petrophysical data in interpreting the results of the observation of different geophysical fields, processes and phenomena, solution of exploration geophysics and mining problems, planning and realization of the extra-deep borehole drifts and mines etc.;

7) synthesis of modern geoinformation, experimental PT-data, respective theoreretic calculations and preparation of summary petrophysical profiles of the lithosphere.

R e f e r e n c e s

/1/ Lebedev T.S.,Problems of thermobaric studies of lithospheric mineral matter properties // Geofiz. Zhurn., Kiev-1985. - 7, No 6.- P. 62-82 (in Russain).

/2/ Lebedev T.S., Korchin V.A., Savenko B.Y. et al. Physical properties of mineral matter under thermobaric conditions of the lithosphere. - Kiev: Naukova Dumka, 1986. - 200 p. (in Russain).

/3/ Lebedev T.S. Model studies of physical properties of mineral matter in high-pressure-temperatures experiments // Physics of the Earth and Planetary Interiors. - 1980. - 23. - P. 292-303.

/4/ Lebedev T.S. Thermobaric petrophysical studies and some prognoses of their development // Geofiz. Zhurn. Kiev-1987. - 9, No 2. - P. 3-19 (in Russain).

HIGH PRESSURE - HIGH TEMPERATURE HISTORY OF PYROXENITES

Hans-Joachim Bautsch [1]

Pyroxenes may be the most abundant rock forming minerals on the earth and the planetary system. They are stable throughout a wide range of thermodynamic conditions. They have a somewhat denser packed crystal structure than olivine and garnet of the same composition. Their crystal structure is flexible in the conformance between the tetrahedral coordinated chains and the [6]- or [8]-coordinated polyhedra of the other cations. Thus the pyroxenes are a solvent of several socalled internal and external normative components, some of them are other rockforming minerals. This pressure and temperature dependent solubility is the basis of most of the pyroxene geothermometers and -barometers. A good review about the knowledge and problems on pyroxene was edited by PREWITT (1980). Since this time there was done some more experimental works on pyroxene systems at high pressures and there could obtained some more petrogenetic experiences about pyroxenitic rocks.
These dates and some thermodynamic statements are useful for petrographers for geobarometric and geothermometric measurements of pyroxenitic rocks with a siple geologic formation. But there are some problems with pyroxenes as xenocrysts in a strange rock matrix or as relics in a polymetamorphic host rock. As deduced from geologic reasons these pyroxenes have a primary high pressure origin.
I will demonstrate and discuss some results and problems of our petrographic studies on
1.) such xenocrysts in kimberlithogenic rocks (Hinterhermsdorf, Elbsandsteingebirge), 2.) xenocrysts in basaltic rocks (Ostritz, Lausitz), and 3.) reli tic coarse-grained pyroxenites in serpentinites of the Saxonian granulite complex (Waldheim).

---

1) Humboldt-Universität zu Berlin, Museum für Naturkunde, Invalidenstr. 43, Berlin, 1040, DDR

1. In a polymict breccia of a tertiary pipe, composed to 3/4 of basaltic lapilli and matter and to 1/4 of fragmental material of sedimentary rocks, xenocrysts of alkali-Ti-pyroxenes and Fe-Ti-oxide minerals are common (BAUTSCH & VOLLSTÄDT 1985). Very rare are separate crystal fragments of quadrilateral clino- and orthopyroxenes with two types of exsolution: i) coarser lamellae or myrmecitic intergrowth of spinell (pleonaste) or garnet (Capyrope), and ii) less than 5 µm thick to submicroscopic lamellae of an other pyroxene phase and /or garnet and/or spinell. An extrapolation of the integral chemical composition for homogenous pyroxenes gives crystallisation conditions with pressures above 20 kbar and temperatures above 1000°C to 30 kbar /1200°C, using the modell of GASPARIK (1984). The pyroxene-garnet correlation (ELLIS & GREEN 1979) gives P/T-dates for the exsolution nearly in the same range. Both are not in accordance with a continental crust geotherm.

2. In the tertiary basalts of the Lausitz the xenocrysts of clinopyroxene, fragments of single crystals up to 7 cm and never found with any other primary mineral in contact, contains µm -sized exsolution domains of anorthite. They can be interpretated as exsolution products of a primary solved Ca-Eskolaspyroxene component. The application of the existing high pressure, high temperature experimental works in the system di-Catspx-Caespx gives contradicting results. The rocks from the few known localities with plagioclase exsolution lamellae in pyroxene are of different petrogenetic origin. We believe, that the exsolution domains of anorthite in the pyroxenes of an early deep situated cumulative stage are reaction products of metasomatic processes and kinetic moments may be more important.

3. The relictic coarse-grained pyroxenites are small layered or lense-like inclusions in tectomorphic chrysolite serpentinites near Waldheim in the Saxonian granulite complex whose geolic origin and evolution has many discussions and questions up to the present time. The pyroxenites may be classified into five petrographic types in regard of thier modal mineral composition and fabric, but it is difficult to correlate them in a petrogenetic sense. The integrating petrochemical compositions gives an average of SiO_2 - 50%, Al_2O_3- 14%, FeO - 5%, MgO - 17%, and CaO - 14%, which is in correspondence with a komatiitic composition. To some rock types it is unambiguous that they primary consisted only of coarse-grained homogenous pyroxenes. With such MgO/CaO-ratios and Al_2O_3 -contents sometimes above 15% they are stable above the miscibility gap diopside enstatite (NICKEL & BREY 1984) and the pyrope(gt) → pyrope(px) transfor-

mation. This is in correspondence with temperatures above 1400°C and pressures above 30 kbar. The pyroxenes now showes a plenty of various transformation and deformation structures, wich we have studied particulary with universal stage microscopy, transmission electron microscopy, and single crystal X-ray diffraction (REICHE & BAUTSCH 1985). Most attractive are the exsolutions of garnets, generally indicating high pressure transformations of the pyroxenes. The transformation character is diffusive by reason of the distribution coefficients in comparison with the surrounding pyroxene matrix. The width of the lamallae reaches from nm-to mm-size. The growth mechanism is sometimes spinodal with coherent interfaces, an interesting crystal chemical phenomena. In most cases the growth mechanisms are with nucleation and lattice defects at the interface. Morphologically the exsolution lamellae are parallel to (100), (010), (001), (120), $(1\bar{1}0)$, (3,11,1), $(3,\bar{1}\bar{1},1)$ and (6,11,1), of the clinopyroxene. All but the first one are new observed orientations in pyroxenes. In some cases the lattice orientations could be defined (first the index of pyroxene): (100) // (110) and [001]// [001], (100) // (100) and [001]// [011] , (010) // $(1\bar{1}2)$ and⊥(100) // [110] , (120) // (101) and [100] // [110]. The geothermobarometric interpretation of the chemical analysis of pyroxene-garnet pairs gives wide spreading results. Most of them can collected to two groups, one with pressures up to 40 kbar and temperatures between 1200 to 1600° C, and the other with pressures of 10 - 15 kbar at temperatures of 820 - 940°C (ELLIS & GREEN 1979). This indicates two stronger metamorphic high pressure, high temperature events. The first event is in correspondence with what we call an eclogite facies. The differences between the pressure and temperature dates of this group may be caused by an inhomogenic distribution of pressure and temperature or by a later influence of the cation distribution. The second event is in correspondence with the granulite facies. In connection with the microscopic pictures it is possible to show the development of the pressure-temperature regime of the granulite metamorphism.
It begins with higher pressures and temperatures (11 - 14 kbar, 940° C) and keeps longer times with falling pressure and temperature down to 8 kbar and 820°C. Some smaller exsolution lamellae in recrystallized pyroxenes gives a temperature of 680° C at nearly 5 kbar. They may affected during conditions of an amphibolite facies, for which are some other petrographic indications. There are numerous other transformation reactions of the pyroxenes: a) exsolutions of spinell, affected during lower conditions of the granulite facies, b) exsolutions of amphibole, resulting from the diffusion of the solved hydrogen, and c) kelyphitic reactions of or between pyroxenes, garnet, and spinell in several phases, some

of conditions of higher pressures and temperatures. The pyroxenitic rocks, particular those of type I and V, showes intensive deformations, often combined with recrystallisation of the pyroxenes.
High pressure plastic deformation gives strong undulations with angles up to 60° C and minimum radius of 1,2 mm. On the contrary other coarse-grained specimens only some meter distant from this in the host serpentinite shows none deformative effects.
The given examples of pyroxenes with a high pressure origin and history shows informations on the pressure and temperature of their formation. On the other hand there remains some questions in the case of more complicated rock genesis and of the influences of kinetic effects on the transformation reactions of pyroxenes.

Literatur:
BAUTSCH, H.-J. VOLLSTÄDT, H. (1985): Freiberger Forschungsh. C 398, 107-112
ELLIS, D.J., GREEN, D.H. (1979): Contrib. Mineral. Petrol. 71, 13-22
GASPARIK, T. (1984): Contrib. Mineral. Petrol. 87, 87-97
NICKEL, K.G. BREY, G. (1984): Contrib. Mineral. Petrol. 87, 35-42
PREWITT, C,T. (Edt.)(1980): Pyroxenes.- Rev. Mineral. 7, 1-525, Miner. Soc. Amer., Washington
REICHE, M., BAUTSCH, H.-J. (1985): Freiberger Forschungsh. C 393, 19-33

Synthesis and Stability of a New Hydrous, High-Pressure Phase with a Pumpellyite Structure in the System $MgO-Al_2O_3-SiO_2-H_2O$

W. Schreyer[1)]

The quaternary system $MgO-Al_2O_3-SiO_2-H_2O$ (MASH) which contains about 35 different crystalline phases (Fig. 1), is of fundamental importance for the understanding of silicate minerals both in the Earth's crust and its mantle. The most important mantle minerals forsterite, enstatite, and pyrope garnet are rich in Mg and lie along or near the MS side of the system; high-pressure phases of the partial system MSH with spinel, perovskite and other structures were omitted in Fig. 1 for simplicity reasons. On the other hand, the more Al-rich portion of the of the MASH-system contains minerals like cordierite or the Al-silicates andalusite etc., that occur predominantly in rocks of crustal origin such as metamorphosed clays and other metasediments.

From a crystal chemical point of view it is of interest that the many different silicate phases of the MASH system known up to now can be grouped with nearly all the structural varieties of silicates, which crystal chemists have learned to distinguish (Liebau 1985). Thus, there are numerous nesosilicates with isolated SiO_4-tetrahedra such as forsterite, pyrope, yoderite, or Mg-chloritoid. There are single-chain silicates represented by enstatite and other pyroxenes, but also by Mg-carpholite. Double-chain (band) silicates appear as the amphiboles anthophyllite and gedrite; phyllosilicates with T_2O_5-sheets occur as the minerals talc, chlorite, kaolinite, and others: Cordierite with six-membered rings of tetrahedra plays a dual role, as these rings are interconnected by more tetrahedra to form a 3-dimensional framework so that a tectosilicate is formed. What has been missing, until our recent discovery (Schreyer et al. 1986), is a quaternary MASH sorosilicate phase, that is one with a structure containing Si_2O_7-groups of double tetrahedra such as in the CaAl-silicate zoisite and all the epidote-minerals.

The missing link has now been synthesized at 50 kbar water pressure, 700°C, in the form of a new high-pressure phase named MgMgAl-pumpellyite in Fig. 1. It lies compositionally close to pyrope and may thus be of significance as a mantle mineral as well.

The new phase is isostructural with the CaMgAlFe silicate mineral pumpellyite, which is a hydrous sorosilicate with Al- and Mg-octahedra and Ca in 7-fold coordination against oxygen. Its general formula may be given as $W_4{}^{[7]}X_2{}^{[6]}Y_4{}^{[6]}Z_6{}^{[4]}O_{20+t}(OH)_{8-t}$, where W = Ca, Mn; X = $(Mg, Fe^{2+}, Mn)_{2-t}(Fe^{3+}, Al)_t$; Y = Fe^{3+}, Al; Z = Si. The number of hydroxyl

---

1) Institut für Mineralogie, Ruhr-Universität, D-4630 Bochum 1, Federal Republic of Germany

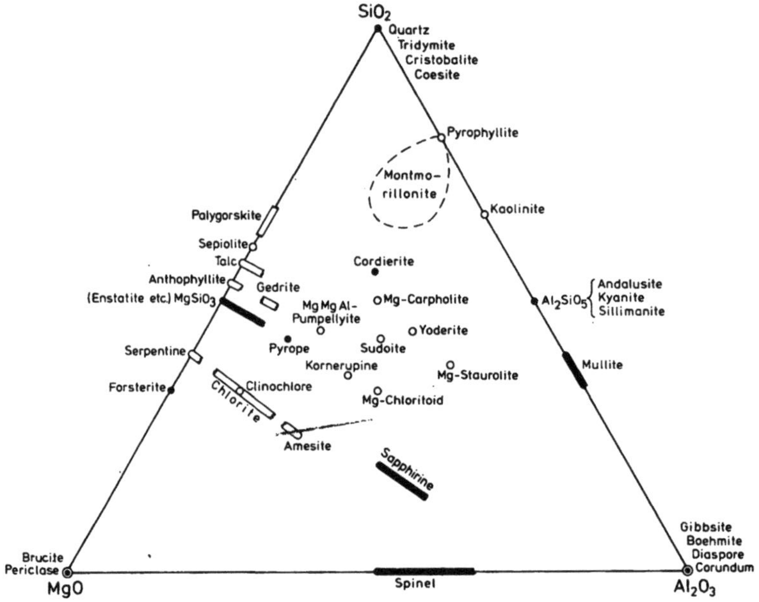

Fig. 1. Crystalline phases of the system $MgO-Al_2O_3-SiO_2-H_2O$ projected from the H_2O apex of the tetrahedron onto the water-free base. Solid dots and bars indicate anhydrous phases, open circles and bars hydrous ones.

groups can vary in conjunction with changing occupancy of the X site following the equation $M^{2+}+H^{+}=M^{3+}$.

A pumpellyite structure in the MASH system is only possible when Mg substitutes for all Ca in the W sites. This is in accordance with the "pressure-coordination rule" which indicates, in the present case, that Mg generally present in silicates in sixfold coordination increases its coordination number with pressure by entering the 7-fold W sites. The high-pressure garnet pyrope with $Mg^{[8]}$ exhibits a comparable example.

Based on the above considerations 3 hypothetical compositions were prepared as gels and run in a piston-cylinder apparatus in the presence of excess H_2O. 100 % yields of MgMgAl-pumpellyite were only obtained with the composition corresponding to the formula $Mg_4(MgAl)Al_4Si_6O_{21}(OH)_7$, that is with a Mg : Al : Si ratio of 5 : 5 : 6. This formula was further corroborated by analytical determination of the H_2O^+ content of the product, which is in excellent agreement with the theoretical value. Thus, MgMgAl-pumpellyite is the Mg-analogue of the synthetic phase $Ca_4MgAl_5Si_6O_{21}(OH)_7$ prepared at low water pressure and named MgAl-pumpellyite by Schiffman and Liou (1980).

Optical and X-ray properties of the new phase MgMgAl-pumpellyite are reported by Schreyer et al. (in press). Its cell volume is smaller by about 9 % than that of the Ca end member. Thus, although the heavier Ca atom is replaced by the lighter Mg, the density increases from 3.19 to 3.27 g/cm^3.

In order to determine the thermodynamic stability of the new phase, bracketing runs were performed under different PT conditions using presynthesized MgMgAl-pumpellyite seeded with various breakdown assemblages. As a result, the minimum water pressure for MgMgAl-pumpellyite was found to be about 37 kbar, and the temperature stability at 50 kbar water pressure ranges from about 400°C to nearly 800°C. On the low-temperature side the new phase breaks down, through reaction with additional H_2O, into the assemblage chlorite+talc+Mg-carpholite (compare Fig. 1); the requisite univariant reaction curve has a negative dP/dT-slope. With increasing temperature, the ensuing breakdown assemblage is talc+Mg-carpholite+Mg-chloritoid, which has the same H_2O content as MgMgAl-pumpellyite. Thus, the related reaction curve has a very flat dP/dT-slope and represents the lower pressure stability limit of the new phase near 37 kbar, 550°C. The upper thermal stability limit of MgMgAl-pumpellyite is given by 3 dehydration reactions with steep positive dP/dT-slopes producing the following assemblages (in the order of increasing temperature and water pressure): talc+kyanite+Mg-chloritoid; talc+kyanite+pyrope; kyanite+pyrope+coesite (see Fig. 1).

As a consequence of this stability field, MgMgAl-pumpellyite could only form within the Earth at a minimum depth of about 12o km and along unusually low geotherms (< 5°C/km). Although such low geotherms do not prevail in the Earth's upper mantle during normal, static conditions, they may occur as transitional stages in subduction zones, that is within downgoing slabs of cold oceanic lithospheric plates. In addition, it is expected that the stability field of MgMgAl-pumpellyite will be influenced by other chemical components present in the mantle, especially by Fe^{2+} and Ca. Even water activity smaller than 1 could shift the stability field towards lower pressures and temperatures and thus perhaps into a range of possible low geotherms prevailing locally inside the dynamic Earth.

References

Liebau, F.: Structural chemistry of silicates. Springer-Verlag Berlin 1985.

Schiffman, P. and Liou, J.G.: Synthesis and stability relations of MgAl-pumpellyite, $Ca_4Al_5MgSi_6O_{21}(OH)_7$. Journal of Petrology 21, 441-474, 1980.

Schreyer, W., Maresch, W.V., Medenbach O., and Baller T.: Calcium-free pumpellyite, a new synthetic hydrous Mg-Al-silicate formed at high pressures. Nature 321, 510-511, 1986.

Schreyer, W., Maresch, W.V., and Baller, T.: A new hydrous, high-pressure phase with a pumpellyite structure in the system $MgO-Al_2O_3-SiO_2-H_2O$. Advances Phys. Geochem., Korzhinskii Volume (in press).

FRACTIONATION OF TRACE ELEMENTS IN MINERALS AS INDICATOR FOR THE GENISIS: IN SITU INVESTIGATION OF THEIR PHYSICOCHEMICAL BEHAVIOUR UNDER HYDROTHERMAL P AND T CONDITIONS

Bilal,B.A.[1)]

Trace elements of similar chemical properties are fractionated in minerals in typical manner,depending on the genetic origin(pegmatetic,hydrothermal or sedimentary) of the mineral. For instance the contents of the lanthanides(normalized on their contents in chondrites) differ in hydrothermally formed fluorites in such a way that a decrease from La to the middle of the series and than a slight increase to Lu is observed [1]. Such fractionation can be used as an indicator to reveal the mineral genesis(pressure p, temperature T and chemical composition X of the system at the mineralization stage). However, a genetic interperatation of the fractionation observed provides the evaluation of a physicochemical model describing it as a function of the parameters(p,T and X) of the mineral formation. Such model has to account for the different behaviour of the considered trace elements in thermodynamic and kinetic processes taking place during the mineralization.

Kinetic processes like different rates of diffusion of the trace elements in the host mineral after formation can be excluded,since the diffusion coefficients are too small to lead to the fractionation in the observed order of magnitude. Different kinetic of the primary coprecipitation of trace elements with crystallizing host mineral are also not observed in model experiments. Thermodynamic processes,in contrary, like complex formation equilibria in the mineral bearing solution as well as distribution equilibria between the mineral bearing solution and the crystallizing mineral seem to be responsible for the fractionation,in so far as 1.) the cocrystallizable „free" ions of the trace elements are differently bound due to complexation with various stability and 2.) the substitutabilty of the cations of the host mineral by these ions is determined by their specific properties (e.g.radius).

The distribution coefficients λ_i and the complex stability constants β_i have to be determined in situ experiments as functions of p,T and X conditions relevant to mineral formation. Diagramms can then be obtained showing calculated fractionation values as a function of λ_i and β_i at different p,T and X. Provided the evaluated model is right,the mineral genesis can be revealed on comparison of the analytically found fractionation with such diagramms.

Bilal and Becker [2] have investigated the distribution of the lanthanides between fluorite and a fluorite bearing model solution of the ionic strength I = 1 mol/kg(NaCl) at pH =1-3 and temperatures up to 250 °c along with the corresponding saturation pressures. No significant variation of λ_i has been found within the series.It seems,therefore that the fractionation of the lanthanides in fluorite is mainly due to their complexation with different degree in the fluorite bearing solution.

It is assumed that fluorite bearing solutions are weak acidic (pH 3-5) and contain the complexing ligands F^- ($c \approx 10^{-3}$ mol/kg),Cl^- ($c \approx 1$ mol/kg),SO_4^{2-} ($c \approx 10^{-2}$ mol/kg)and OH^- ($c \approx 10^{-9}$-10^{-5} mol/kg) depending on the pH and the ion product of water at different temperatures and pressures). Bilal,Hermann and Fleischer [3] and Bilal and Koß [4] studied the complex formation by means of Cl^- and SO_4^{2-} at 25 °c and 1 bar. Within the range of the mentioned concentrations,no significant complexation by these two ligands could be found. Bilal and Baumann [5] studied this aspect by means of measurement of the fluorescence intensity and fluorescence life time of Terbium at T up to 200 °c and p up to 0.5 kbar in absence and in presence of the two ligands. They found similar results. Consequently,it seems that F^- and OH^- play the important role in the complex formation of the lanthanides in fluorite bearing solution.

Recently Bilal and Langer [6] have studied the complexation of five representative lanthanides ,Ce,Nd,Tb,Er and Lu by means of F^- and OH^- in fluorite bearing model solution at the temperatures up to 200 °c and pressures up to 1 kbar. The initial pH of the solution was 3.6 and the ionic strength was held constant by NaCl at I = 1 mol/kg. The activity of the F^- was measured by means of a fluoride membrane electrode developed by Bilal,Langer and Müller [7] for operation at high temperature and high pressure.The activity of OH^- was indirectly determined by measuring the pH by means of high pressure high temperature cell having Pt/H_2 electrodes developed by Becker and Bilal [8] and using the values of the ion product of water determined by Becker and Bilal [9] as a function of temperature and pressure in 1 m NaCl solution.

1) Hahn-Meitner-Institut Berlin: Glienicker Str. 100, D-1000 Berlin 39

Although, the pH of the solution investigated by Bilal and Langer [6] increased slightly from 3.6 at 25 °C and 583 bar to 3.65 at 200 °C and 1020 bar, the corresponding OH^- concentration changed drasticaly due to the rapid increase of the ion product of water from 2.10^{-14} to $6.5 \cdot 10^{-11}$ mol^2/kg^2 and also because of the increased formation constant of HF from 814 to 7787 mol/kg. Fig.1 shows the pOH values of solutions having pF ≈3(= approximative value of saturated fluorite bearing solution) at different T,p and pH.

Bilal and Koß [10] showed that at pF = 3 the OH^- participates soon at pOH ≈9 in the complexation of the lanthanides. According to fig.1, this pOH value is established in solutions with pH =3 at ≈ 90°C , in those with pH =4 at ≈ 45°C and in solutions with pH = 5 at ≈ 20°C. Hydroxo and fluorohydroxo complexes have , therefore, to be considered in hydrothermal systems bearing fluorite, even if they are weak acidic.

Fig.2 shows by the way of example the percentage distribution of Tb species present in the the solution investigated by Bilal and Langer (pH = 3.6, pF = 3) which is calculated according to

$$\frac{[\mathrm{Tb}\ F_i OH_j]}{c_{Tb}}\ \% = \frac{100 \cdot \beta_{i,j}[F]^i [OH]^j}{1 + \sum_{i,j} \beta_{i,j}[F]^i [OH]^j} \qquad (1)$$

($\beta_{o,o} = 1$, c_{Tb} = total concentration of Tb)

using $\beta_{i,j}$ determined by Bilal and Langer as function of p and T. The formation of hydroxo and fluorohydroxo complexes begins at relatively low temperature of ≈40°C. At 200°C the mixed ligand complex becomes the main specie (72%) in the solution. In solutions with pH = 5 this complex is formed soon at room temperature (≈ 10%).

Bilal, Becker, Koß and Nies [11] studied the typ of distribution which controls the coprecipitation of the lanthanides (Ln_i) in fluorite. They found a logarithmic one at normal p and T conditions and pH = 7, whereas at 200 °C, 50 bar and pH = 1 the function

$$\left(\frac{Ln_i}{Ca}\right)_{solid} = D_i \left(\frac{[Ln_i]}{[Ca]}\right)_{solution} \qquad (2)$$

was obtained, which indicates a homogeneous distribution. D_i is the linear distribution coefficient that was found by Bilal and Becker to be approximately the same for all lanthanides. In fluorite bearing solutions with pH 3-5 $[Ln_i]$ is therefore the only term controlling the fractionation, since [Ca] equals always the saturation concentration of CaF_2. The percentage parts of the „free" ions of two Lanthanides Ln_1 and Ln_2 are determined by the complex formation according to equation (1). The ratio of such two parts equals the fractionation degree between the two lanthanides. It is given by

$$\left(\frac{Ln_1\ \%}{Ln_2\ \%}\right)_{solid} = \frac{[Ln_1]\ \%}{[Ln_2]\ \%} = \frac{(1 + \sum_{i,j} \beta_{i,j}[F]^i [OH]^j)_{Ln_2}}{(1 + \sum_{i,j} \beta_{i,j}[F]^i [OH]^j)_{Ln_1}} \qquad (3)$$

Fig.3 shows the variation of the percentage parts of the „free" ions of the lanthanides as function of p and T at pH 4 and 5. The fractionation is represented by the ratio of values on lines parallel to the ordinate. These vary in distinctive way in the two graphs with decreasing temperature and pressure within the lanthanide series. The sequence of the fractionation found in hydrothermally formed fluorite (Ce > Nd > Tb > Er < Lu) is represented in graph ((A), pH 5) at T > 100 °C and p > 675 bar, whereas in graph ((B), pH 4) values of T > 200 °C and p > 1000 bar are necessary to obtain it. At low temperatures and pressures a different sequence, namely Er > Lu > Nd > Tb > Ce would be expected according to the graphs in fig.3 .

This simple model shows in principal how to find out the parameters of mineral formation by means of the fractionation of trace elements in mineral samples. However, one has to consider further aspects in this model, such as the probable presence of other ions which may be also complexed by the same Anion forming the mineral and thereby change the mineral solubility and as a consequence the saturation concentration of the complexing ligands. In that case the equilibrium concentration of the ligands must be calculated from the data of the simultaneous equilibria system.

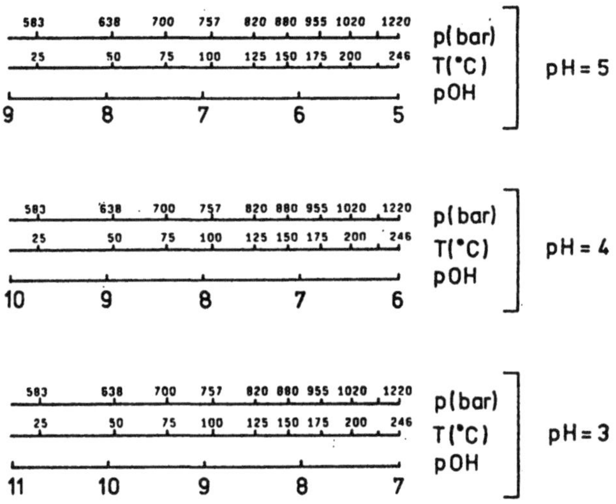

Fig.1 . pOH as a function of T and p at different pH

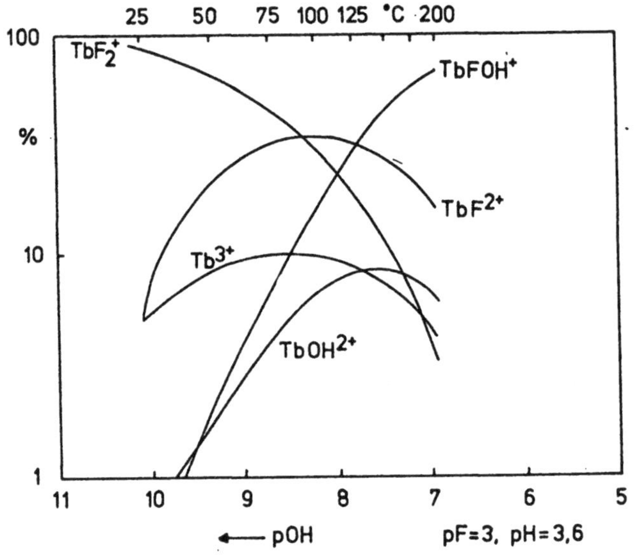

Fig.2. Percentage distribution of Tb species as a function of T at pF=3 and pH=3.6

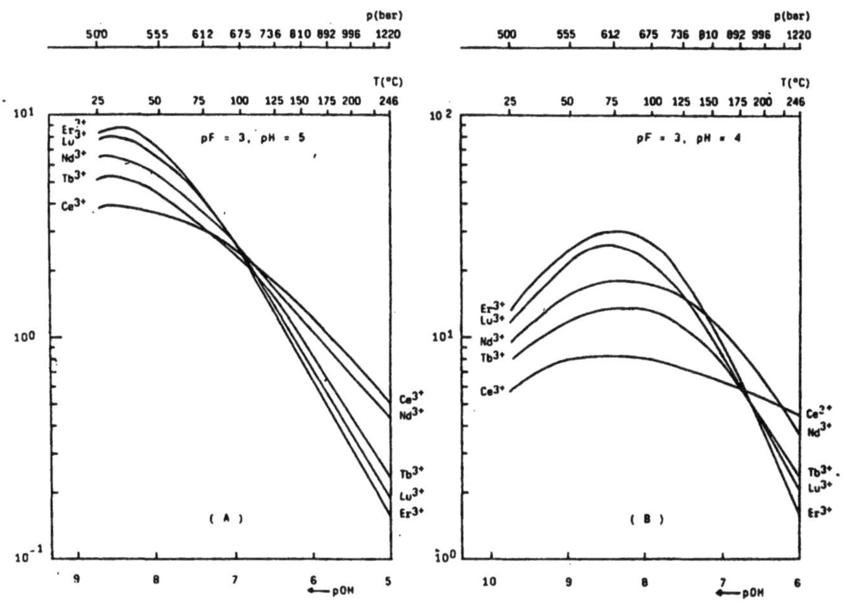

Fig.3. Variation of the percentage parts of the „free" ions of the Lanthanides as function of p,T at different pH

Literature

/1/ Möller,P.: In Systematics and the Properties of the Lanthanides,p.561,Copyright 1983 by Reidel Publishing Company

/2/ Bilal,B.A.and P.Becker : J.Less Common Metals 112 (1985),221

/3/ Bilal,B.A.,F.Hermann and W.Fleischer : J.Inorg.Nucl.Chem. 41 (1979),347

/4/ Bilal.B.A. and V.Koß : J.Inorg.Nucl.Chem. 42 (1980),1065

/5/ Bilal.B.A. and M.Baumann : unpublished work

/6/ Bilal.B.A. and P.Langer : unpublished work

/7/ Bilal.B.A.,P.Langer and E.Müller : Application filed with German Patent Office Aug.19887

/8/ Becker,P. and B.A.Bilal : Fresenius Z.Anal.Chem. 317 (1984),118

/9/ Becker,P. and B.A.Bilal : J.Solution Chem. 14 (1985),367

/10/ Bilal,B.A. and V.Koß : Polyhedron 1 (1982),239

/11/ Bilal,B.A.,P.Becker,V.Koß and H.Nies : In the Rare Earths in Modern Science and Technology,Vol 3,p.323,edited by McCarthy, Silber and Rhyne,Plenum Publishing Corporation

NEW CONCEPT OF THE DIAMOND GROWTH IN THE EARTH CRUST

Alicja Niedbalska, Andrzej Szymański 1/

In geology, the years 1960-1980 brought important achievements in the field of research on natural diamonds. Among them, the major ones were the considerable increase in the number of diamond-bearing rock types and the separation of new structural varieties of diamond /e.g. lonsdaleite/. Natural diamond occurs in kimberlites and the related ultrabasic xenoliths and basic plutonic rocks, in meteorites, impactites and secondary deposits of old and contemporary scatterings. Recently, diamond have also been found "in situ" in certain nonkimberlite ultrabasic and basic rocks and in alkaline basalts /4/. The origin of diamonds has so far not been determined unambigously, and in discussions on it, three views are held, namely:

1 - that diamonds crystallized from magma "in situ", where they occured;
2 - that diamonds were an original component of deeper lying eclogite rocks which were molten by kimberlite moving upwards, and in this way diamonds which used to be deep inside the earth, came closer to the surface;
3 - that diamonds are products of crystallization of magma, which moved them from deep inside the earth to its surface. Recently, this theory has had most folowers.

All the three theories on the formation of diamonds in nature agree on that this mineral can arise only from magma and only in the conditions of very high temperatures /about 1600 K/ and very high pressures /of the order of 10.000 MPa/. So high pressures occur only at depths exceeding 100 km and it is only very seldom that magma reaches the surface of the earth from such depths.

Research on kimberlites suggest that as the depth increases, the content of diamonds, and additionally that of such elements as Ni, Co and Cr, decreases, but the amounts of Fe and Ti increase /3/.

The earliest studies on the isotopic composition of carbon in diamond and kimberlite carbonates indicated that there is an essential - genetic - relation between diamonds and bitumens saturating kimberlites, and also carbonates of intrusive rocks /7,9/. Recent studies /3,5/ on the isotopic composition of carbon /^{12}C and ^{13}C/ indicated very large variations in $\delta^{13}C$ between 0 and -10°/oo, and less often between +2,5 and -34°/oo which does not demonstrate unambiguously the plutonic origin of carbon. At the same time, structural studies an organic molecules indicate that the distances between carbon atoms are different from those of diamond and graphite-layers by about 0,5% whereas the energy of C-C bonds differs just by 1-3% /1/.

Since organic compounds are enriched by lighter carbon ^{12}C compared with carbon compounds of nonorganic origin /resulting from the process of photosynthesis/ and since isotopic studies, mentioned above, do not determine unambiguously the enrichment of diamonds with heavier carbon, taking also atomic distances into consideration, one can assume that these compounds may be the initial "raw material" for structures of crystalline carbon: graphite and diamond which form in nature. This applies to both hydrocarbons and different varieties of kerogen.

Hydrocarbons of organic /and partly also of magmatic/ origin migrating in the deep regions of the earth's crust often gather in rock crevices /as a rule, kimberlites are impregnated by them/, where they undergo pressure condensation. A consequence of this may be the later polymerization consisting in their combination into spatially arranged hydrocarbon chains which maintain under pressure a state close to solid state. The disintegration of such polycondensates or semi-solid products, connected with the escape of hydrogen, may, with the simultaneous increase in the content of the element C in the remainder, be accelerated by ionizing radiation /cosmic radiation or the presence of radioactive elements/, or by highly active free radicals, e.g.: H^+, CH_3^+, $C_2H_5^+$ /6/, which form "in situ". The transfer of such partly or completely carbonized polycondensates, along with the bearing rock mass, close to molten basic magma /or conversely/ causes a slow increase in temperature with unchanged, or increasing, pressure of the system in question. In such conditions, the further disintegration and structural reconstruction of hydrocarbons may take according to the reaction

$$C_nH_m \longrightarrow C \text{ /diamond or graphite/ } + 1/2\ H_2$$

Analyses of admixtures in natural diamonds showed the presence, e.g., of methane, bitumens and nitrogen, therefore organic compounds must have reached kimberlite magma before diamond crystallized. The origin of nitrogen can be explained by the slow transition of radiocarbon ^{14}C to ^{14}N, as result of isotopic degradation, according to the reaction

$$^{14}_{6}C \longrightarrow \beta \longrightarrow {}^{14}_{7}N$$ /with disintegration period of 5730years/

Such a conception of the formation of diamonds in the nature explains their presence in the upper parts of kimberlites and discrepancies in the isotopic composition of carbon. In the upper parts of kimberlites there were much lower pressures and temperatures, at which, according to previous theories, diamonds cannot have formed. On the other hand, it should be assumed that much less energy is need for the reconstruction of an organic structure containing free radicals - the best catalysts for reactions - than for breaking up the existing structure and its reconstruction into another /e.g. graphite into diamond/. Since the process of the formation of diamond from hydrocarbons would have been slowed down by a decrease in energy /and thus in temperature too/ by 436 kJ per each mole of escaping molecules of H_2, this suggest much lower pressures and temperatures necessary for diamond to form. Thus, this precious mineral need not at all have formed and depths exceeding 100 km, as it has so far been assumed in geology.

If carbon were of plutonic origin, studies of the isotopic composition of carbon in diamond would not show so large differences. These discrepancies indicate the organic origin of carbon /5/, even if assumed that most diamonds are enriched by ^{13}C and only slightly by ^{12}C. Its increased concentration in diamond, despite its lower concentration in organic compounds, can be explained by its easier - selective - entry into a new crystalline structure due to its reactivity which is much greater than that for carbon ^{12}C. It operates on the principle of self-catalyst and first builds itself into crystals of diamond /8/. It is only than that it builds itself into the structure of diamond carbon ^{12}C. The discovery by, space chemistry of the presence of hydrocarbons in meteorites, particularly in carbonaceous ones, and in lunar rocks is evidence for the organic origin meteorite diamonds.

The hypothesis that diamond in nature formed from organic compounds is now verifield experimentally. The authors assume that vitreous carbon as obtained from a slow process of carbonization of phenol-formaldehyde resin, which is a single-element vitrifield polymer with cluster structure /2/, used as carbon material in a typical industrial diamond synthesis, can greatly lower the pressure range used previously. Samples were prepared in two version: Vitreous carbon in the form of solid plates in a typical charge for synthesis /Fig.1a/, and in the form of powder mixed with a metallic catalyst /Fig.1b/. The material prepared in this way underwent high-pressure synthesis on QRD type press manufactured by the ASEA Company, at pressures of the order of 6000 MPa and temperatures of 1770 K /normal conditions of industrial synthesis/, and also at half as low pressure of about 3800 MPa with the same temperature. Each time, the sinthesis time /a full cycle from the chamber loading to its unloading/ was 22 minutes.

After the synthesis, the material was tested by X-ray powder method to find the presence of diamond crystallites. For the presence of diamond crystallites in the carbon material /in contact with a metallic solvent/ confirms the validity the present thesis of the organic origin of carbon building crystals of diamond and graphite, and their growth within carbonized polycondensates of alifatic hydrocarbons /diamond/ or aromatic ones /graphite/. In this case, a further growth of crystals on the basis of such a crystalline nucleus, in natural conditions, is related to the long-term formation of the system under high pressures in the region with high temperature.

In samples synthetised under industrial conditions /about 6000 MPa/ vitreous carbon did not transform to crystalline structure. Diffraction patterns show only a broadening difusion halo near 2,057 peak, with total transformation of rest glassy carbon into graphite /3,35 peak highest intensity on the pattern/. This does not exclude the possibility of the presence of extremaly small crystallites with diameters of less than 10 nm in the material, but does not confirm, either, the possibility of obtaining diamond nuclei in this pressure-temperature system in the course of a short-term process. On the other hand, apart transformation of basic mass of glassy carbon into graphite structure /3,35 peak/ diffraction patterns for samples synthetised at lower pressure show distinct lines representing a slight, but nevertheless distinct, amount of small crystallites of diamond /2,057 and 1,260 peaks with relative intensities respectively 5 to 10 and 1 to 2/. Maximal intensity of diamond peaks we observe for 15% higher temperature, of about 2040 K, and 36,7% lower presssure, about 3800 MPa.

This experimental results confirm the possibility of the formation of natural diamond from hydroarbons as the initial material, in the conditions of much lower presssures, and accordingly, closer to the earth's surface.

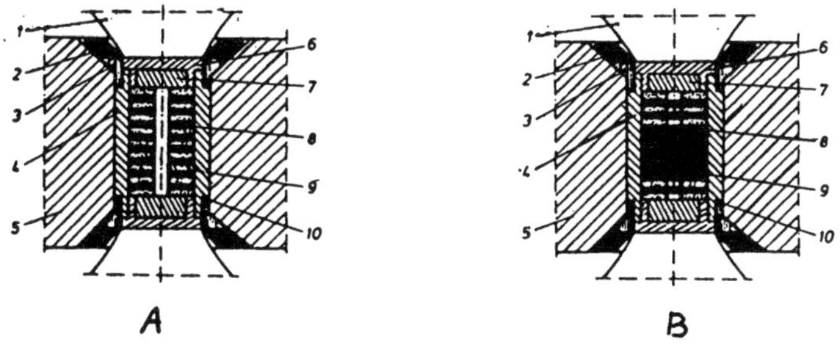

Fig.1. Diagram of charge used in a Belt-type chamber /cross-section/:
A - Vitreous carbon in the form of plates,
B - Vitreous carbon in the form of powder,
1- tungsten carbide piston, 2- rubber ring, 3- raw pyrophyllite fired at 1248 K, 5- tungsten carbide matrix, 6- steel contact rings supporting pyrophyllitite pastilles, 7- fixed pyrophyllite pastilles, 8- central steel mandrel, 9- cobalt plates, 10- graphite plates.

Literature:

[1] Bokij G.W.: Kristallokhimiya, Izd.UGU, 1960

[2] Goodmann C.H.L.: A new way of looking at glass, Glass Technol., 28, /1/ /1987/, 19-29.

[3] Kostrovitskij S.I.: Geokhimicheskiye osobiennostii mineralov kimberlitov, Izd. Nauka, Novosibirsk, 1986.

[4] Kvasnitsa B.N.: Mielkiye almazy, Naukova Dumka, Kiev, 1985.

[5] Polanski A.: General and organic geochemistry /in Polish/, Warsaw University Press, 1986, 120-127.

[6] Stolarczyk L., Stolarczyk U.: Free radicals /in Polish/, Warsaw, 1960

[7] Vinogradov A.P. et. al.: Vozmoznye istochniki ugleroda almazov po izotopnym dannym $^{12}C/^{13}C$, Geokhimya 6 /1965/, 643-651.

[8] Wadas R.: privat com. /1987/.

[9] Wickman F.E.: The cycle of carbon and the stable carbon isotopes, Geochim.Acta, 9 /1956/, 136-153.

1/ Institute of Electronic Materials Technology, Wolczynska 133, Poland

A CERTAIN PROBLEM OF CRYSTALLIZATION

F.Egerer 1)

Nowadays a very important and interesting problem of crystallization is being investigated in mineralogy. This problem can be solved by finding the answer to the following guestion:

Why does it occur under natural circumstances that exactly the greater part of minerals and those of different grain size crystallyze from the supersaturated solution of a closed system?

This subject has already been investigated by Zoltai, Stout, Newham, Wojnarovics, Egerer and many other specialists.

It was pointed out in a paper published earlier (Egerer, 1985.), that the distribution of elements changes on the grain boundaries of fully crystallized rocks like granites.

Concentration of elements which cannot even be found or are only scattered in the neighbouring grains can be observed on the grain boundarics. Fig. 1. shows a micro-probe line scain taken at the grain boundaries of orthoclase and quartz relating to different elements. Fig. 1. clearly represents, that the quantity of K, Si, Al dicreases on the boundary (grains of orthoclase can be found on the left side and those of quartz are on the right side). At the same time the quantity of Cr and Fe elements increases suddenly, so the fact that the grains do not decompose is significantly caused by the Cr and Fe elements. The general problem of this paper was given by this recognition. Namely, finding the reason for the phenomenon why crystals of different grain size form, when the formation of crystals starts in place of the crystal nucleus in a supersaturated solution. That is, why does formation of a crystal finish while another one or more crystals continue to grow?

Experiments and results

In order to make the above mentioned problem clear, the calcite rhombohedron was chosen as an ideal mineral for studying the distribution of

---

1) Technical University for Heavy Industry, Department of Mineralogy and Petrography, 3515 Miskolc, Hungary

elements on the $(10\bar{1}1)$ crystal face (Fig. 2/a.) and on the surface parallel to it inside the crystal (Fig. 2/b.). The photo of the regional spectrum taken of the crystal boundary from the $(10\bar{1}1)$ face by the electron mikro-probe can be seen on Fig. 3. This figure clearly illustrates, that besides the big quantity of Ca cations, Si, S, Cl and K appears on the surface of the calcite crystal. At the same time - as it is illustrated in Fig. 4. - only Ca cations can be found inside the crystal, but odher elements are missing.

Conclusions

Our results show, that so called "extraneons" elements, like Cr and Fe, Which cannot be found inside the mineral, infiltrate into the edge surface. It means that the distribution of charges on the surface of the mineral is different from the distribution inside the mineral. That is, the infiltrating "extraneons" ions may have a decisive effect on completing the process of crystallization; on formation of the final crystal face.

We imagine this process as follous:

The infiltrating "extraneons" ions (in our case the Si, S, Cl and K ions) change the charges in the developing crystal face and as a result, they change the necessary potential for the separation of Ca and CO_3 ions in the way that it is impossible for other Ca and CO_3 ions to get separated from the developed crystal face. This puts an end to the growth of the crystal face as well as the crystal. At the same time it is possible for the crystals starting from the crystal nucleus to continue growing if "extraneons" ions did not get separated from the boundary faces.

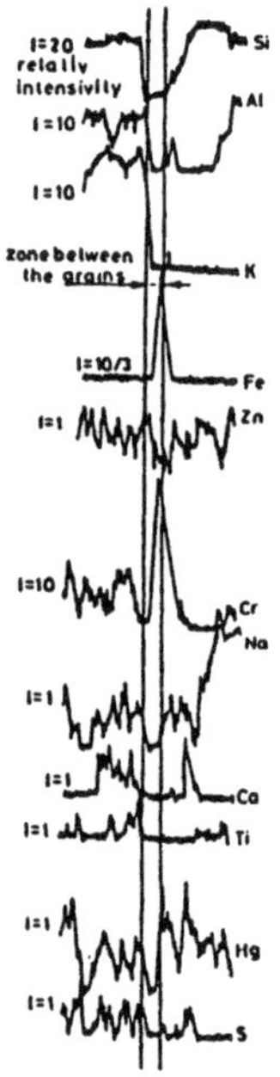

Fig.: 1

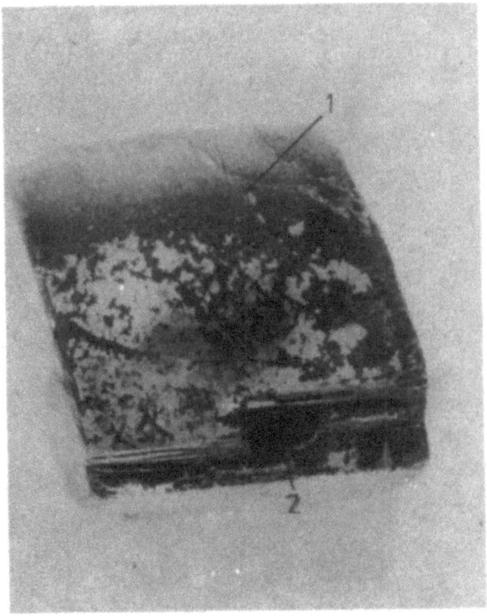

Fig.: 2

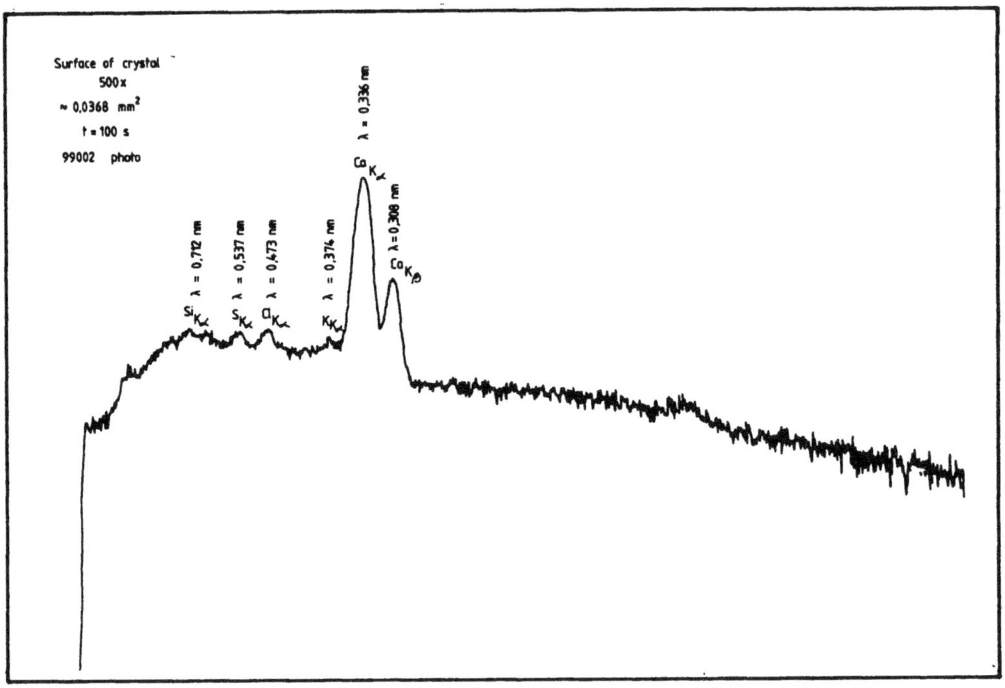

Fig.: 3

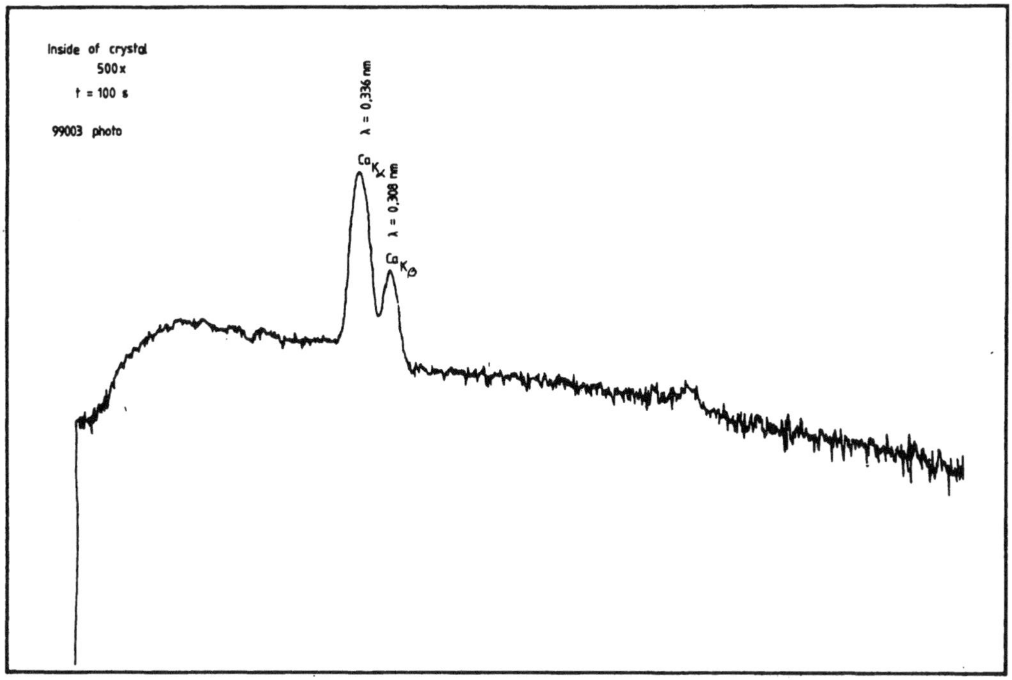

Fig.: 4

Abstract

The contact of mineral grains is one of the most interesting practical aspects of mineralogy. In this work the author investigated the distribution of charges and elements of the crystal boundary. He points out, that the anomalous distribution of charges on the surface may play a decisive role in finishing the process of crystallization.

RELATION BETWEEN THE ELASTIC PROPERTIES AND THE STRUCTURAL BEHAVIOUR OF HIGH- METAMORPHIC ROCKS UNDER THE THERMODYNAMIC CONDITIONS OF THE UPPER MANTLE

Mueller, H.-J.; Schmaedicke, E. 1)

I N T R O D U C T I O N

Seismic investigations have a special importance for the exploration of the deeper crust and the upper mantle. High pressure / high temperature investigations of the elastic properties of granulite facies rocks are a valuable contribution for the systematic interpretation of seismic velocity - depth - functions. Accordingly, the main factor is to discover the interpretation between the variation of the physical properties and the material - structural parameters.

E X P E R I M E N T A L   P R O C E D U R E S

A 2.0 GPa - gas pressure chamber was employed for the experiments. An internal heating provided temperatures up to 1,000 °C in a sample space of 13 mm diameter and 20 mm length.
The determination of the ultrasonic compressional and shear wave velocities Vp and Vs was carried out with the aid of buffers. Separate pairs of transducers were used for the generation and the recording of Vp and Vs. After attaining the maximum pressure the temperature was slowly rised stepwise. The duration of the temperature influence amounted to about 30 minutes. In order to determine any relation between the elastic properties and the phase behaviour optical analyses of the rock samples have been performed using standard optical and powder - diffraction techniques. After quenching the results were compared with the initial composition.

S A M P L E S

The rock samples from the Saxonian Granulite Massif are high - metamorphic, granulite facies ones, such as granulites, pyriclasites and pyroxenites.
The chemical composition of one representative of each rock kind is listed in Table 1.

----

1) Central Institute for Physics of the Earth
Academy of Sciences of the GDR, Potsdam, 1561

Table 1: chemical analyses (percentage by weight)

	granulite	pyriclasite	pyroxenite
SiO_2	75.1	49.0	47.4
TiO_2	0.16	3.36	0.081
Al_2O_3	11.7	13.6	10.4
Fe_2O_3	0.4	4.2	0.9
FeO	1.0	8.42	4.76
MnO	0.02	0.18	0.18
MgO	0.5	7.5	16.6
CaO	0.7	9.2	15.9
Na_2O	3.2	2.4	0.5
K_2O	4.9	0.3	0.2
H_2O	1.1	0.9	1.9
P_2O_5	0.12	0.071	0.018
sum	98.1	99.1	98.8

EXPERIMENTAL RESULTS

granulite

The Vp- and Vs- values are given in Fig. 1. For the initial rock composition and that after the high pressure / high temperature experiments see Table 2. The increasing influence of the temperature on the values of Vp and Vs starts at 600 °C. The dropping values are mainly caused by the beginning of partial melting at about 600 °C. Concerning the results of STERN et WYLLIE (1981) the conditions of 1 - 2 GPa and 600 - 650 °C are close to the solidus. In our experiments the melt fraction increases up to 50 % by volume at 1,000 °C. The melt seems to contain the whole content of garnet and plagioclase and a great part of the potash feldspars and less quartz. By passing the area from the solidus to the liquidus the minerals disappear in the following order: potash feldspar, garnet (or biotite), plagioclase and quartz (according to HUANG et WYLLIE, 1986).

pyriclasite

The velocity data are shown in Fig . 2. The results of the optical phase determinations are summarized in Table 2. The increasing temperature influence on the velocity begins at about 750 °C. According to LAMBERT et WYLLIE (1970) the solidus for basaltic systems is between 800 °C and 900 °C (2 GPa) but it can be modified by the specifical basaltic composition and the available water amount. Quartz and plagioclase disappear somewhat above the solidus. Sporadically the investigated samples show narrow melt films at prefered grain boundaries only at 1,000 °C. The start of melting is only a partial explanation for the dropping velocity data above 750 °C. Further, the modified optical properties of plagioclase could be taken as an argument for its instability probably causing the described elastical behaviour. Above 1.5 GPa plagioclase should be replaced by the coexisting phases quartz and jadeite.

p y r o x e n i t e

The data of the elastical properties are represented in Fig. 3. Contrary to granulite and pyriclasite the velocity values demonstrate an almost linear decrease with increasing temperature. The mineral assemblage garnet - pyroxene should be relatively stable under the described conditions regarding the known phase relationships.

Table 2: initial phase compositions of the samples and those caused by the experiments at 2 GPa (percentage by volume)

	initial state	600 °C		750 °C	1,000 °C
<u>granulite</u>					
quartz	43	42	disk quartz	30	27
potash feldspar	44	42	quartz/feldspars	29	20
plagioclase	8	7		2	-
garnet	3	3		3	-
glass (colourless)	-	4		34	47
glass (brown)	-	-		-	4
<u>pyriclasite</u>					
pyroxenes	39				39
plagioclase	57				57
accessory minerals	3				3
glass (colourless)	-				<1
<u>pyroxenite</u>					
pyroxenes	95				95
garnet	4				4

C O N C L U S I O N S

Generally the results show that a partial melting essentially influences the elastic properties whereas phase instabilities connected with transitions under sub- solidus conditions act somewhat weaker on the elastic behaviour.

On the contrary, a stable assemblage does not show comparable phenomena. The decreasing data are probably only due to the deterioration of the grain contact which is increasing along with the temperature.

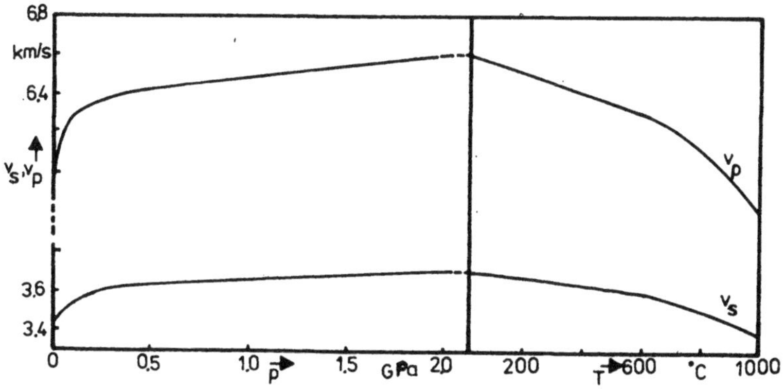

Fig. 1

granulite

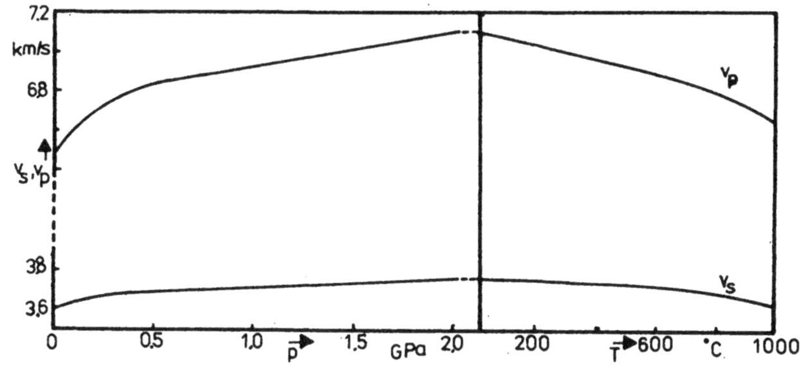

Fig. 2

pyriclasite

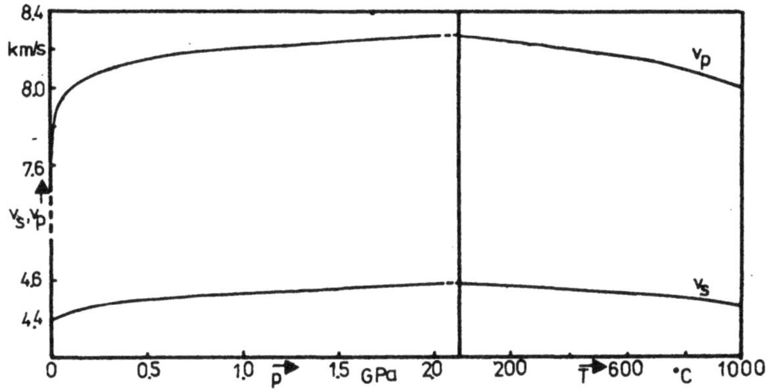

Fig. 3

pyroxenite

Fig. 1-3: elastic wave velocities Vp and Vs in dependence on pressure and temperature up to 2 GPa and 1,000 °C, respectively

r e f e r e n c e s

[1] STERN, C.R.; WYLLIE, P.J.: Phase relationships of I-type granite with H_2O to 35 kb. J. Geophys. Res., 86 (1981), 10412-10422

[2] HUANG, W.-L.; WYLLIE, P.J.: Phase relationships of gabbro-tonalite-granite-water at 15 kb with applications to differentiation and anatexis. Amer. Miner., 71 (1986), 301-316

[3] LAMBERT, I.B.; WYLLIE, P.J.: Melting in the deep crust and upper mantle and the nature of the low velocity layer. Phys. Earth Planet. Int., 3 (1970), 316-322

THE MEASUREMENT OF ELECTRICAL RESISTIVITY OF ROCKS UNDER SIMULATED CRUSTAL CONDITIONS

Glover, P.W.J.[1], Ross, R.G.[+], Vine, F.J.[1], Turner, G.[*]

ABSTRACT

The observed field determinations of crustal conductivity depth profiles show anomalously high conductivities in the lower crust. This has yet to be explained by a consistent theory and interpretation of field results is difficult due to the lack of relevant laboratory-based conductivity measurements. Although there are laboratory measurements of rock conductivity of the upper crustal regime[4] no laboratory measurements on saturated rocks are available at lower crustal temperatures and pressures. Measurements of saturated rock cores at temperatures of up to 800°C and pressures up to 1 GPa while enabling the variation of pore pressure independently of confining pressure poses considerable problems in designing an experimental cell. An initial design of experimental cell is presented. Its design will be discussed with special attention to problems obtaining an insulated two terminal conductivity measuring electrode arrangement at high temperatures while retaining a jacketed sample by the use of a guard-ring electrode arrangement. The design of seals will also be discussed.

INTRODUCTION

Field studies commonly reveal a zone of high electrical conductivity within the Earth's lower crust. Various attempts have been made at explaining such behaviour but all are hampered by the lack of knowledge of the electrical behaviour of rock types found at such depths under high pressures and temperatures. Direct measurements of the electrical properties of rocks in situ is not possible so laboratory experiments under artificially created pressure temperature conditions are of great value. One of the possible explanations of the lower crustal high conductivity layer involves the presence of an aqueous electrolyte. It is therefore important that such measurements are possible on saturated rocks with the ability to vary the pore pressure independently of the overburden pressure. Up until now such work has been carried out up to the upper crustal conditions of 0.4 GPa and 250°C[1,2,3,4]. The value of temperature and pressure in the lower crust is higher than this in most areas. The range of temperatures and pressures needs to be extended to 800°C and 1 GPa respectively. This is the primary aim of this work and demands the design of a measurement cell that will withstand such an extreme environment.

EXPERIMENTAL CELL DESIGN

An initial measurement cell has been made (Fig. 1). Its design has been dictated by the high pressures and especially temperatures at which it is planned to be used. The ability to vary the pore pressure of a saturated rock sample independently of the

---

[1]School of Environmental Sciences, University of East Anglia, Norwich, Norfolk, England
[+]School of Mathematics and Physics, University of East Anglia, Norwich, Norfolk, England
[*]School of Information Systems, University of East Anglia, Norwich, Norfolk, England

confining pressure requires that the rock is surrounded by a sleeve that will separate the external pressure transmitting medium from the internally pressurised electrolyte solution. This sleeve must be sufficiently deformable to transmit the confining pressure to the rock sample. For temperatures up to around 250°C a PTFE sleeve has been used[3]. The electrical insulating properties of the PTFE enables the end pieces of the cell to act as electrodes. Extending the range of temperatures at which such a conductivity cell can operate entails the replacement of the PTFE sleeve with another deformable material that can withstand high temperatures. Unfortunately there is no such material that is an electrical insulator as well as fulfilling these requirements. A copper cylinder, 250 μm thick, has been chosen to sleeve the sample. The use of an electrically conducting copper sleeve raises two major problems that need to be solved independently:

a) Electrodes placed at each end of a cylindrical sample need to be set in a suitable electrical insulator that can be sealed to the sleeve and electrode leadthroughs.
b) Leakage currents shunted around the sample by the copper sleeve give rise to a measurement of apparent conductivity that is considerably greater than the true value of the rock conductivity.

The next two sections deal with these two problems in this order.

MATERIALS AND SEALING CONSIDERATIONS

The use of a copper sleeve has dictated many of the requirements for other parts of the cell. Reference to Figure 1 shows the need for setting the electrodes in a block insulator at each end of the cell. Such electrode holders are required to have high electrical resistance, zero connective porosity, high compressive strength, be refractory to 850°C and have an ability to be sealed to the copper sleeve and electrode leadthroughs. Pure recrystalised alumina manufactured by isostatic pressing by Degussa fulfils these requirements and has been tested to 0.4 GPa. A computer model has indicated that no failure should occur due to compressive or tangential stresses at confining pressures up to 1 GPa. Tests on helium leak detector indicate zero porosity for alumina made by isostatic pressing but not for alumina that is slip-cast. The electrode leadthroughs need to be sealed to the alumina to isolate the confining pressure medium from the pore pressure transmitting aqueous NaCl solution. Niobium was chosen as a suitable material for the leadthroughs as it has almost exactly the same linear expansivity as the alumina. This enables the use of a simple cone seal. The initial seal is made by diamond lapping the niobium into a self-locking taper that has been diamond ground into the alumina. Since the confining pressure is always greater than or equal to the pore pressure the seal remains tight at high pressures. The similar and small linear expansivities of both alumina and the niobium ensures the seal remains tight at high temperatures. Copper slip is used as a gasketting material to improve sealing. Electrodes are made from stainless steel to withstand attack by the pore pressure solution, although they may have to be fabricated from another material to withstand attack by hot electrolytes. The electrodes are machined slightly too small to accommodate relative thermal expansion with respect to the alumina. They are held against the sample by a light spring, which, with the presence of a surface layer of electrolyte on the sample, in the case of saturated samples, provides good electrical connection. The alumina/copper sleeve seal shown in Figure 1 is a modification of the design of H. Jolly, School of Mathematics and Physics, University of East Anglia. The seal is initially made by swaging the copper sleeve into a groove in the alumina using a rotating wheel tool. A circlet of copper wire is laid in the copper groove and held in place by four jaws. The four stainless steel jaws are tightened together by sliding a tapered collar over them by the use of an assembly tool. The seal will improve at high temperatures if the collar is made from a material of less linear expansivity than the stainless steel and copper because the expansion of the copper sleeve/wire and steel jaws is constrained between the less expansive collar and alumina. Leak tests on this seal show no helium leakage when tested on a helium leak detector and no penetration of high pressure transmitting oil when held at 0.4 GPa for 24 hours in a hydrostatic oil pressure vessel.

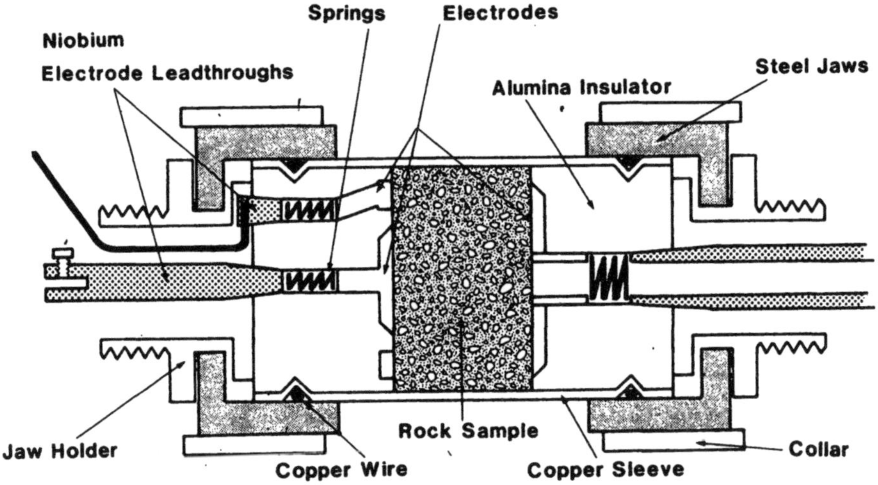

FIG. 1 The Copper Sleeved Guard-ring Experimental Cell

COPPER SLEEVE LEAKAGE CURRENTS AND THE USE OF A GUARD-RING

The electrical conductivity of the copper sleeve gives rise to large leakage currents through it. This shunts most of the current around the sample. It also distorts the current flow around the periphery of the sample near to the sleeve due to the boundary conditions that the conducting sleeve imposes on the system. The total effect of the copper sleeve is to increase the apparent conductivity over 300% compared with the unsleeved case. The problems caused by the copper sleeve can be removed by the use of a guard-ring. An annular guard-ring electrode that surrounds the collector electrode and is held at the same instantaneous potential as the collector electrode with respect to the emitter electrode, collects the leakage currents that flow through, or are distorted by the copper sleeve and drain them to ground without them being measured. The result is that the conductivity measuring bridge measures a central cylinder of rock where the flow of current is dependent upon the conductivity of the rock alone and the edge effects caused by the copper sleeve do not affect the measurement. The guard-ring potential is generated by active operational amplifier circuitry configured as a voltage follower. It can provide a guard-ring potential that differs less than 0.1% from the collector potential, isolate the guard-ring from the collector and drain the required leakage current. This is due to its characteristics of a very high input impedance and low output impedance at the conductivity bridge operation frequency of 1592 Hz. High slew rate versions ($2x10^{+7}$ V s^{-1}) of the commonly available 741 op-amps were used to enable close potential mapping at high bridge potential outputs. Other specifications of the op-amp enable the full range of unsaturated rock conductivities to saturated rock conductivities to be measured. The measurement set up is shown in Figure 2. A Wayne-Kerr B642 Autobalance conductivity bridge has been used. The op-amp circuitry has been applied to a simple array of resistors modelling the behaviour of a loaded cell with an arranged leakage current 3.41 times the 'sample' resistor current. The measured apparent resistance of the 'sample' resistor was within 0.5% of its true resistance with the guard-ring in operation whereas it fell to less than 62% of its true value with the guard-ring inoperative. Tests of the guard-ring circuitry on a proper model of the copper sleeved cell configuration containing a saturated rock sample has given consistent results that systematically differ from the PTFE sleeved case by less than 4%. Comparing this with the 300% plus increase in apparent conductivity with the guard-ring circuitry inoperative shows how effective this method is. It is hoped that the 4% error of this technique will be able to be reduced especially if used with unsaturated rocks.

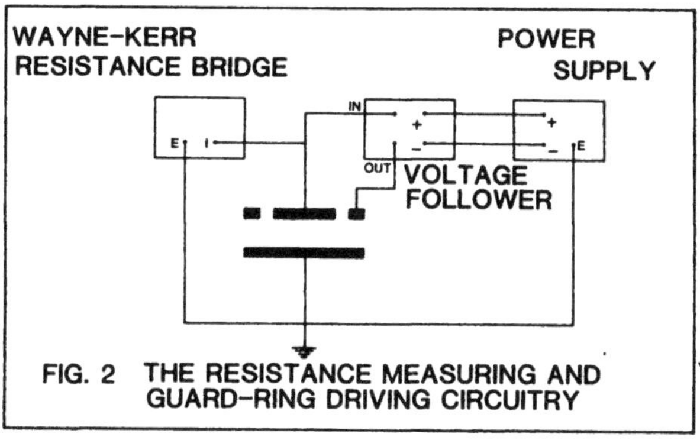

FIG. 2 THE RESISTANCE MEASURING AND GUARD-RING DRIVING CIRCUITRY

CONCLUSIONS

Results from tests on the new measurement cell and its guard-ring driving circuitry show how well they seem suited to carrying out the task for which they were designed. It is hoped that the coming year will produce saturated rock conductivity measurements at higher temperatures and pressures.

REFERENCES

[1] Evans, C.J. The seismic and electrical properties of crystalline rocks under simulated crustal conditions. Unpubl. Ph.D. Thesis, University of East Anglia. 1980.

[2] Lee, C.D. Electrical conductivity of rocks subjected to upper crustal conditions. Unpubl. Ph.D. Thesis, University of East Anglia. 1977.

[3] Lee, C.D. et al. Geophysics J.R. astr. Soc. (1983), 72, 353-371.

[4] Tousaint-Jackson, J. Laboratory and theoretical studies of electrical conduction in crystalline rocks. Unpubl. Ph.D. Thesis, University of East Anglia. 1984.

PRESSURE-DEPENDENT ULTRA-LOW PERMEABILITY DETERMINED BY MEASUREMENTS OF ELECTRICAL PROPERTIES

D.Freund 1); J.Iffland 2); R.Boettger 1); F.Theilig 3)

INTRODUCTION

The isolation of natural gas fields by very low porosity cap rocks depends among other properties on their ultra-low permeability. For a suite of such rocks it will be tried to get informations about the behaviour of their permeability reduction resulting from electrical measurements under hydrostatical pressure.

THEORETICAL RELATIONS

On the basis of DARCY's and ARCHIE's laws permeability and electrical conductivity are related (BRACE [1]) by

$$k=(m^2 / b)\ F^{-1.5} \qquad (1)$$

where k=permeability, m=hydraulic radius(the volume of interconnected pores referred to surface area), b=shape factor, and F=formation factor (ratio of conductivity of an electrolytic saturant to the conductivity of the fully saturated sample: σ_e / σ_r). DACHNOV's [2] version of ARCHIE's law combines conductivity with porosity (PHI) and is consistent in this power-law form with the empirical KOZENY-KARMAN equation

$$k=c\ \frac{\Phi^3}{S_o^2} \qquad (2).$$

Here S_o =specific surface, $c \approx 0.2$(WONG, KOPLIK, TOMANIC [3].
This capillary bundle model also represents the starting point of a network model after SCHOPPER et al.[4] and relates k to F and PHI by

$$k= \frac{1}{F} \cdot \frac{1}{2\,S^{*\,2}_{por}} \qquad (3),$$

where S^*_{por} = capillary surface(Pape et al.[5,6,7]. Relation(3) is identical

---

1) Akademie der Wissenschaften der DDR, Zentralinstitut fuer Physik der Erde , Telegrafenberg, DDR-1561 Potsdam

2) VEB Geologische Forschung und Erkundung Halle, Betriebsteil Schwerin, Schleifmuehlenweg 6, DDR-2786 Schwerin

3) VEB Erdoel-Erdgas Gommern, Stammbetrieb, Magdeburger Chaussee , DDR-3304 Gommern

with

$$m=(bkF)^{1/2} \tag{4}$$

from the equivalent channel model after WALSH and BRACE [8], if b=2 and

$$m=\frac{1}{S^{*}_{por}} \tag{5}.$$

SAMPLES AND EXPERIMENTS

On siltstones and fine-grained sandstones sampled from a well core with porosities ranging 0.9-2.3, and clay contents from 11.0 to 34.0 VOL.% (Table 1.) permeabilities have been measured at steps of 5, 20, 40, , and 80 MPa by a standard gas technique (Iffland [9]). As shown in Fig.1. permeabilities are reduced about 3 orders of the magnitude range.

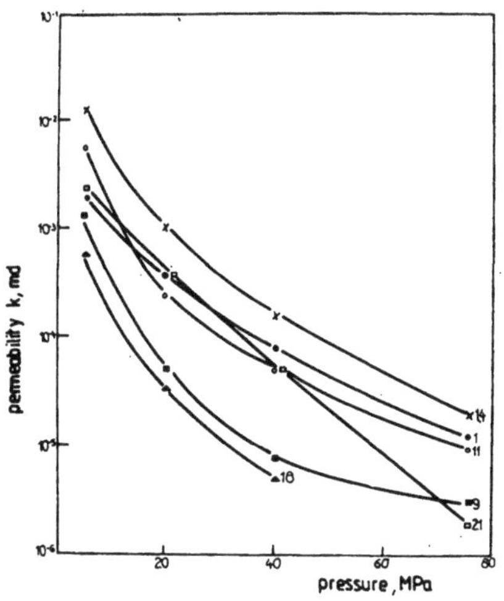

Electrical data were obtained in a second pressure run after an apropriate relaxing time in a sample and pressure arrangement similiar to BRACE [9]. The specimens were saturated with 0.2m NaCl. The pressure steps were identical with the permeability test points up to 80 MPa.

Fig.1. Permeability, k (md) versus pressure, p(MPa). (Numbers signs the samples)

Table 1. Sample Characterization.

Sample	Mineral composition (%) by X-ray qu	pl	K-fsp	anh	calc	dol	clay	Initial Por.(%)	Slope r in (6)	Slope m in (8)
1	48	6	2	4	5	11	24	1.7	5.0	2.19
9	43	7	3	14	4	11	18	1.2	4.4	2.41
11	38	7	2	9	8	2	34	2.3	7.0	2.23
14	54	5	4	12	2	6	17	2.3	8.5	2.94
18	42	7	2	19	6	-	24	1.5	8.3	2.83
21	47	8	3	22	5	4	11	0.9	4.4	2.57

R E S U L T S

The results of electrical measurements up to 200 MPa(No.9.up to 300 MPa) are given as formation factor F in Fig.2.

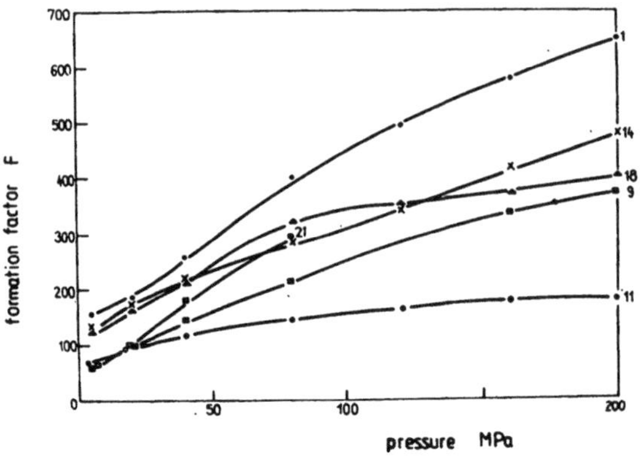

Fig.2. Formationfactor F ,versus pressure, p (MPa).

The representation of k and F in a double logarithmic scale (Fig.3, upper part) follows for all samples a straight line according to

$$\log k \propto -r \log F \tag{6}$$

with a correlation coefficient of 0.97 and the exponent r ranging from 4.4 to 8.5 (see Table 1.).

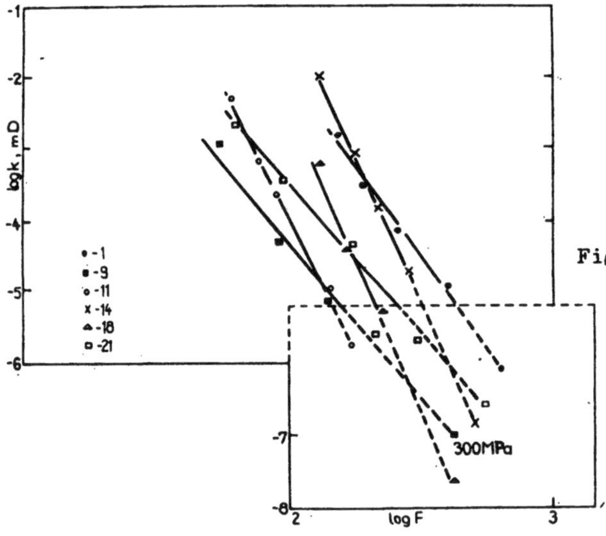

Fig.3. Relation $k \propto F$ Experimental data and extrapolated k-values from electrical measurements The dotted line indicates the lower limit of the available technique for the permeability determination.

The strong interdependence of permeability and electrical properties in the lower pressure range (Fig. 3., upper part) permits to extrapolate ultra low permeability values (dashed lines) into the high pressure field (Fig. 3., lower part).
This result indicates that permeability can derived from electrical measurements under hydrostatic stress for low-porosity sedimentary rocks, too.
The now available k-values give the possibility to plot k versus p. These data can be fitted fairly closely (R=-0.98) by a log - log plot according to

$$k = a_o \, p_e^{-m} \qquad (7).$$

For comparing the different samples we calculate the ratio k/k_0 , where k_0 is the permeability on the pressure p_0 (0.1 MPa) at the surface of the Earth. When p_e (effective pressure at depth) in (7) is replaced by p_0 we get the empirical relation

$$k/k_o = (p_e/p_o)^{-m} \qquad (8).$$

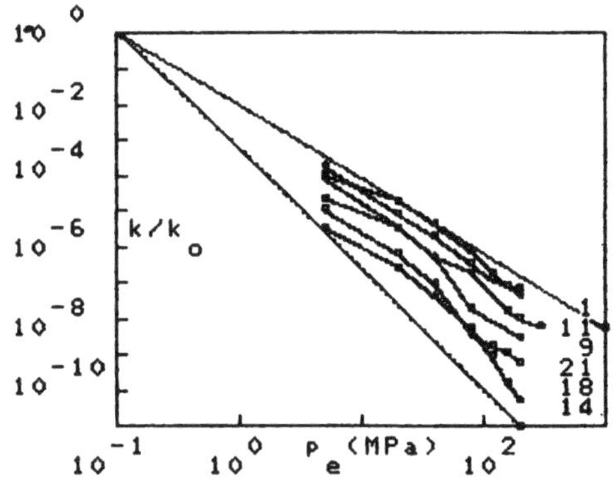

Fig. 4. Experimental and extrapolated permeability data (points connected by lines) versus effective pressure. The two lines envelop, respectively, the best fitting of the upper boundary, $k/k_0 = (p_e/p_0)^{-2.1}$, and the lower boundary, $k/k_0 = (p_e/p_0)^{-3.1}$.

This permeability-depth function shows Fig. 4. The slope m for the investigated samples is given in Table 1.

R E F E R E N C E S

[1] BRACE, W. F.: Permeability from resistivity and pore shape. J.G.R. 82 (1977) 3343-3349

[2] DACHNOV, W. N.: Promyslovaya Geofizika; Moscow (1959); in Russian

[3] WONG, P., J. KOPLIK, J.P. TOMANIC: Conductivity and permeability of rocks Physical Review B, 30 (1984) 6606-6614

[4] SCHOPPER, J.R.: A theoretical investigation on the formation factor permeability/porosity relationship using a network model

Geophys.prospecting 14(1966)301-341

[5] PAPE,H.,L.RIEPE, J.R.SCHOPPER: Calculating permeability from surface area measurements. 7th Europ. Logging Symp. Trans.; Paris (1981)

[6] see [5] A pigeon-hole model for relating permeability to specific surface. Log Analyst 23 (1982) 5-13

[7] see [5] Permeability of porous rocks derived from internal surface effects. Preprint 47th EAEG Meeting, Budapest, 4.-7.6.85

[8] WALSH,J.B., W.F. BRACE: The effect of pressure on porosity and the transport properties of rock. J.G.R.89 (1984) 9425-9434

[9] IFFLAND, J.: Ein Beitrag zur Bewertung von Speichern geringer Porositaet und Permeabilitaet. Report 84/81 VD B-GII/53/84 KEE, Gommern (1984)

VELOCITY-POROSITY AND VELOCITY-CLAY CONTENT RELATIONSHIPS FOR DRY CLASTIC SEDIMENTARY ROCKS FROM ULTRASONIC MEASUREMENTS UNDER PRESSURE

D. Freund 1)

I N T R O D U C T I O N

Crystalline rocks are mainly investigated with respect to their velocities of elastic waves in the dry state while experiments on sedimentary rocks are commonly performed with saturated samples. Obviously, there is a lack of investigations on dry sedimentary rocks at high pressure. But it is this kind of experiments which can serve as a simulation for the gas-filled state of sedimentary reservoir rocks.

S A M P L E S   A N D   B A S I C   D A T A

For about 88 clastic silicate sedimentary rocks sampled from a deep well core, velocity data of elastic waves were obtained at pressures of 8,24, 40,60,120,160,200,250, and 300 MPa using the pulse-transmission technique and a hydrostatic pressure equipment. Compressional and shear wave velocities are determined by the transit time of ultrasonic pulses(1MHz). The samples used in this study are sandstones (56), siltstones (22), and shales (10). The porosities ranged from 0.5 to 15.0 % and were determined by standard techniques. Clay contents were obtained by thin section point-counting and X-ray analysis and varied from less than 1 % to 88 % by volume.

R E S U L T S   A N D   D I S C U S S I O N

The velocities of compressional (v_p) and shear waves (v_s) increase with pressure. In the lower pressure range (<150 MPa) the rate of increase is large and greater for v_p than for v_s. From ≈150 MPa up to 300 MPa velocities grow nearly linearly.
It is known that porosity (PHI) lowers the velocities of elastic waves, but a cross-plot of these data for all samples shows no connection. Other parameters than porosity change the velocity, too. In first order the various mineral contents of the individual samples, e.g. the clay content, diminishes the velocities of compressional and shear waves (TOSAYA [1982], CASTAGNA et al. [1985], HAN et al. [1986]).

---

1) Central Institute for Physics of the Earth
Academy of Sciences of the GDR, Potsdam, 1561

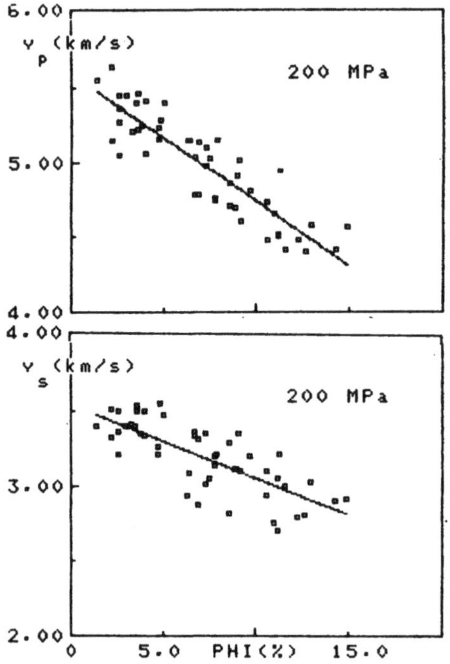

Fig. 1. Velocities of elastic waves v_p and v_s [km/s] versus porosity, PHI [%] at a single pressure value of 200 MPa.

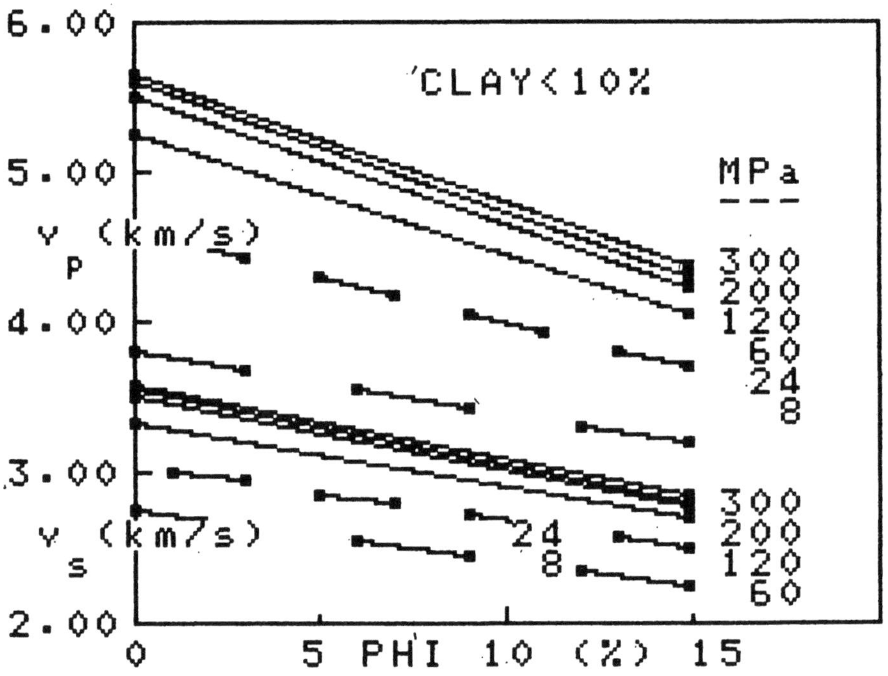

Fig. 2. Straight lines obtained by least-squares regression for the velocity-porosity-equation for 50 sandstones at different pressures.

For getting the effect of porosity on the velocities, samples with less than 10% clay content were selected from the complete suite. The velocities of compressional and shear waves as a function of porosity of these 50 samples are shown in Fig. 1 at a single value of pressure (200 MPa). A strong linear dependence between velocities and PHI is demonstrated. Fig. 2 presents the cross-plots according to the straight lines obtained by least-squares regression (Table 2) for various pressures resulting from measurements on 50 well-consolidated sandstones (the dashed lines refer to fits with low correlation). The correlation coefficient r rises with increasing pressure.

Table 1: Linear equations and correlations coefficient r for the plots at Fig. 2 at different pressures [MPa]

v_P [km/s] =	A*PHI[%] +	B	r	v_S [km/s] =	A*PHI[%] +	B	r
8 MPa	- 0.041	3.81	0.38	8 MPa	- 0.032	2.75	0.48
24 MPa	- 0.062	4.62	0.65	24 MPa	- 0.037	3.06	0.66
60 MPa	- 0.081	5.26	0.84	60 MPa	- 0.042	3.34	0.69
120 MPa	- 0.086	5.51	0.90	120 MPa	- 0.048	3.50	0.76
200 MPa	- 0.086	5.60	0.89	200 MPa	- 0.050	3.56	0.77
300 MPa	- 0.085	5.65	0.89	300 Mpa	- 0.049	3.58	0.76

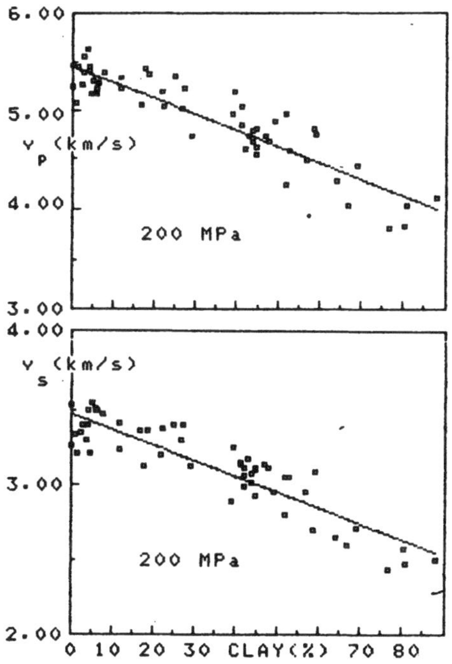

Fig. 3. v_P and v_S [km/s] versus total clay content [%] at 200 MPa for 56 samples (20 sandstones, 19 siltstones, and 10 shales) with <5 % porosity.

In a second step 56 samples with low porosities (<5%) were used to analyse the effect of clay content on the compressional and shear wave velocity. As an example, the plots of both the velocities versus clay content at 200 MPa is given in Fig. 3. Fits by regression analysis for the different values of pressure demonstrate the close connection between these parameters (Fig. 4) in the higher pressure range as follows from Table 2.

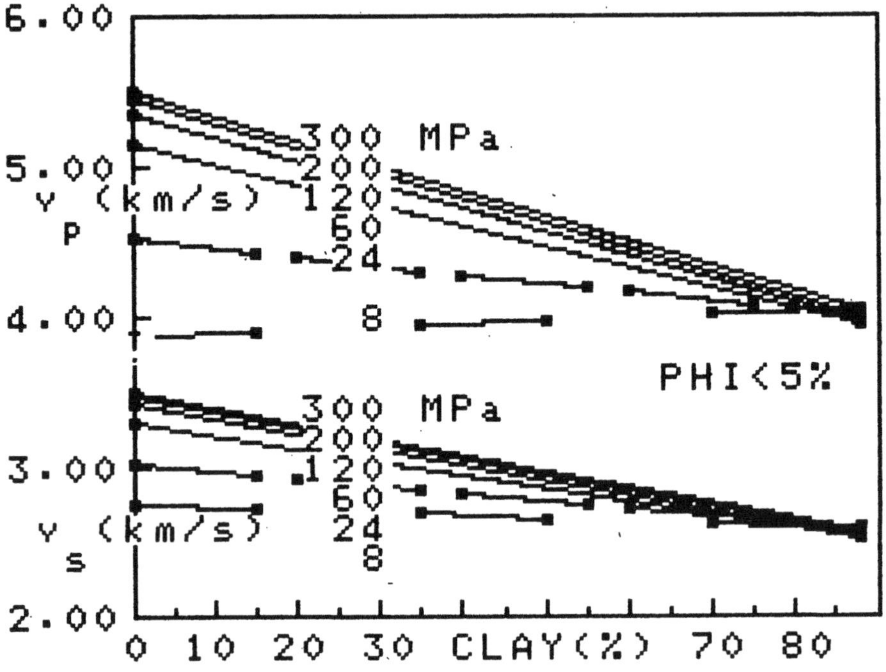

Fig. 4. Plots of velocities of elastic waves v_p and v_s versus clay content for a selected suite (see Fig. 3.) at various pressures.

Table 2: Linear equations and correlation coefficients r for the straight lines given at Fig. 4.

v_p [km/s] = A*CLAY[%] + B	A	B	r	v_s [km/s] = A*CLAY[%] + B	A	B	r
8 MPa	+0.002	3.89	0.12	8 MPa	-0.002	2.77	0.19
24 MPa	-0.006	4.53	0.44	24 MPa	-0.005	3.04	0.59
60 MPa	-0.013	5.16	0.82	60 MPa	-0.009	3.32	0.86
120 MPa	-0.016	5.37	0.89	120 MPa	-0.010	3.44	0.89
200 MPa	-0.017	5.46	0.90	200 MPa	-0.011	3.49	0.90
300 MPa	-0.017	5.52	0.91	300 MPa	-0.011	3.52	0.91

The absence of a good correlation between v_p and v_s versus PHI[%] and CLAY[%] in the lower pressure range is in contrast to the results of investigations on saturated clastic sedimentary rock samples. For the dry state at low pressure the influence of cracks on the velocities is significant and covers the effect of porosity and clay contents. At higher pressure steps ($\gtrsim$ >120 MPa) both velocities correlate linearly with porosity and clay content. The effect of porosity in reducing velocities is about 5.0 as great as the clay content for v_p and 4.5 times as great for v_s .

R E F E R E N C E S :

[1] TOSAYA, C., A. NUR : Effect of diagenesis and clays on compressional velocities in rocks. Geophys. Res. Letters, 9 (1982) 5-8.

[2] CASTAGNA, J. P., M. L. BATZLE, R. L. EASTWOOD : Relationships between compressional-wave and shear-wave velocities in clastic silicate rocks. Geophysics, 50 (1985) 571-581.

[3] HAN, D., A. NUR, D. MORGAN : Effects of porosity and clay content on wave velocities in sandstones. Geophysics, 51 (1986) 2093-2107.

REGULARITIES OF BEHAVIOUR OF ELASTIC WAVES VELOCITIES IN SEDIMENTARY ROCKS AT HIGH PRESSURES

G.I.Petkevitsh[1)]

Due to the increase of depth survey and wide use of the seismic methods of investigstion, the elastic wave velocity becomes a basic parameter for prediction of properties and matter composition on high depths. The prediction procedure is based on simulation of elastic wave velocities behaviour in rocks at high P,T-parameters, corresponding to the depths under investigation, comparison of the laboratory date and other methods of wave velocity determination and construction of prediction relations. Laboratory measurements are of a special importance for sedimentary rocks, which are strongly inhomogeneous in composition, structure and stress state in real conditions.

For a numbers of years, beginning from the 60-es, determination of elastic wave velocities in the conditions, simulating thouse of seams, for sedimentary rocks of the Carpathian region of the Ukrainian SSR was carried out by the author [1-2]. The seam conditions are obtained in special chambers with controlled internal and external pressure, providing effective loading up to 80 - 100 MPa, thus it is possible to imitate the depths up to 5 - 7 km. The effective pressure value was estimated according formula

$$P_{eff} = P_{ext} - \eta P_{int} \quad (\eta = 0.85).$$

The temperature effect was not considered, since the temperature up to 150°C may by neglected. The sonic test method was used in a wide range of frequences. At the same time determination of longitudinal v_l and transverse v_s waves velocities, the characteristics of waves damping, electric resistance, longitudinal specimen deformation and variation of pores space volume was studied. It allowed calculation of the system of necessary petrophysical characteristics of the sedimentary rocks.

Our experience in theoretical, laboratory and well investigations on stress state effect on elastic properties of sedimentary rocks allowed to formulated a number of regularities of elastic wave velocities behaviour [1-2].

1. Elastic wave velocity is a fundamental parameter in the sciences of the Earth, since it is highly informative of matter composition of the rocks, their structure, stress state, fluid content and others

----

1) Institute of Geology and Geochemistry of Combustible Minerals of the Ukrainian Akademy of Sciences, 290 047 Lvov, Nautchnaja street 3-a.

factors and has the greatest possibilities for diagnostics of geological media. Together with mineral composition (analogue - velocity in the solid phase V_1 or density δ_1), porosity (coefficient of porosity K_p), connectivity (pores compressibility β_p), effect of pressure P - are the main factors, determining elastic wave velocity. Consequently, $V_P = f(V_1, K_P, \beta_P, P_{eff})$. The evaluation of the role of pressure factor, obtained in statistical analysis of the experimental date for the Carpathian sandstones, are comprise 0.1-0.3 (in the part of a unit). For the uniform rocks effect of the others factors decreases and the pressure influence increases. Thus, prediction ability of the velocity parameter may be realised either at the presence of the independent date on other parameters or for the narrow class of the uniform rocks.

2. The character of variation of the velocity under pressure is studied rather well. Usually, it is a $V_P = aP^{1/n}$ $(n \simeq 6)$ dependency. However, for the Carpathian rocks $n \simeq$ 10-13, proving their higher lithification. The detailed study of the $V = f(P)$ dependency shows, that it is close by its shape to the curve of crack porosity reduction. The existance of such relationship is explaned by subsequent cracks and pore voids closure under external pressure.

In laboratory conditions, simulating the depths parameters, usually only reverse (elastic) changes of rocks are reproduced. A recurrence of the phenomena under repeated "loading - unloading" cycles was studied, so it is possible to speak about different elastic "stability" of rocks. For rocks proferties prediction on great depths only the values of the velocities at pressure P_H, corresponding to the depths of their bedding, may be used. Just those values V_{P_H} are the true values of the velocity on corresponding depths. According to these values the curve $V = f(H)$ may be reproduced for the specimens picked out from different depths ($V_{P_{H_1}}, V_{P_{H_2}} \dots V_{P_{H_i}}$). The velocity values at overloading ($P_i > P_H$) are not typical for the depth conditions and may be used for velocities extrapolation only at high elastic stability of rocks (Fig.1.).

3. Relative increase of velocity $\frac{V_P - V_0}{V_0}$ at the pressures, corresponding to the depths of rocks bedding, achives 25-40% and depends of the solid phase elasticity, porosity, voids shapes, filler properties and so on. Thus, $\Delta V/V_0$ increases with high K_P. For a number of samples these dependencies were expressed by correlation equations like: $V_P = f(V_0, K_{P_0}, P)$, where V_0, K_{P_0} - "initial" parameter value. It follows from the above mentioned that the uniform rocks have the character of velocity variation under pressure formed in the initial parameters at conventional (normal) conditions.

Thus, a strong inhomogeneity of the sedimentary rocks and comparatively high sensitivity to pressure reduse the experimental results and ability to predict other properties. Still, comparatively low effective

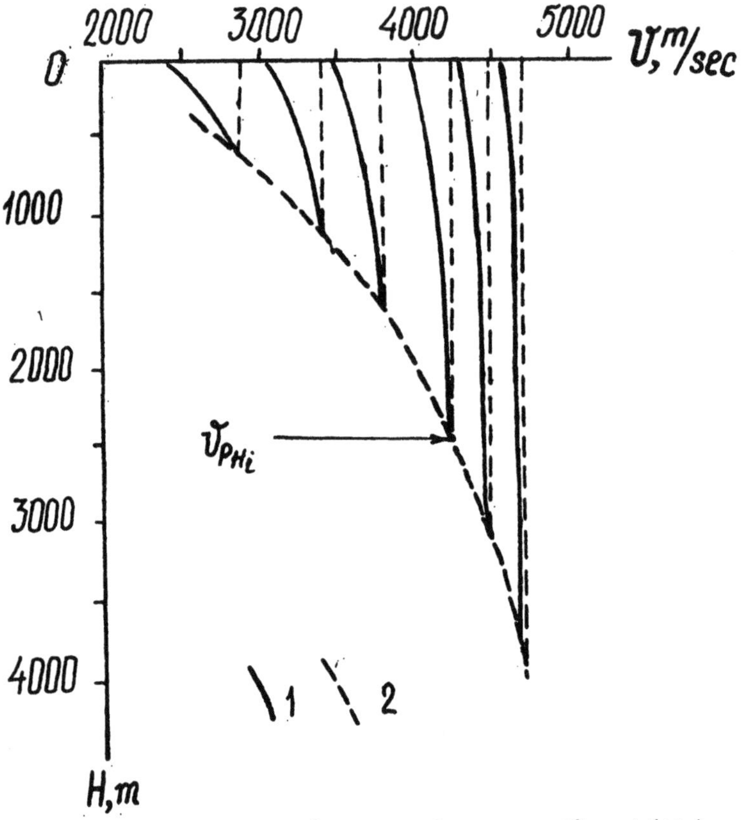

Fig.1. Construction of prognosis curve $V_p = f(H_i)$.
1 - laboratory date $V = f(P)$; 2 - values V_{PH_i};
3 - curve $V_p = f(H_i)$.

pressures values to the depths of 5 - 7 km allow to simulate deformation of sedimentary rocks predominantely in an elastic part, compraising the interval of natural loading, and to obtain satisfactory results.

4. Comparison of the laboratory velocity date and well-logging measurements allows to estimate the effect of different loads in definite rocks formations (Fig.2). Thus, the velocity on some depth may be presented as a sume of four components: some "initial" velocity for unconsolidated sediment, component of unreversible rocks variation for period of geological history, component of reversible (elastic) static compression, component of reversible geodynamic compression, i.e.

$$V_p = \Delta V_0 + \Delta V_{unrev} + \Delta V_{rev.stat} + \Delta V_{rev.dyn}$$

"Dynamic" component is determined as a difference between "well" velocity and "laboratory" velocity for a given depths. "Static" component - is an increase of velocity, measured in laboratory conditions due to application of pressure on the given depth. The "unreversible"

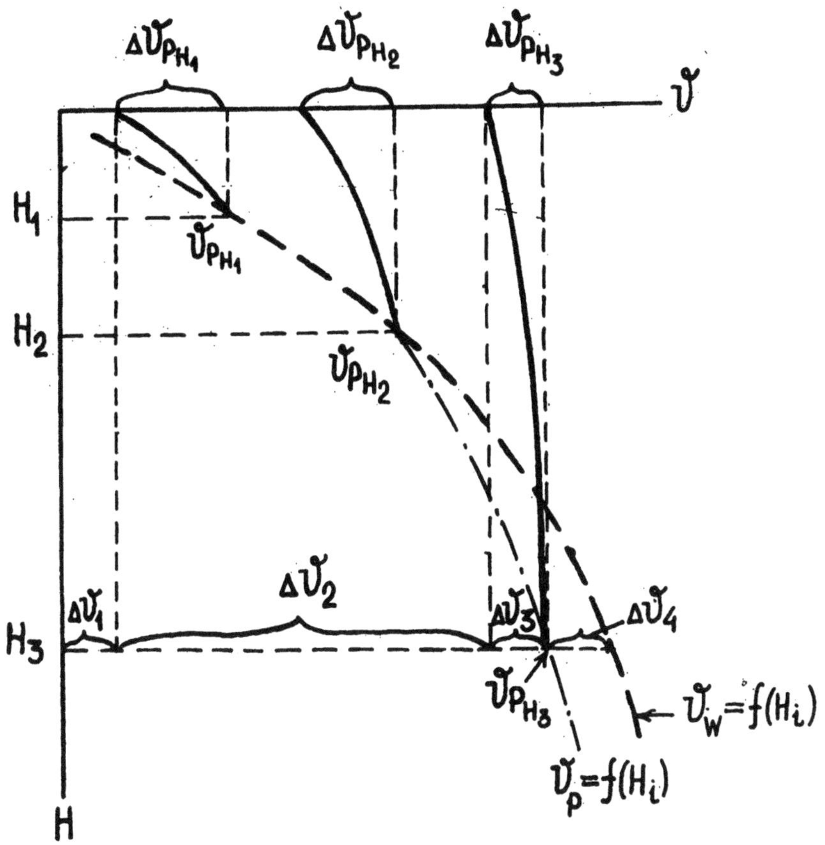

Fig.2. Schematic diagram of evaluation the effect of different loads on rocks velocities.
$\Delta\vartheta_1$ - "initial" value in sediment; $\Delta\vartheta_2$ - unreverse variation (paleovalue); $\Delta\vartheta_3$ - reverse static variation $\Delta\vartheta_4$ - reverse geodynamic variation.

component is obtaind in case if we residue dynamic and static components as well as initial velocity for a sediment of the given composition from "well" velocity value.

When speaking about applied field, the mentioned regularities were used in prediction of velocities in sedimentary rocks on great depths, particularly, in near-well spase, as well as for evaluation of coefficient of porosity on depth via velocity (interval time) according to dependency $K_P = f(\Delta T, \lg H)$, where ΔT is date of acoustic logging.

Literature

[I] Петкевич Г.И. Информативность акустических характеристик неоднородных геологических сред, Киев, Наукова думка, 1976. -214 с.

[2] Петкевич Г.И., Шеремета О.В., Притулко Г.И. Методика петрофизического изучения коллекторов нефти и газа в условиях, моделирующих пластовые, Киев, Наукова думка, 1979. -128 с.

THE SOLUBILITY OF CARBON AND CARBON DIOXIDE IN THE MAGMATIC MELTS AND CRYSTALS UNDER HIGH PRESSURE AND TEMPERATURES.

A.A.Kadik and S.N.Shilobreeva[1)]

Volatile compounds of the system C-O-H such as H_2O, CO_2, CO are important volatile components taking part in the magmatic processes. Investigation of interaction between these volatile components as well as carbon and silicate melts and crystals permit to estimate the role of magmatic processes on the distribution of the volatile components in the upper mantle and crust, on the composition of the volatile components rising from the interior to the Earth surface and formation of planet's atmosphere and hydrosphere.

Up to now vide experimental material is accumulated on the saturation conditions of H_2O and CO_2 in silicate melts under high pressure and temperature. These data permit to consider regularities of the saturation of the magmatic melts with these volatiles up to pressure from 0.5 kbar to 20-30 kbar corresponding to the magma formation region in low and upper crust and upper mantle.

In this paper we will pay the special attention to the influence of such factors as pressure, composition of melts and presence of water in the system on the saturation of melts with CO_2. These factors have to influence the magma degassing during their evolution and uprising to the surface of the Earth.

The conducted analyse have shown, that the solubility of CO_2 in the melts increases with pressure. No extremum was discovered in these systems under the examined pressure. It was stated that solubility CO_2 in pressure interval of 0.5-5 kbar in the melts of basic and acid compositions under the dependence of pressure could be described by the equation of linear type (Fig.1).

The solubility of CO_2 in silicate melts is characterized by the complicated temperature dependence, which is still under discussion However, the dependence of temperature on CO_2 solubility is negligible in the temperature interval which characterizes magmatic eruptions. The minimum was established on the dependence of CO_2 solubility from the basity of melt (Fig.2). This minimum could be explained by two CO_2 solution reactions in silicate melts. First is the reaction in the mole cule form:

$CO_{2\ (vapor)} = CO_{2\ (melt)}$, the second is the interaction between CO_2 and melt oxygen:

$$CO_{2\ (vapor)} + O^{-}\ (melt) = CO_3^{2-}\ (melt)$$

[1)] Present address: Vernadsky Institute of Geochemistry and Analytical Chemistry of the USSR Academy of Sciences, Moscow

On the dependence curve of CO_2 and H_2O solubility from the composition of vapor phases the maximum was established for high temperature region 1400-1650°C [4] as well as the lower one 850-1250° [5] corresponding to the real magma temperature. Thus the effect of water influence on the solubility CO_2 which was firstly established by Eggler [1] could have the petrological meaning.

The conducted thermodynamic analyses based on the experimental data in albite-H_2O-CO_2 system had shown that the existing models which describe the equalibria in the systems melt-H_2O and melt-CO_2 don't describe the experimental data on the joint solubility of H_2O and CO_2 in silicate melt in wide region of fluid composition, temperature and pressure. The empirical dependence is proposed for the description of the system silicate melt-H_2O-CO_2. It describes the composition of vapor and melt in wide region of T and P. Our analysis of CO_2 and H_2O solubility regularities and empiric dependence permit to construct the surface of the saturation magmatic melts with H_2O and CO_2 under high pressure and temperature.

On the base of these surface we can point the next regularities in the changing in H_2O and CO_2 content during decompression.

1. The negligible variations in H_2O concentration in melt in wide pressure range is a distinct feature of the degassing of melts containing H_2O and CO_2. The water content remains practically constant till rather low pressure values (Fig.3). Only in the pressure range less than 1 kbar this process leads to a rapid decrease in the water concentration in the melt. This fact means that it is difficult to find out the initial H_2O concentration in magmas with the help of the quenching products, formed near the Earth's surface. The content of water in these products can differ from the initial in deep magma fraction.
2. Unlike water the melt degassing in the pressure range from 20 kbar up to the lowest leads to uniform decrease in CO_2 concentration. The initial ratio CO_2 to (CO_2+H_2O) in system negligibly affected the CO_2 concentration under the given pressure (Fig.4).
3. The strong prevalence of CO_2 in the fluid phase is a remarkable feature of fluid phase composition of magmas in wide pressure range (Fig.5) The CO_2 concentration is equal to 95-98 mol.%. The degassing of the basic magmas can lead to formation of fluids with high water concentration only in the pressure range lower than 500 bar. Thus the main compo nent of basalt degassing on depth is CO_2.

Recently in the last years the possibility of carbon entrance in crystalline structure of oxides and silicate was under the discussion. Freund et al. [2] had proposed that carbon may be in atomic state in it. The possibility of carbon entrance in the crystals of Earth mantle may be of important in the evolution of fluid and redox regime of the Earth. Academician Vinogradov [6] had proposed that the volatiles of C-O-H system are mainly the reaction products between water and carbon

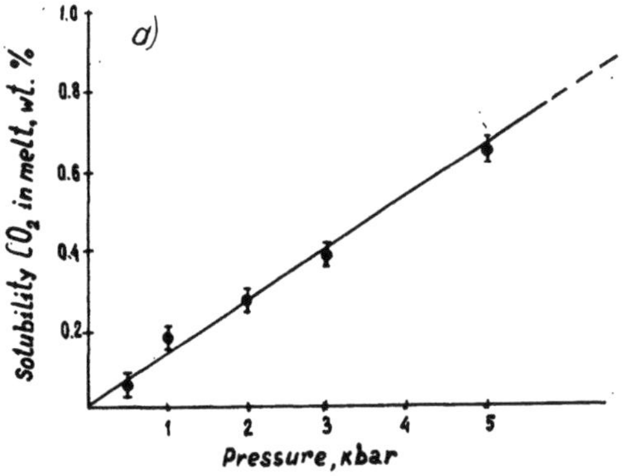

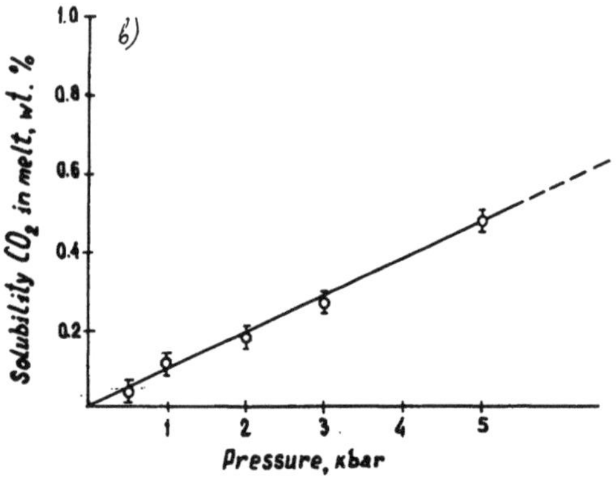

Fig. 1a: The dependence of CO solubility from pressure in melt of basalt

Fig. 1b: The dependence of CO solubility from pressure in melt of andesite

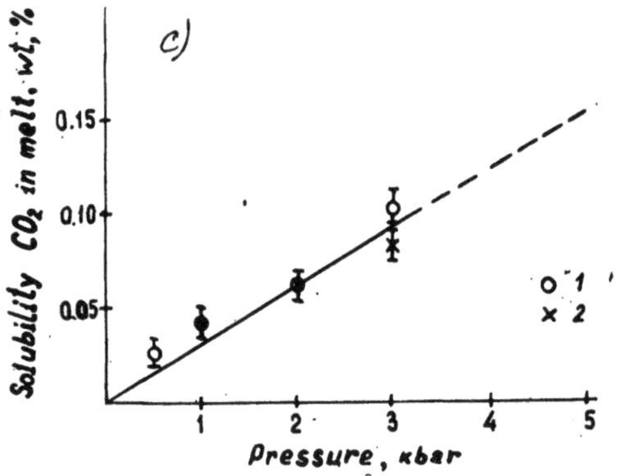

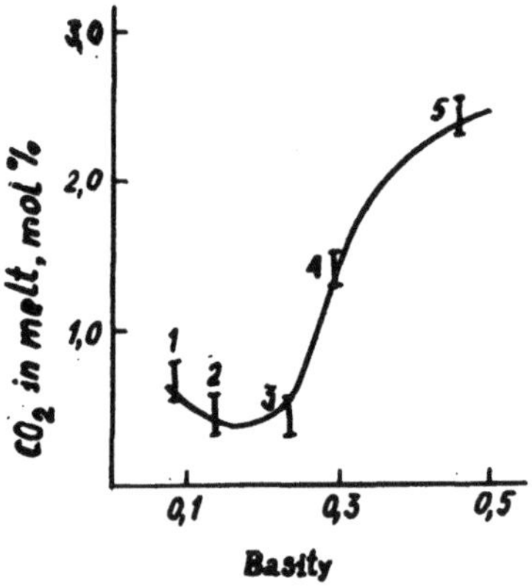

Fig. 1c: The dependence of CO solubility from pressure in melts of riolite (1) and dacite (2) under 1250 C.

Fig. 2: The dependence of CO solubility from basity of melts under pressure 3 kbar and temperature 1250 C. The composition of melts:
1- albite-SiO ; 2- granite;
3- dacite; 4- andesite;
5- basalt

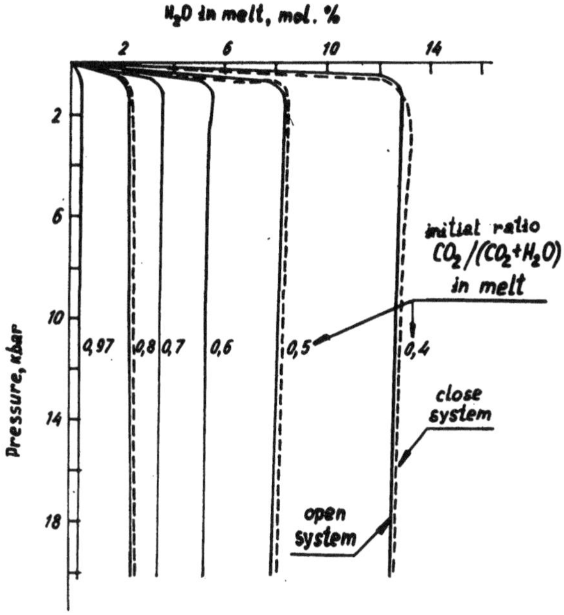

Fig. 3: Variation of concentration of H O in melt during the degassing.

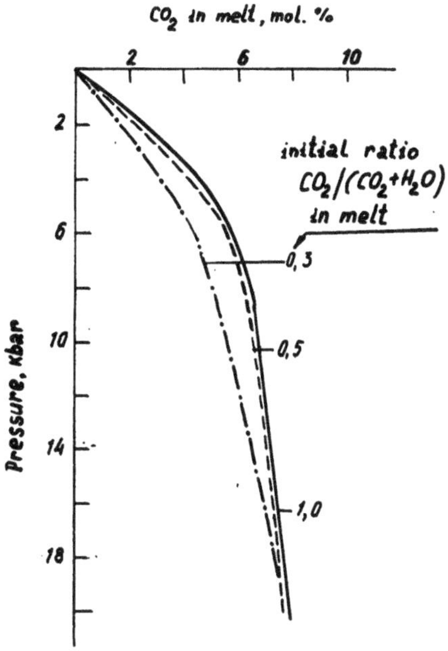

Fig. 4: Variation of concentration of CO in melt during the gas release.

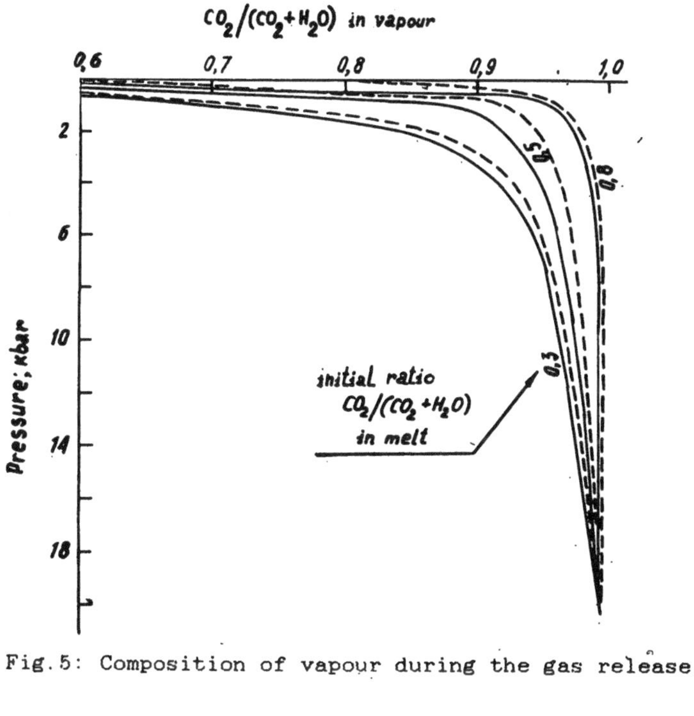

Fig. 5: Composition of vapour during the gas release.

dispergate in the Earth mantle.

Thus in the geochemistry of mantle processes the most important problem is the possibility of carbon solubility in crystals and silicate melts under high temperature and pressure.

We have investigated with this purpose the equilibria forsterite--graphite and basaltic melts-graphite under 1600-1800°C and 25-50 kbars.

The method of autoradiography was applied for the determination of carbon content and its distribution in silicates. We used radioactive carbon - 14 isotope.

The homogeneous distribution of carbon in the melt was established, This permitted to estimate its content equal 30 ppm under 25 kbar and temperature 1600°C [3]. The carbon distribution in forsterite crystals after experiments under 25 kbar and 50 kabr isn't homogeneous. That is possibly connected with solution mechanism of carbon in crystals. The next experimental and theoretical investigation are required to solve this problem.

Literature

[1] Eggler,D.H.:Role of CO_2 in melting processes in the mantle.Carnegie Inst.Washington Year Book, 72 (1973), 457-467.

[2] Freund,F, H.Kathrein,H.Wengeler, and R.Knobel: Carbon in solid solution in forsterite - a key to the untraktable nature of reduced carbon in terristrial and cosmogenic rocks.Geochim.Cosmochim.Acta. 44 (1980) 9, 1319-1333.

[3] Kadik,A.A.,S.N.Shilobreeva, et.al.:Experimental investigation of carbon in basaltic melt and forsterite crystals under high temperature and pressure. Geochimia, in press (inRussian).

[4] Mysen,B.O.:The role of volatiles in silicate melts: solubility of carbon dioxide and water in feldspar, pyroxene and feldspathoid melts to 30 kbar and 1625°C.Amer.J.Sci. 276 (1976) 8, 969-996.

[5] Shilobreeva S.N.:The solubility of carbon dioxide in melts of basic and acid composition and conditions its release from magmas in the volcanic regions.Ph.D.Thesis.Vernadsky Institute of Geochemistry and Analytical Chemistry of The USSR Academy of Sciences,Moscow 1984 (in Russian).

[6] Vinogradov,A.P.: The gas regime of the Earth.Science,Moscow 1964 (in Russian)

THE APPARATUS FOR INVESTIGATION OF ROCKS AND MINERALS UNDER THE QUICK-VARIABLE UNAXIAL STRESSES

R. Wiśniewski, A.J. Rostocki, J. Rogowski

Abstract

The paper presents the device for investigation of mechanical properties of rocks and minerals under condition of quick-variable compression stresses in an unaxial state of stress. The device provides rise of stress at constant rate within the range from MPa/s to few GPa/s. The size of samples can very up to 5x5x5 cm^3. The device has enabled to test mechanical characteristics of several mineral materials including coal, sandstone, marble, granite, etc. An influence of the stress rate on the increase of dynamic strenght has been noticed. It was possible to analyze various types of destruction process in various samples. The device seems to be usefull for any geological research where the quick- variable loads are essential.

Introduction

Investigations of different properties of rocks and minerals under conditions of quick-variable stresses are very important for geological research, seismic problems and for coal-mining industry /what is connected with a roof-scrump phenomenon/. The apparatus for testing of rocks in unaxial compressive stress states, at constant stress rate is presented below. The stress rate can be changed from about 1 MPa/s to about 10 GPa/s. These values are in middle, between those which can to be obtained using classical compression testing machines and those which can be obtained using shock-wave generating methods. Dimentions of rock specimens would not be bigger then 5x5x5 cm^3. In order to determine the influence of initial stress state on behaviour of investigated rocks under quick-variable load, the specimens can be affected to the initial static stress equal approximatelly 1/2 of maximum static value. Using described here apparatus there is possible to record stress-time, strain-time and stress-strain dependences. For recording of above mentioned dependences a common method with oscillograph equipment and photographic apparatus was used. Independent of oscilographic recording an electronic system of measurement of maximal values of investigated parameters was used. That enabled to increase an accuracy of measurement.

Description of the apparatus

A cross-section of a main part of the discussed apparatus is show in fig. 1a. The apparatus is based on the same principles of pressure pulse

Institute of Physics, Warsaw University of Technology

generation as it was described in [1]. Similary to [1], the simple and very convenient in use a pile-driver system as impulse source of energy was employed. Falling down loads (1) from definite heights /to 5m/ strike throught unelastic plate (2), the ridigly connected together piston (3) and cylinder (4) and compress the liquid in cylinder (7). In the paper [1], there is pointed out that if the relation $m = 0.5\ p_{max} A_{ef}/g$, (where m - means the faling down masses, g - an earth acceleration, p_{max} - an maximum generated pressure in a chamber, A_{ef} - an effective area of piston-cylinder system) will be fullfiled a generating pressure is approximatelly linear function of compresing time.

The pressure is measured by means of pressure gauges (6) and (8). The high pressure electrical leads (5) allows to make an measurement using not very complicated electronic system. In Fig. 1b there are shown two pressure- time dependences recorded using upper (6) and lower (8) pressure gauges.

b.

c.

a.

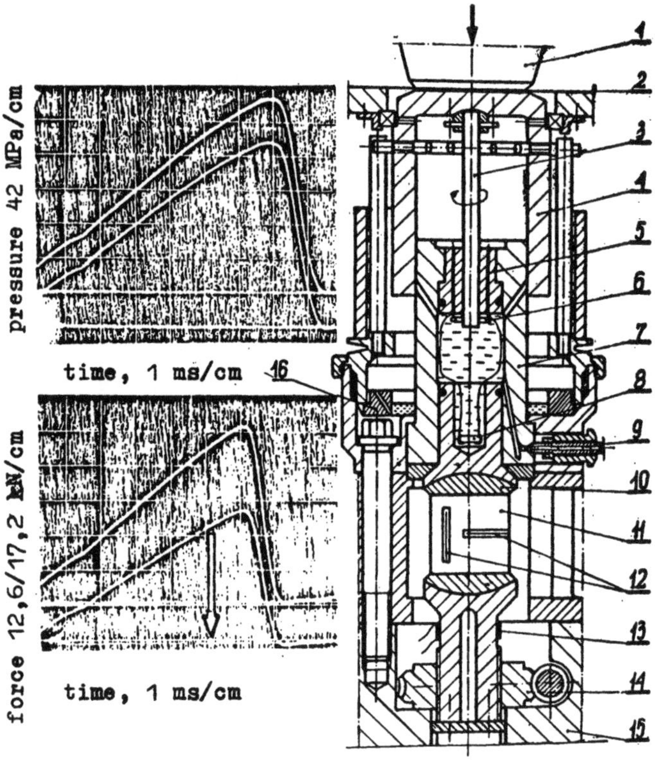

Fig. 1 a. Cross section of quick- variable stress generator /for meaning of symbols see text/; b. - pressure-time dependences recorded using (6) and (8) pressure tranducers, c.-force-time dependences recorded using (13) and (12) strain transducers

The way of compression is determined by the distance between a special hole system in the cylinder (7) and the hole system in the cylinder (4). This distance can be regulated from 0 to 80 mm so the pressure maximum generated in this apparatus may be regulated from zero up to maximum value / in this case up to 400 MPa /. The pressure is dropped when mentioned above hole system begin to cover each other.

Pressure in chamber (7) acting on a low inertial moveable upper anvil (10) generates in specimen (11) linear - dependence on time stress pulse. The reaction force of a down non - moveable anvil is measured by means of resistance strain gauges (13) bounded to the anvil. In Fig. 1c there is shown force - time dependence recorded by the strain gauge (13) and force - time dependence recorded by dynamometr during testing of apparatus. One can see the full likeness of it. The arrow on that figure shows in which moment the investigated specimens were destroyed. So, the observed during such investigations rapid fall of force, acting on being destroing specimens is adequate to their behaviour exactly and is not the specific property of apparatus. On investigated specimens the strain gauges, as it seems, have so many adventages / for example they have not practically any inertion / so they are the best among other gauges as inductive or capacitive ones.

The initial stresses in the specimen can be generated by means of a worm system (14) placed into the frame of the apparatus (15).

The repetition of generated stress pulse is easy thanks to additional system of refiling of the chamber (7) by capilar (9). When the compression process is finished, the velocity of falling down masses is approximatelly the same as the begining of the compression process. So, the very effective dissipation of kinetic energy of falling down masses then have to take place. In our solution of that problem a special pumping system of viscous liquid have been applied (16).

Experimental results

On the Fig. 2a,b,c, there are shown oscillograph registrations of compressive ultimate strength /CUS/ test for finegrained sandstone specimens: a - represents CUS for slow-variable stresses /S-VS/,i.e. $\dot{\sigma}$ = 1 MPa/s; b - represents dependence of stress on time for quick-variable stresses /Q-VS with $\dot{\sigma}$ = 5 GPa/s and c - stress dependence on time for Q-VS with $\dot{\sigma}$ = 8 GPa/s.
If we compare the those results one can see that CUS of sandstone specimens increases with increasing stress velocity and that dependence is not linear one. Dependences strain-stress for sanstone specimens shown on Fig. 2d,e,f,g are widely described in [2].Here we can say

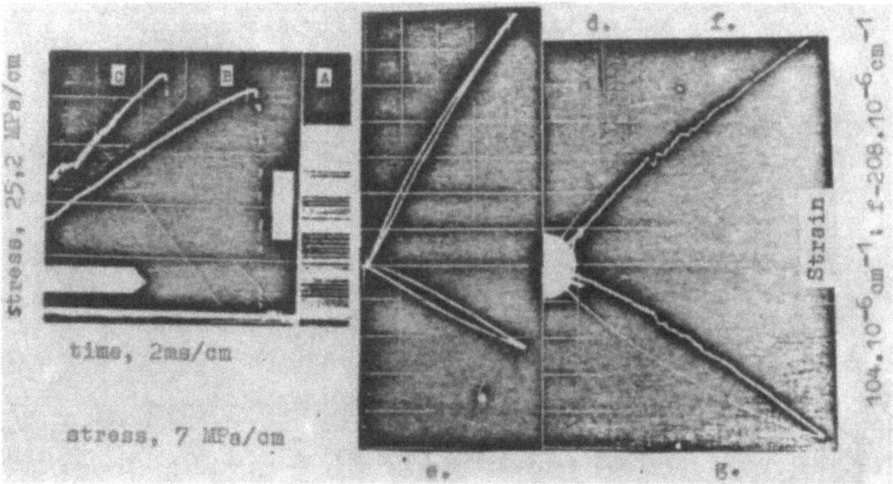

Fig. 2 Stress-time dependences for fine-grained sandstone under slow-variable load /A/ quick-variable load /B/ and /C/. Elastic hysteresis in coarce-grained sandstone under slow-variable load /d/,/e/ and quick-variable load /f/,/g/.

that changes of Young modulus and Poisson ratio were not so big as CUS at the some conditions of experiment.

On the Fig. 3 a,b,c,d there are shown oscillograph record of CUS test for coal specimens a,b - represents CUS data for Q-VS states, c - for Q-VS state with initial stresses and d - for S-VS states observed in eight specimens. As we can see here the disintegrating process is approximatelly four times longer than that for sandstone specimens. Behind this is the case of coal we can see some oscillations observed especially just before disintegrating process. Also some stage character of

Fig. 3 Stress-time dependence for coal under quick-variable load /a/ and /b/, under quick-variable load with initial stress /c/ and under slow-variable load /d/ for eight specimends

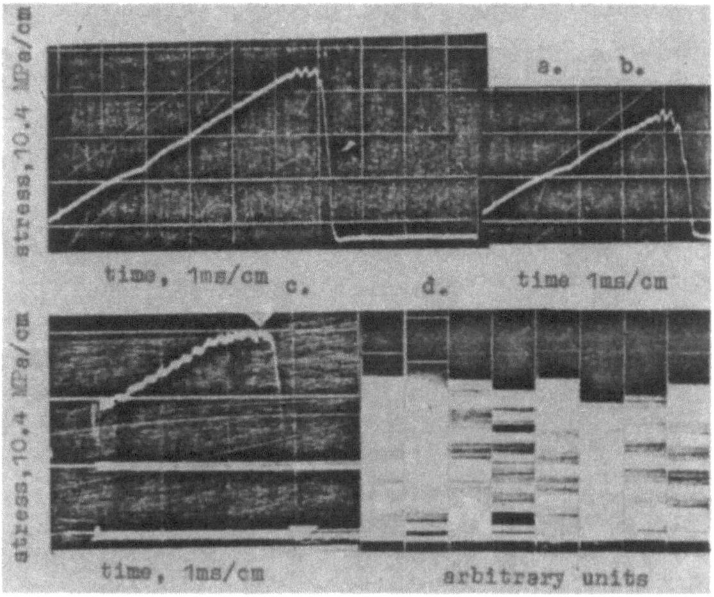

Fig.4 Stress-time dependence for granite /Michałowice, Poland/ under different stress rate and for marble /Sławniowice,Poland/. Arrow shows maximal stress /mean value/ under slow-variable load for granite

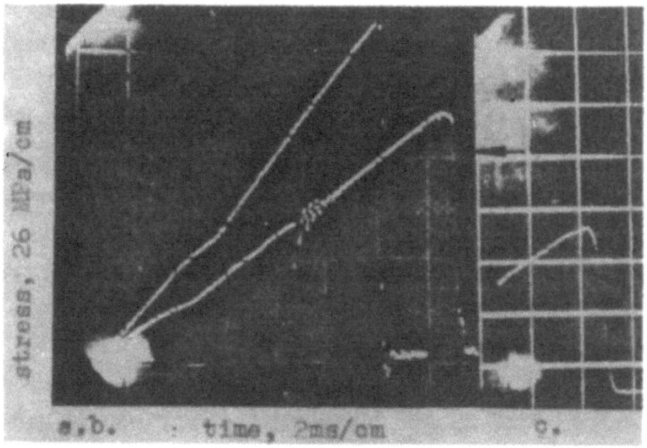

this process is observed too.
Because of laminar structure of coal and large anisotropic properties two directions of applied stresses were chosen. One of it was parallel to layers and second one perpendicular, so we have UCS// and UCS⊥.
We have obtained [3] : UCS// = 33,6 ± 10 MPa and UCS⊥= 35,4 ± 0,7 MPa for S-VS states and UCS//= 36,8 ± 1,9 MPa and UCS⊥ = 41,4 ± 3,1 MPa for Q-VS states.
Fig. 4 a,b,c represents results obtained for granite/a/,/b/and marble/c/ specimens. There are shown dependences for Q-VS states with $\dot{\sigma}$= 10 GPa/s and for $\dot{\sigma}$=18 GPa/s. The value of CUS obtained for S-VS condition is market by arrows /mean value/. One can see a strong influence of stress rate on CUS of granite specimen. It was also possible to notice that the time of disintegration for the marble specimens were much longer than those for granite ones.

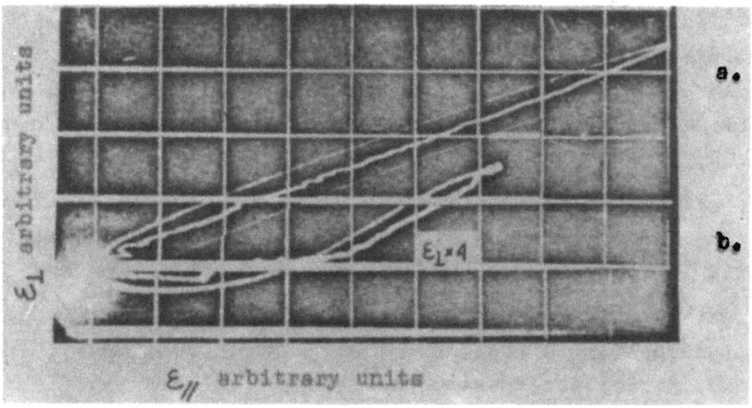

Fig. 5 Transverse strain $\varepsilon_\perp$ versus longitudinal strain $\varepsilon_{//}$ for granite /a/ and marble specimens

In Fig. 5 there are shown dependences of transverse strain $\varepsilon_{\perp}$ versus longitudinal strain $\varepsilon_{\parallel}$ for granite and marble specimens under Q-VS states. From these dependences we have obtained informations about dynamical values for Poisson's ratios. For marble specimens Poisson's ratio is equal about 0,3 and is nearly constant. For granite specimens Poisson's ratio is very small at the begining and increases with increasing longitudinal strain.
For further demonstration of our work in Fig. 6 a,b there are shown strain- stress dependences for marble /a/ and granite /b/ specimens under Q-VS states.

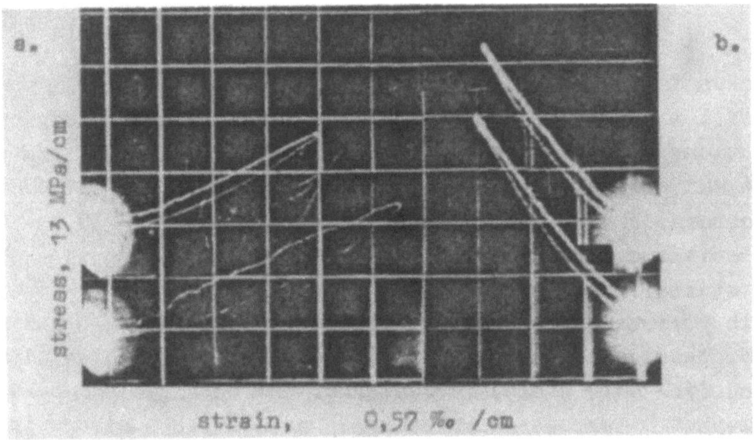

Fig. 6 Stress / σ / - strain / ε / dependences for marble /a/ and granite /b/ samples under different quick-variable stress states

References:

1. Wiśniewski R, Rostocki A, Bock W.
Generator of Triangular Pressure Pulse up to 500 MPa High Pressure Science and Technology, Pergamon Press, V.1 /1980/,176-180.

2. Wiśniewski R. Trombik M.
Apparatus for investigations of mechanic properties of coal and rocks under condition of quick-variable load at uniaxial stress state. /in polish/, Przegląd Górniczy Nr 7-8, Tom XL /1984/,254-6

3. Wiśniewski R, Trombik M.
Investigations of coal specimens under conditions of dynamic load. Raport for Research Institute of coalmining, Katowice, Poland /1982/, Unpublished.

HIGH PRESSURE PROPERTIES OF THE ALKALINE EARTH METALS

H. Olijnyk[1] and W.B. Holzapfel[2]

Abstract: The experimental results on structural phase transitions, pressure-volume relations and superconductivity for the alkaline earth metals under high pressure are reviewed. The relation between these material properties and electronic structure is discussed.

Introduction

Recent high pressure investigations on the alkaline earth metals (AEM) have revealed an unforeseen manifold of structural phase transitions in this group of elements [1] . At the same time theoretical studies have focused on the relation between electronic structure and phase stability. It is now recognised that many of these phase transitions are primarily driven by the degree of s--d transfer in the conduction bands [2,3] and that these metals become divalent d-band metals at sufficiently high pressure [4] . However, the details of this transition for the individual metals are not extensively investigated and the properties of these metals above the s--d transition are as yet largely unknown. A comparison of available high pressure data for the AEM, such as crystal structures, P-V-relations and superconducting properties may therefore stimulate further experimental and theoretical investigations.

Structural behaviour -- Experimental results

A summary of the presently known phase transitions in the AEM at room temperature and high pressure [1] is given in Table I, together with corresponding results for the divalent lanthanides Yb and Eu [5]. Not all high pressure structures for Ba and Sr have as yet been identified. For Ba-IV Takemura et al [6] propose a bct lattice (2 atoms per unit cell, c/a = 1.27). However, Ba-IV appears to remain an unsolved problem because this structural assignment cannot account for all observed diffraction lines. On the other hand our x-ray diffraction data [1] indicate a hcp lattice for Ba-V, which becomes stable around 46 GPa with increasing pressure. This structural assignment is supported by the good agreement between observed and calculated d-values, and gives also reasonable values for the atomic volume (Table II).

[1]Bayerisches Forschungsinstitut für experimentelle Geochemie und Geophysik, Universität Bayreuth, Postfach 101251, 8580 Bayreuth, FRG.

[2]Fachbereich Physik, Universität - GH - Paderborn, Postfach 1621, 4790 Paderborn, FRG.

Be:	hcp $\xrightarrow{11}$ orthor.
Mg:	hcp $\xrightarrow{50}$ bcc
Ca:	fcc $\xrightarrow{19.5}$ bcc $\xrightarrow{32}$ sc
Sr:	fcc $\xrightarrow{3.5}$ bcc $\xrightarrow{26}$ Sr-III $\xrightarrow{36}$ Sr-IV $\xrightarrow{46}$ "Ba-IV"
Ba:	bcc $\xrightarrow{5.5}$ hcp $\xrightarrow{7.5}$ Ba-III $\xrightarrow{12}$ Ba-IV $\xrightarrow{46}$ Ba-V
Eu:	bcc $\xrightarrow{13}$ hcp $\xrightarrow{32}$ Eu-III
Yb:	fcc $\xrightarrow{4}$ bcc $\xrightarrow{35}$ hcp

Table I. Phase transitions and transition pressures for the AEM [1] and Eu and Yb [5].

d_{obs}	I_{obs}	hkl	d_{calc}
2.780	6	-	-
2.669	17	100	2.669
2.427	2	002	2.427
2.338	100	101	2.338
2.102	2	-	-
1.793	8	102	1.793
1.540	10	110	1.540
1.407	2	-	-
1.381	4	103	1.381
1.333	1	200	1.334
1.300	8	112	1.300
1.290	6	201	1.286
1.254	7	-	-
1.169	2	202	1.169
1.029	1	203	1.028
1.006	1	210	1.009
0.988	2	211	0.988
0.951	1	114	0.952
0.931	1	212	0.931

Table II. Observed d-values and relative intensities for Ba-V at 54 GPa and calculated d-values for hcp indexing with a=3.082 Å and c=4.840 Å, c/a = 1.57.

Structural behaviour -- Comparison with theoretical results

A comparison of the various metals becomes more transparent if one scales the atomic or Wigner - Seitz radii (R_A) by the ionic radii (R_I) [7]. In this representation it becomes evident that the observed structural sequence for the heavier AEM (Ca, Sr, Ba) may be summarized as fcc--bcc--complex--Ba-IV--hcp. Corresponding phase transitions for the individual metals occur at almost the same critical radius ratios (Fig.1). The sequence also applies in part to the divalent lanthanides Yb and Eu, however, at higher compression a deviatoric behaviour is observed for Eu. Be and Mg cannot be arranged into the systematic of their heavier homologues and show no similarities to each other.
Calculations of the structural stability in one electron band theory indicate very clearly that for the heavier AEM most of these transitions are driven by s--d transfer within the conduction bands [2]. Within this context the present representation shows that the relative contributions of ns,(n-1)d electrons to the conduction bands may depend more or less on the radius ratio R_A/R_I. The appearance of the bcc structure in Mg at high pressure is also driven by d-electron effects, but the nature of the d-band in this case is sufficiently different, that the correlation is less significant [3]. The lack of d-bands near the Fermi-level in Be explains easily within this framework the different structural behaviour of this light divalent metal [8].
The s--d transfer is terminated with the total depletion of the s-bands as the bottom of the s-band moves above the Fermi-level, resulting in a new class of divalent d-band metals. For Ba this occurs at a relative volume of 0.4 - 0.5V_0 [9], which corresponds to the stability range of Ba-IV. The reappearance of a close packed structure, hcp, in Ba-V

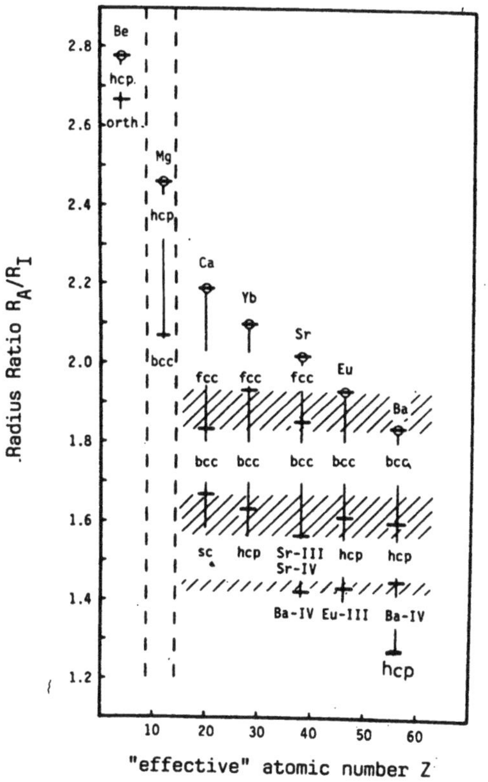

Fig.1: Systematics of phase transitions for the AEM and Eu and Yb.

and the observed stiffening in the P-V relation in this compression range (Fig.2) may indicate the increasing dominance of ion-core repulsion [10].

Pressure - volume relations

Room temperature isotherms [11] are compared in Fig.2, which shows the dependence of the relative volume V/V_0 on the reduced pressure P/B_0.B_0 denotes the initial bulk modulus. In this representation the compression curves with the same B_0' ,which is a measure for the initial rate of increase of the bulk modulus B, coincide. The heavier metals tend to follow the same curve for $V \gtrsim 0.6V_0$, and are more compressible than Mg. For the heavier metals the s--d transfer leads to a decrease in the electronic contribution to the total pressure, as compared to the case without s--d transfer, and thus gives rise to the observed softening in the equation of state [12]. The observed stiffening in the compression curve of Ba for $V \lesssim 0.5V_0$ may indicate the loss of the softening mechanism by termination of s--d transfer and the beginning dominance of core-core repulsion. In this treatment of the data Yb becomes rather incompressible for $V \lesssim 0.5V_0$,which may suggest an anomalous behaviour of Yb at high pressures, as compared to the heavier AEM.

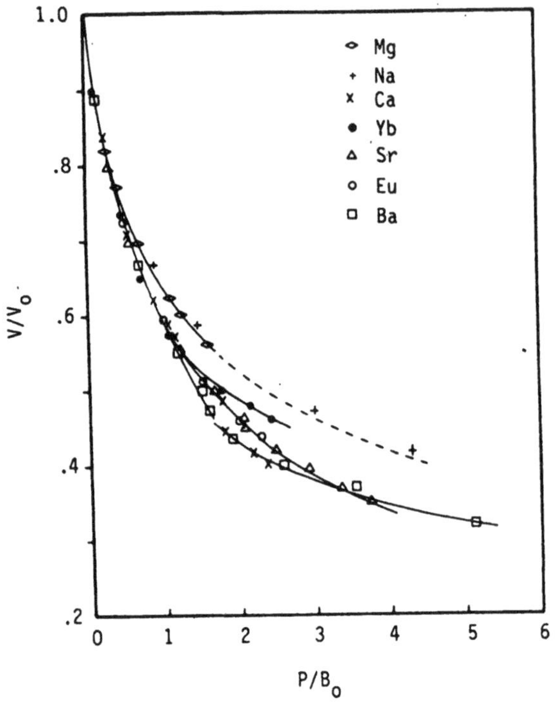

Fig.2: Reduced pressure-volume curves for the AEM [11] and Yb and Eu [5].

Superconductivity

Under pressure Sr and Ba become superconductors [13,14]. For both metals superconductivity even starts in the bcc - phase. From Fig.3 it can be noted that the onset of superconductivity occurs at slightly different values of the radius ratio R_A/R_I for Ba and Sr. This may suggest that these critical radius ratios shift to smaller values if one proceeds from heavier to lighter AEM. For Ca no superconductivity above 1.2 K has been detected in the pressure range up to 50 GPa (R_A/R_I = 1.56) [15]. A strong increase of T_c is observed in the range of considerable s--d transfer, however, the data for Ba indicate a decrease of T_c in the d-band limit. Yb and Eu do not show superconductivity up to the highest pressures of 56 GPa and 40 GPa, respectively [14,15].

Conclusion

Within the present results the "complex" region in the structural sequence of the heavier AEM remains to be solved. It has been suggested that soft phonon modes, caused by intervals of especially rapid s--d transfer may be responsible for the "complex" region [4]. In this context the superconducting properties of Ca might be quite interesting. Another difficulty concerns the treatment of Yb and Eu as divalent metals. There is experimental evidence that in the considered pressure range both metals may undergo a transition into an intermediate valence state [5]. The observed deviations from "regular" divalent behaviour may be related to this transition.

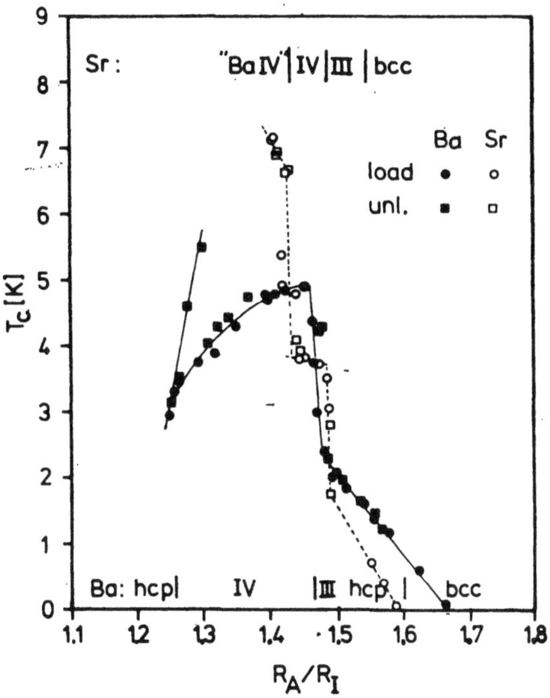

Fig.3: Superconductivity in Sr and Ba [15] as a function of radius ratio. Note that Ba-V can be recovered at lower pressures.

References

1 Olijnyk, H. and Holzapfel, W.B.: J. de Phys. C8 (1984) 157
2 Skriver, H.: Phys. Rev. Lett. 49 (1982) 1768
3 McMahan, A.K. and Moriarty, J.A.: Phys. Rev. B37 (1983) 3235
4 McMahan, A.K., Physica 139 & 140B (1986) 31
5 Takemura, K. and Syassen, K.: J. Phys. F15 (1985) 543
6 Takemura, K., Yagi, T. and Syassen, K.: 27th Japanese High Pressure Research Groups Meeting (1986)
7 Olijnyk, H. and Holzapfel, W.B.: Phys. Lett. 99A (1983) 381
8 Chou, M.Y., Lam, P.K. and Cohen M.L.: Phys. Rev. B28 (1983) 4179
9 Duthie, J.C. and Pettifor, D.G.: Sol. St. Comm. 27 (1978) 613
10 McMahan, A.K.: Phys. Rev. B29 (1984) 5982
11 Olijnyk, H.: Ph.D. Thesis, Universität Paderborn (1985)
12 Ross, M.: Rep. Prog. Phys. 48 (1985) 1
13 Wittig, J. and Matthias, B.T.: Phys. Rev. Lett. 22 (1969) 634
14 Bundy, F.B. and Dunn, K.J. in: Physics Of Solids Under Pressure, eds. J.S. Schilling & R.N. Shelton (1981), North Holland, Amsterdam, p.401
15 Bireckoven, B.: Ph.D. Thesis, Jül-report 2129 (1987)

EFFECT OF PRESSURE ON THE STRUCTURE OF THE MOLYBDENUM TRIOXIDE

*M. Malinowski, K. Łukaszewicz
**L. Kihlborg, S. Åsbrink

MoO_3 is a member of a whole family of molybdenum oxides investigated in detail by Magneli and Kihlborg [1].

The crystal structure of MoO_3 was determined by Bräkken [2] and Wooster [3], and later reinvestigated by Anderson [4] and Kihlborg [5] in space group Pbnm, $a = 3.9628(7)$ Å, $b = 13.855(3)$ Å, $c = 3.6964(6)$ Å, Z = 4. The crystal structure of molybdenum trioxide may be described as built up of MoO_6 octahedra joined by common edges and corners, in the a direction by corners only. The octahedra are strongly deformed with four shorter Mo-O bonds and remaining two bonds considerably longer as shown in Fig. 1. It was interesting to check how the anisotropy of bond lenghts corresponds with the anisotropy of compressibility.

Experimental

A diamond-anvil high-pressure cell of a new design [6] was used in this experiment. The new cell allows two kinds of diffraction geometry (transmission and transverse) to be available simultaneously, and due to that at least 85% of the Ewald sphere is measurable. The MoO_3 single crystal was of the size of about 150x100x50 μm^3. The sample and CsCl single crystal as internal pressure standard [7] were placed in a cylindrically drilled hole of 0.21 mm in diameter in a gasket 15x15x0.35 mm in size. The hardened Inconel alloy X-750 was applied as a gasket material. The pressure fluid was methanol-ethanol in a ratio 4:1 [8]. The high-pressure cell was mounted on a Philips PW 1100 four-circle diffractometer, and MoKα radiation was obtained from an X-ray fine-focus tube operated at 60 kV and 35 mA, with a graphite monochromator. The temperature inside the cell oscillated between 296 and 297 K, which is within the limits of accuracy of lattice parameter measurements. For the determination of the unit-cell parameters at different pressures 11 independent reflections were used. To reduce the zero-point errors of 2Θ and ω circles, each reflection was measured several times at +2Θ, +ω and -2Θ, -ω and the Bragg angles were averaged. The accuracy in the pressure determination was better than 0.2 kbar.

---

* Department of Crystallography, Institute for Low-Temperature and Structure Research, Polish Academy of Sciences, 50-950 Wrocław, POLAND

**Inorganic Chemistry Department, Arrhenius Laboratory, University of Stockholm, S-10691 Stockholm, SWEDEN

Results

The lattice parameters of MoO_3, unit-cell volume and axial ratio *a/c* versus pressure are listed in Table 1 and presented on Fig. 2.

Tab. 1. The lattice parameters, unit-cell volume and ratio *a/c* versus pressure.

P Kbar	*a* Å	*b* Å	*c* Å	V Å³	*a/c*
0.001	3.9667(8)	14.04(9)	3.7029(9)	206.3(13)	1.0712(3)
1.0	3.9642(8)	13.99(10)	3.7037(10)	205.5(15)	1.0703(4)
4.9	3.9605(7)	13.80(7)	3.7014(7)	202.2(10)	1.0700(3)
9.3	3.9538(5)	13.59(5)	3.7014(6)	198.9(7)	1.0682(2)
15.4	3.9456(5)	13.45(5)	3.7011(6)	196.4(7)	1.0661(2)
20.4	3.9391(8)	13.39(8)	3.6993(10)	195.1(11)	1.0648(4)
24.8	3.9331(8)	13.33(8)	3.6991(10)	193.9(11)	1.0633(4)
28.5	3.9279(10)	13.25(9)	3.6999(12)	192.6(14)	1.0616(4)
34.2	3.9216(6)	13.18(5)	3.6982(7)	191.2(7)	1.0604(3)
38.8	3.9169(8)	13.10(7)	3.6990(9)	189.7(10)	1.0589(3)
44.3	3.9100(7)	13.05(7)	3.6971(9)	188.6(10)	1.0576(3)
46.1	3.9068(10)	13.01(10)	3.6964(12)	187.8(14)	1.0569(4)
48.1	3.9060(7)	13.02(6)	3.6969(9)	188.0(9)	1.0566(3)
51.8	3.9005(10)	12.90(7)	3.6963(6)	186.0(10)	1.0552(3)
57.5	3.8918(9)	12.84(7)	3.6965(8)	184.7(10)	1.0528(3)
61.1	3.8882(9)	12.81(9)	3.6945(11)	184.0(13)	1.0524(4)
68.3	3.8804(12)	12.67(10)	3.6939(12)	181.6(14)	1.0505(5)
71.0	3.8752(9)	12.64(9)	3.6951(10)	181.0(13)	1.0487(4)
74.1	3.8714(7)	12.62(7)	3.6930(9)	180.4(11)	1.0483(3)

The cell parameters *a*, *c* and ratio *a/c* show a strictly linear decrease for a pressure increasing from normal up to about 75 kbar, whereas the decrease of *b* axis and unit-cell volume in the pressure range 0.001 - 15 kbar is clearly nonlinear. For the best fit to these data and mean compressibility calculation [9], polynomial regression analysis was applied, using the POLREG program [10].

The mean compressibility of lattice parameters *a*, *c* in the whole range and lattice parameter *b* as well as unit-cell volume in the range 20.4 - 74.1 kbar are following:

$$\beta_a \ (0.001 - 74.1 \text{ kbar}) = 3.23(2) \times 10^{-3} (\text{kbar})^{-1}$$
$$\beta_c \ (0.001 - 74.1 \text{ kbar}) = 3.35(5) \times 10^{-4} (\text{kbar})^{-1}$$
$$\beta_b \ (20.4 - 74.1 \text{ kbar}) = 1.11(15) \times 10^{-2} (\text{kbar})^{-1}$$
$$\beta_V \ (20.4 - 74.1 \text{ kbar}) = 1.46(19) \times 10^{-2} (\text{kbar})^{-1}$$

The compressibilities of *b* and V in the range 0.001 - 20.4 kbar are presented in Table 2.

Table 2. The compressibility of lattice parameter *b* and unit-cell volume.

Pressure range kbar	β_b x 10^{-2} (kbar)$^{-1}$	β_V x 10^{-2} (kbar)$^{-1}$
0.001 - 1	4.32(16)	4.86(21)
1 - 4.9	4.14(14)	4.67(18)
4.9 - 9.3	3.40(16)	3.86(18)
9.3 - 15.4	2.53(15)	2.91(19)
15.4 - 20.4	1.28(17)	1.52(18)

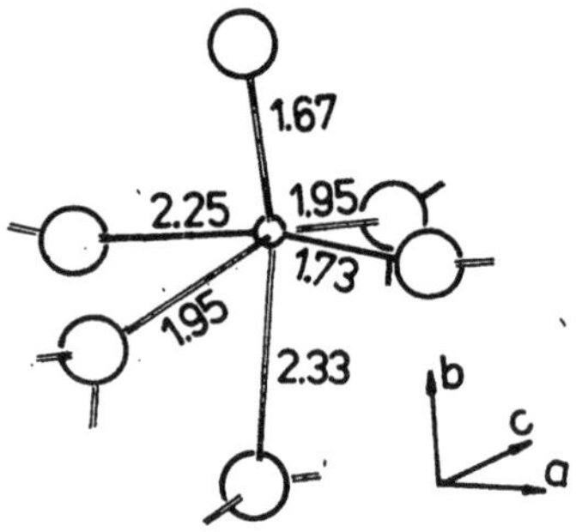

Fig. 1. The coordination of oxygen atoms (large spheres) around the molybdenum atom (small sphere).

The differences between β_a, β_b and β_c clearly indicate that the compressibility of MO_3 depends mainly on the Mo - O bond lengths and shape of the MoO_6 octahedra and to smaller extend on the way the octahedra are linked together. The compression along *c*-axis is determined mainly by the two short bonds Mo - O with length equal to 1.95 Å (see Fig. 1). The weaker bond Mo - O with distance equal 2.25 Å is nearly parallel to *a* axis. The compressibility of this axis is therefore higher then *c*. The bond distance Mo - O with length equal to 2.33 Å is responsible for greatest and nonlinear compressibility along *b* axis.

From anisotropy of compressibility it follows that the strongly deformed MoO_6 octahedra at high pressure tend to become more regular. This conjecture will be checked after completing the crystal structure determination of MoO_3 uder high pressure which is in progress.

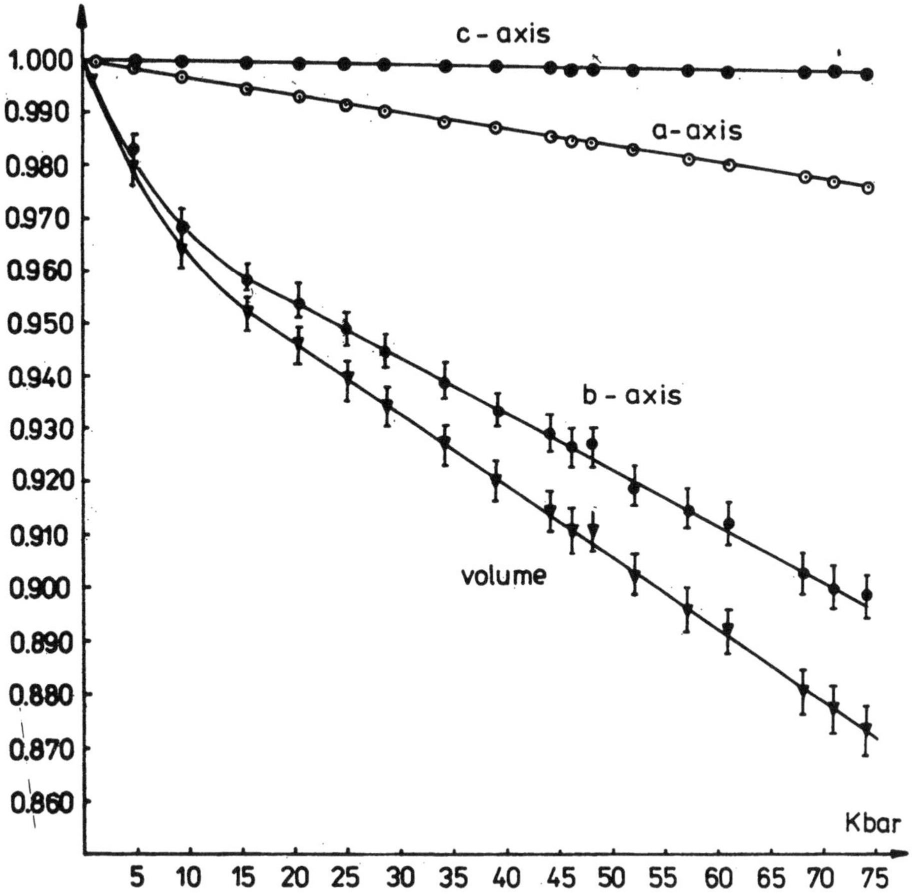

Fig. 2. Lattice parameters and unit-cell volume of MoO_3 as functions of pressure at room temperature. For parameters *a* and *c* e.s.d.'s are equal to the radii of the circles.

[1] A. Magneli and L. Kihlborg, Acta Chem. Scand. 9 (1955) 471
[2] H. Bräkken, Z. krist. 78 (1931) 484
[3] N. Wooster, Z. krist. 80 (1931) 504
[4] G. Andersson and A. Magneli, Acta Chem. Scand 4 (1950) 793
[5] L. Kihlborg, Arkiv Kemi 34 (1963) 357
[6] M. Malinowski, Submitted to J. Appl. Cryst. (1987)
[7] D.L. Decker, J. Appl. Phys. 42 (1971) 3241
[8] G.J. Piermarini, S. Block and J.D. Barnett, J. Appl. Phys. 44 (1973) 5377
[9] R.M. Hazen, L. Finger, Comparative Crystal Chemistry (1982) New York: John Wiley
[10] K.J. Johnson, Numerical Methods in Chemistry. New York, Basel: Marcel Dekker.

THERMODYNAMIC PROPERTIES AND PLASMA PHASE TRANSITION OF HYDROGEN AND RARE GAS PLASMAS

A. Förster, W. Ebeling, W. Richert [1)]

1. Introduction

Hydrogen as well as noble gas plasmas show unique features with respect to their thermodynamic properties. In particular, this is due to their relatively high ionisation energy. A direct consequence is the possibility of two critical points which are well separated. The data of the first critical point which is connected with the gas-liquid transition are well known. They are located at about 10^1 .... 10^2 K and 10^5 .... 10^7 Pa. The second critical point is expected to be located in the strongly coupled plasma region at about 10^4 .... 10^5 K and 10^9 .... 10^{11} Pa. The corresponding phase transition is of first order and accompanied by an abrupt change in mass density and in the degree of ionisation. This plasma phase transition is so to say a high-temperature modification of the well-known insulator-metal transition.

The existence of a phase diagram of this type was discussed for the first time in 1943 by Landau and Zeldovich /1/. First calculations of the plasma phase transition are due to Norman and Starostin /2,3/ and to Ebeling and Sändig /4,5/. In spite of quantitative corrections with regard to the exact location of the critical point and the coexistence line, there are several quantitative statements which are independent of the nature of the approximations:

(i) In the p-T diagram, the pressure region 10^0 .... 10^3 GPa includes a line of a first order phase transition connected with delocalisation of electrons.

(ii) At lower temperatures this transition corresponds to a insulator-metal transition whereas at higher temperatures $T \gtrsim 10^3$ K the plasma phase transition occures.

(iii) The transition has the tendency to occur at lower pressure with increasing temperature.

(iv) Above a critical temperature $T_{C,2} \sim$ 10 000 .... 20 000 K, instead of a sharp transition, a soft transition to full ionisation is observed with increasing pressure.

For the purpose of this paper we restrict ourselves to the particular case of xenon. Calculations for hydrogen and other plasmas yield qualitatively similar results and have been performed /6-11/ within the same scheme. Xenon is the leading candidate for the plasma phase transi-

1) Humboldt-Universität zu Berlin, Sektion Physik, Invalidenstrasse 42. Berlin, DDR-1040, German Democratic Republic

tion among the rare gases. Static compression experiments at lower temperature indicate, that the insulator-metal transition occures at pressures above 100 GPa. Ross and McMahan /13/ presented calculations of this transition and predict that the transition occur at a pressure of about 130 GPa. The exact phase diagram of xenon at higher temperatures is yet not clear and is the subject of many discussions.

2. Calculation of the Thermodynamic Functions

As the basic thermodynamic function we choose the free energy density $f = F/V$. Once an appropriate expression for the free energy density is obtained the other thermodynamic quantities, as the chemical potential μ or the pressure p, could be calculated numerically.

In our model the plasma consists of atoms as well as of electrons and single positively charged ions. Under high pressure there are, in addition to the electrical interaction, strong effects originating from the spatial extension of the plasma particles. Furthermore, effects of polarisation and Van-der-Waals attraction must be taken into accound. So the free energy density is constructed as follows:

$$f = f_{id} + f_{Coul} + f_{hc} + f_{VdW} + f_{pol}$$

In the ideal part f_{id} the atoms and ions are treated as Boltzmann particles whereas the electrons are considered as obeying the Fermi distribution with an arbitrary degree of degeneration. The contribution of the electrical interaction, f_{Coul}, is approximated by a Padé formula. This formula /6,7,10,11/ is an elaborated mathematical construction which show a correct behaviour in the limit cases of high and low temperature and density. In these cases it yields the known analytical expressions for the free energy density of a nonideal two-component plasma. In the median region the formula has been fitted to numerical data. Effects of excluded volume are considered by a hard-core ansatz, f_{hc}. Modelling the plasma particles as hard spheres we take advantage of the general formula for a mixture of different species of hard spheres reported by Mansoori et al. /14/. According to our problem we should deal with a system of three different species: atoms with the radius R_a, ions with the radius R_i and electrons with the radius R_e equal to zero. For the atoms we chose the temperature dependent radius proposed by Barker and Henderson /15/. The radii of the rare gas ions are not very well investigated. On the basis of scattering data we made the choise $R_i = 0.9\ R_a$. Van-der-Waals attraction and polarisation effects are taken into consideration using a pair approximation by means of the second virial coefficient. The attractive interaction of the atoms is modelled by a Lennard-Jones potential /16/. A convenient approximation for the polarisation potential is the totally screened Debye potential obtained by Redmer and Röpke /17/.

3. Results and Discussion

Let us first of all underline that the given approximation for the free energy is well elaborated with respect to the electrical interaction but is only a rough approximation with respect to the atom-atom and atom-charge interaction.

The ionisation equilibrium is given by the condition of the chemical equilibrium: $\mu_a = (\partial f/\partial n_a)_{n_e,T} = \mu_{ei} = (\partial f/\partial n_i)_{n_a,T} + (\partial f/\partial n_e)_{n_a,T} = \mu$. The phase transition is obtained by considering the change of the pressure as a function of the density, i.e. by considering the criterion of stability $(\partial p/\partial V)_T < 0$.

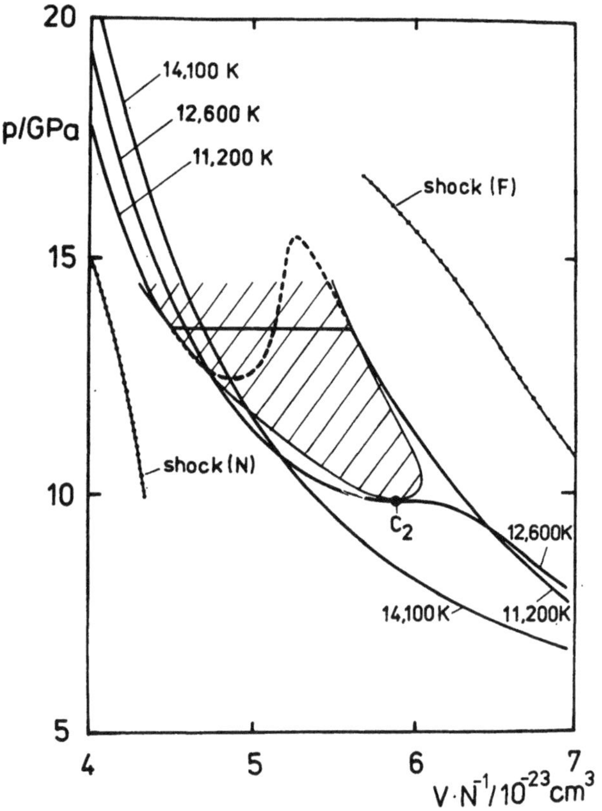

Figure 1

Isotherms (thick lines) for three different temperatures including the critical, $T_{c,2}$ = 12 600 K, on the pressure-volume plane. The thin lines represent the border of the coexistence area (area hatched). Within this area the equilibrium pressure remains constant. For undercritical temperatures, $T < T_{c,2}$, there is a plasma with a relatively low degree of ionisation right of the coexistence area, a plasma with a relatively high degree of ionisation left of the coexistence area, whereas within this area we expect the coexistence of two fluid phases differing in mass density and in the degree of ionisation. C_2 - critical point of the plasma phase transition. The dotted lines labelled "shock" refer to shock-wave experiments (Hugoniots) done in Livermore (N) and Moscow (F) (see text). The temperature increases along these curves with increasing pressure.

In Fig. 1 isotherms corresponding to different temperatures are given on the p-V plane. It can be seen that below a critical temperature $T_{C,2}$ the typical Van-der-Waals loop arises indicating the occurence of a thermodynamic instability. For overcritical temperatures the pressure rises monotonically with increasing density. The critical point of xenon is located at: temperature - $T_{C,2}$ = 12 600 K, pressure - $p_{C,2}$ = 9.9 GPa, heavy particle density - $n_{C,2} = 1.7 \cdot 10^{22}$ cm^{-3}, mass density -

$\varrho_{C,2}$ = 3.7 g cm^{-3}, free electron density - $(n_e)_{C,2}$ = $5.6 \cdot 10^{21}$ cm^{-3}, degree of ionisation - $\alpha_{C,2}$ = 0.33. By means of a Maxwell construction based on both the data for the pressure p and the chemical potential μ we obtain the equilibrium pressure for the coexistence area and the border of this area.

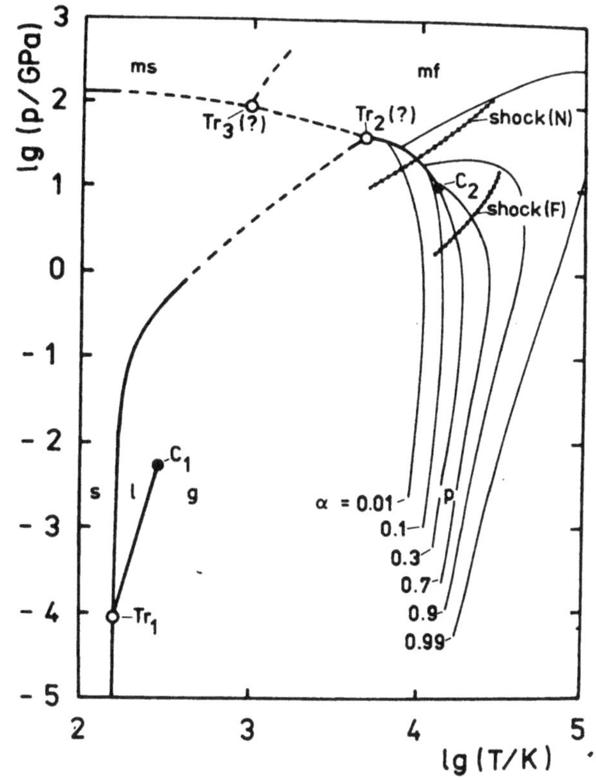

Figure 2

The fluid phases of xenon on the pressure-temperature plane. Thick lines represent the first order phase transitions. Tr_1 - triple point between the solid (s), liquid (l) and gaseous (g) phase. C_1 - critical point of the gas-liquid transition. For temperatures $T \gtrsim 10^3$ K the plasma (p) occurs. C_2 - critical point of the plasma phase transition. Dashed lines show a possible supplement of the phase diagram: Tr_2 - a triple point between the dielectric solid, the dielectric fluid and the metallic fluid (mf) phase (the dense and strongly ionised plasma). Tr_3 - a triple point between the metallic solid (ms), the metallic fluid and the dielectric solid phase. The thin lines correspond to a constant degree of ionisation α. The metallisation at lower temperatures is given according to Ross and McMahan /13/. The dotted lines correspond to shock wave experiments (see Fig. 1 and text).

Figs. 2 and 3 present the plasma phase transition together with the gas-liquid transition. It can be observed that around the second critical point C_2 the mass density and the degree of ionisation change drastically. On the p-T plane (Fig. 2) this is illustrated by lines of a constant degree of ionisation. The phase diagram in the temperature region 10^3 K $\lesssim$ T $\lesssim 10^4$ K is yet not clear. A possible situation is shown in Fig. 2: the melting line and the coexistence line of the plasma phase transition intersect in the triple point Tr_2. Fig. 4 gives the corresponding picture of the phase areas fitted together on the n-T plane.

The shock-wave data shown in Figs. 2 and 3 are from different experiments starting from the liquid (N) and from the gaseous (F) state. The data (N) from Nellis et al. /18,19/ cross the instability region whereas the data from Fortov et al. /20,21/ are outside what correspond to the initially lower density. The thermodynamic data on xenon

Figure 3

The fluid phases of xenon on the density-temperature plane. Denotation of the areas: s-g - sublimation area, l-g - gas-liquid coexistence area, s-f - melting area, s-solid, l-liquid, g-gas, f-fluid, p-plasma, PPT-coexistence area of the plasma phase transition, mf-metallic fluid, ms-metallic solid. Isobars crossing particulary the coexistence area are indicated by thin lines (units: lg(p/GPa)). The arrow at high density refers to the density of the transition to the metallic state at T = 0 K according to Ross and Mc Mahan /13/. The dotted lines correspond to shock-wave experiments (see Fig. 2 and text).

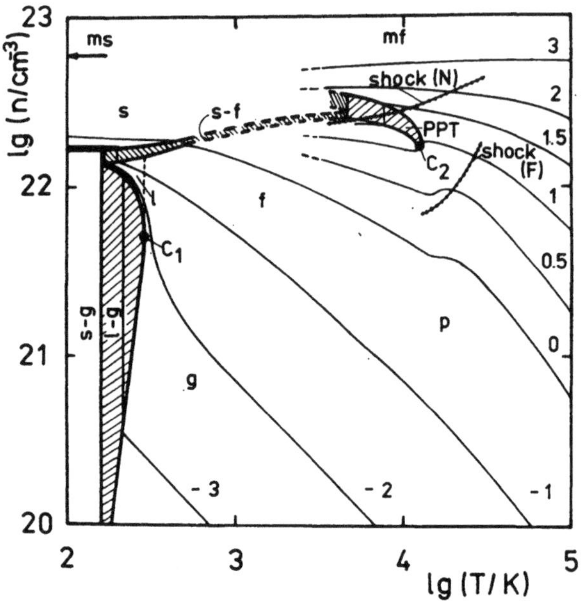

published so far give no clear reference to a phase transition. Careful measurements of the electrical conductivity or of the reflectivity should show whether there is really such a plasma phase transition or not.

In order to get an idea about the impact of the different interactions on the location of the critical point we have made several runs differing in consideration of the interactions /9/. We have found that Van-der-Waals attraction has a negligible impact on the location of the critical point. This is due to the weakness of the attracting interatomic potential. The impact of polarisation is considerable. However, the most dominant contribution comes from the hard-core interaction. We have found that by manipulating the atomic and ionic radii in a range of about 25 p.c. the critical point can be shifted to every pressure between 10^0 and 10^2 GPa. The critical temperature will still remain in the interval between 10 000 and 20 000 K.

Taking the existence of the plasma phase transition for granted, we conclude that its exact localisation will be still a subject of further experimental and theoretical investigations, the letter based on a more elaborate model for the fluid.

References

1. L.D. Landau, Ya.B. Zeldovitch, Acta Physico-Chimica U.R.S.S. **18**(1943)194
2. G.E. Norman, A.N. Starostin, Teplofiz. Vys. Temp. **6**(1968)410
3. G.E. Norman, A.N. Starostin, Teplofiz. Vys. Temp. **8**(1970)413
4. W. Ebeling, phys. stat. sol. (b) **46**(1971)243
5. W. Ebeling, R. Sändig, Ann. Physik (Leipzig) **28**(1973)289
6. W. Ebeling, W. Richert, phys. stat. sol. (b) **128**(1985)467
7. W. Ebeling, W. Richert, Phys. Lett. **A108**(1985)80
8. W. Ebeling, A. Förster, H. Hess, W. Richert, ICTP-preprint, Miramare-Trieste, 1986
9. W. Ebeling, A. Förster, H. Hess, W. Richert, Physika A (submitted)
10. W. Ebeling, A. Förster, R. Redmer, T. Rother, M. Schlanges, ICPIG - XVIII. Invited Papers, Swansea(UK), 1987
11. W.-D. Kraeft, D. Kremp, W. Ebeling, G. Röpke, Quantum Statistics of Charged Particle Systems, Akademie-Verlag Berlin, 1985, Plenum-Press New York, 1985, Mir Publishers Moscow (in Russian), 1987
12. V.E. Fortov, I.T. Yakubov, Physics of Nonideal Plasma (in Russian), Chernogolovka, 1984
13. M. Ross, A.K. McMahan, Phys. Rev. **B21**(1980)1658
14. G.A. Mansoori, N.F. Carnahan, K.E. Starling, T.W. Leland, J. Chem. Phys. **54**(1971)1523
15. J.A. Barker, D. Henderson, J. Chem. Phys. **47**(1967)4714
16. J.A. Barker, in: Rare Gas Solids, ed. by M.L. Klein and J.A. J.A. Venables, Academic-Press, New York, 1976, vol. 1
17. R. Redmer, G. Röpke, Physika **A130**(1985)523
18. R.N. Keeler, M. van Thiel, B.J. Alder, Physika **31**(1965)1437
19. W.J. Nellis, M. van Thiel, A.C. Mitchel, Phys. Rev. Lett. **48**(1982)816
20. V.B. Minzev, V.E. Fortov, Pis'ma v ZhETF **30**(1979)401
21. Yu.B. Zaporozhez, V.B. Minzev, V.E. Fortov, O.M. Batovskij, Pis'ma v ZhTF **10**(1984)133

PHASE TRANSITIONS IN XENON

T. Kahlbaum and H. Hess [1)]

1. Introduction

The subject of our contribution are the thermodynamical properties of xenon on the fluid side of the melting-line in the pressure-temperature (p-T) plane (fig. 2). Especially, a theoretical model is set up for the description of two separate phase transitions occuring in this region in a unified manner: the ordinary liquid-gas transition (LGT) and the so-called plasma phase transition (PPT) from a more or less ionised plasma to a highly conducting metallic state. Several more general aspects and numerical results for the latter one have already been published /1/. Here we present some details of the calculation of composition and equation of state (EOS) for a partially ionised xenon plasma consisting of n^* free electrons (e), n^* ions (i), and n_a atoms (a) per volume unit at temperature T. Similar calculations with a somewhat different approach were reported in the foregoing paper by Ebeling et al.

2. Interactions and Chemical Potential

Choosing the chemical potential to be the basic quantity we start with an expression for the charged particles:

$$\mu_{e/i} = \mu^{id}_{e/i} + \mu^{Coul}_{e/i} + \mu^{hc}_{e/i}. \tag{1}$$

The ideal parts

$$\mu^{id}_{e/i} = k_B T\{J^{-1}_{1/2}(n^*\Lambda^3_e/Z_e) + \ln(n^*\Lambda^3_i/Z_i)\} \tag{2}$$

are Fermi-like for the electrons ($J^{-1}_{1/2}$ - inverse of the Fermi function) and Boltzmann-like for the ions due to their large mass with the thermal DeBroglie wavelength $\Lambda_k = h(2\pi m_k k_B T)^{-1/2}$ and the internal partition function Z_k (see below) of species k = e,i,a. For the contributions due to Coulomb interaction we use a symmetrical variant of Padé approximants /2/ which interpolate between analytically well-known limiting cases:

$$\mu^{Coul}_{e/i} = -8(\pi n^*/k_B T)^{1/2} \longrightarrow -5.83 n^{*1/3}, \tag{3}$$

namely the quantum-corrected Debye law $\sim n^{*1/2}$ for low density and high temperature and the case of high degeneration of the electrons (Gell-Mann and Brueckner, Wigner) and strong nonideality of the ions (Hansen, DeWitt), respectively, $\sim n^{*1/3}$. Originally, these formulae have been worked out only for point particles. However for the treatment of more

---

1) GDR Academy of Sciences, Central Institute for Physics of the Earth Telegrafenberg, Potsdam, 1561, DDR

complex ions it is necessary to take into account at least their spatial extension. This can be done in a first approximation simply by adding a hard core interaction term for a mixture of three components /3/

$$\mu_{e/i}^{hc} = k_B T \Big\{ \sum_{m=0}^{3} c_m(n^*, n_a)(1 - f)^{-m} + c_4(n^*, n_a)\ln(1 - f) \Big\} \qquad (4)$$

as a function of the total packing parameter $f = \pi(n^* l_i^3 + n_a l_a^3)/6 < 1$ (l_k - diameter of species k; $l_e = 0$, $l_i = 2.37$ Å) where the last part represents the excluded volume effect: The electrons are not allowed to penetrate into the volume which is filled by the ions and atoms.
On the other hand we have to know the chemical potential of the atoms. In our model which needs still further improvement here it is given by

$$\mu_a = \mu_a^{id} + \mu_a^{attr} + \mu_a^{hc} \qquad (5)$$

with an ideal part in the usual Boltzmann form

$$\mu_a^{id} = k_B T \ln(n_a \Lambda_a^3 / Z_a) - I \qquad (6)$$

(I - atomic ionisation energy) and a hard core contribution μ_a^{hc} similar to eq. (4). The expression due to attractive van der Waals-like forces

$$\mu_a^{attr} = 2 n_a k_B T B_2(T) \qquad (7)$$

was calculated in terms of the classical second virial coefficient /4/

$$B_2(T) = 2\pi \int_{l_a}^{\infty} dr r^2 (1 - e^{-V_{a-a}(r)/k_B T}) = -2\pi l_a^3 \sum_{m=1}^{\infty} (8e\,\epsilon/k_B T)^m / (8m-3)^{m+1} \qquad (8)$$

for the simple interaction potential

$$V_{a-a}(\bar{r}) = -8e\,\epsilon\,\bar{r}^{-8}\ln\bar{r}, \quad \bar{r} = r/l_a \geq 1 \qquad (9)$$

where potential depth $\epsilon/k_B = 236$ K and atomic diameter $l_a = 3.68$ Å were fitted to reproduce the critical temperature $T_c^{(1)} = 289.9$ K and density $n_c^{(1)} = 5.0 \times 10^{21}$ cm^{-3} of the LGT. All polarisation interactions between atoms and charged particles were neglected at this stage of work.

3. Mass Action Law, Partition Functions, and Equation of State

Now, the condition of chemical equilibrium between electrons, ions, and atoms may be expressed in terms of the chemical potentials (1) and (5) of these species

$$\mu_{e/i}(n^*, n_a) = \mu_a(n_a, n^*) \stackrel{!}{=} \mu_{pl}(n) \qquad (10)$$

which defines the plasma potential μ_{pl} as a function of the total density $n = n^* + n_a$. This mass action law (MAL) provides an implicit relation between n^* and n_a (plasma composition) and can be rewritten in the form of a Saha equation

$$n_a = n^*(Z_a/Z_i)\exp\{(\mu_e^{id} + I - \Delta I)/k_B T\} \qquad (11)$$

with the lowering of the continuum edge

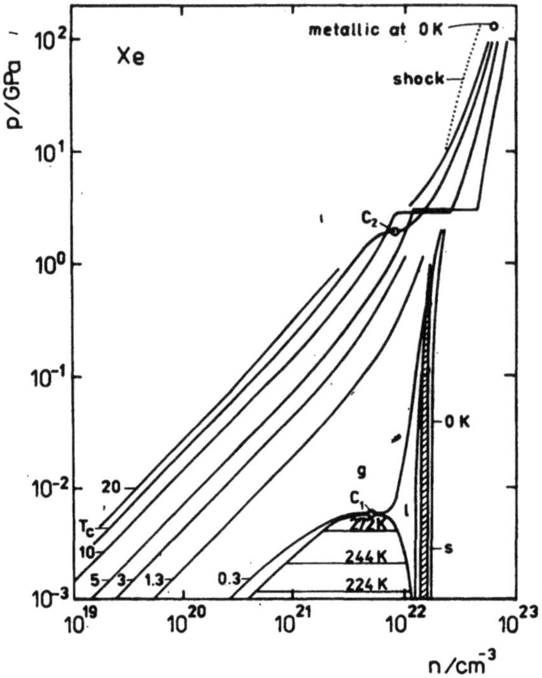

Fig. 1: p-n diagram. Isotherms with $T/10^3$ K; g - gas, l - liquid, s - solid; $C_{1,2}$ - critical point of the LGT (1) and the PPT (2)

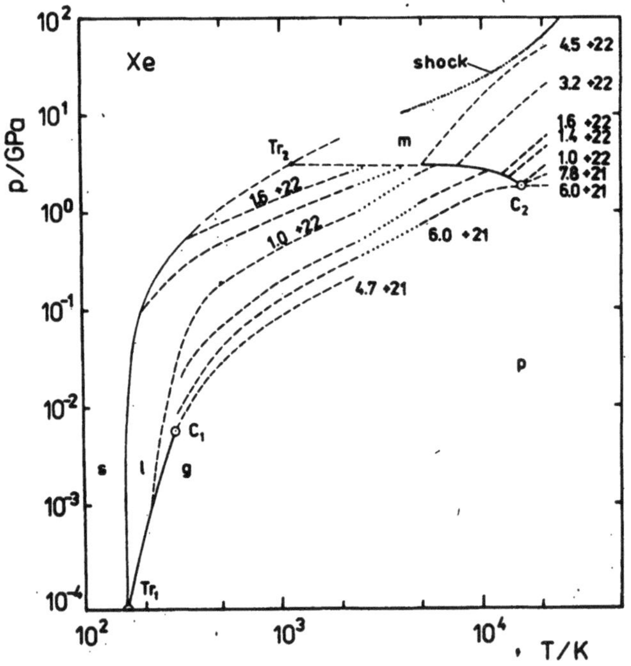

Fig. 2: p-T diagram. Isochores with n/cm^{-3}; p - plasma, m - metal; $Tr_{1,2}$ - triple point of the LGT (1) and the PPT (2)

$$\Delta I(n^*, n_a) = -\mu_{e/i}^{Coul} - \mu_{e/i}^{hc} + \mu_a^{attr} + \mu_a^{hc}. \qquad (12)$$

To calculate the atomic partition function Z_a, we restrict ourselves to the contributions of discrete bound state levels E_m ($\varepsilon_m = E_m + \Delta I$) with degeneracy g_m according to the Planck-Brillouin-Larkin convention /5/

$$Z_a = \exp\{-(I - \Delta I)/k_B T\} \sum_m^{\varepsilon_m < 0} g_m \{\exp(-\varepsilon_m/k_B T) - 1 + \varepsilon_m/k_B T\}. \qquad (13)$$

Since the PPT is closely connected with the vanishing of bound states due to Coulomb screening at densities given by the condition $\Delta I = -E_m$ (soft Mott transition) the number of terms in the sum has to be limited by a cut-off procedure with ΔI resulting in a continuously decreasing function of n. Retaining the structure of Z_a but replacing I, E_m, and g_m by the corresponding values for the ion we cancel ΔI and arrive at Z_i which is therefore independent of n. $Z_e = 2$ is due to spin degeneracy. Finally, the total pressure (EOS) was evaluated by numerical integration of μ_{pl} with respect to n

$$p(n) = \int_0^n dn' \{\mu_{pl}(n) - \mu_{pl}(n')\}. \qquad (14)$$

4. Results

Calculating isotherms for μ_{pl} and p we find two regions where the stability conditions $\partial\mu_{pl}/\partial n$, $\partial p/\partial n \geq 0$ are not fulfilled for $T < T_c^{(1,2)}$ indicating phase transitions of first order. The application of Maxwell constructions between gas or plasma (n_1) and liquid or metal (n_2)

$$\mu_{pl}(n_1) = \mu_{pl}(n_2), \; p(n_1) = p(n_2) \qquad (15)$$

yields the corresponding regions and lines of coexistence (fig. 1, 2). At low temperature we obtain the LGT with a usual sequence of isotherms, a critical point C_1 at the top of the sinusoidal coexistence region, and an increasing coexistence line ending up at $p_c^{(1)} = 7.2$ MPa (experimental value 5.84 MPa). At higher temperature the PPT shows up with crossing isotherms, a critical point C_2 at the bottom of the sickle-like coexistence region, and a decreasing coexistence line p(T). C_2 is located at $T_c^{(2)} = 16{,}000$ K, $n_c^{(2)} = 8 \times 10^{21}$ cm^{-3}, and $p_c^{(2)} = 2$ GPa /6/.

5. References

/1/ Hess, H., and T. Kahlbaum, Proc. IXth AIRAPT, Kiev 1987, in press
/2/ Ebeling, W., and W. Richert, phys. stat. sol. (b) 128 (1985) 467
/3/ Mansoori, G. A., N. F. Carnahan, K. E. Starling, and T. W. Leland, Jr., J. Chem. Phys. 54 (1971) 1523
/4/ Hirschfelder, J. O., C. F. Curtiss, and R. B. Bird, Molecular Theory of Gases and Liquids, Wiley, 1954, p. 233
/5/ Zimmermann, R., and H. Stolz, phys. stat. sol. (b) 131 (1985) 151
/6/ Hess, H., and W. Ebeling, in: Strongly Coupled Plasma Physics, ed. by F. J. Rogers and H. E. Dewitt, Plenum, 1987, p. 185 - 198

Preparation and Measurement of Superconducting PdH(D) in a Diamond Anvil Cell

H.Hemmes[1], A. Driessen[1], R. Griessen[1], J. Rector [1], J. Caro [2] and S. Radelaar [2]

Introduction

One of the most interesting properties of the Pd-H(D) system is that it is one of the few metalhydrides with a superconducting T_c that increases with increasing hydrogen content. After the discovery of superconductivity in $PdH_x(D_x)$ in 1972 [1] it took several years before a quantitative theory emerged for the observed "inverse isotope effect" in T_c between PdH_x and PdD_x [2]. Several mechanisms have been proposed so far to explain the isotope effect [3] but theoretical and experimental work is still needed to identify the relevant contributions to this intriguing effect [3-7]. As theoretical work on the electronic contributions to the superconductivity in $PdH_x(D_x)$ (especially those based on bandstructure calculations) are mainly concerned with stoichiometric PdH(D) (i.e. x=1) experimental data on stoichiometric PdH and PdD are needed for direct comparison with theory. To this end we synthesized Pd-hydrides (deuterides) in a diamond anvil cell and measured for the first time the pressure dependence of T_c of stoichiometric PdH (and PdD). As the preparation of PdH(D) and measurement of T_c in situ in a diamond anvil cell is quite complex we shall focus our attention in this paper mainly on the experimental techniques. A comprehensive report on both experimental and theroretical results for superconducting PdH and PdD will be given elsewhere [8].

Experimental

The experiments are performed in a Diamond Anvil Cell (DAC) which is made completely of a hardened Berylium Copper alloy. With this DAC we can apply a maximum force of $3 \cdot 10^4$N on the diamond anvils. This allows us to use relatively large diamonds and still reach high pressures. As described below the use of large diamonds is necessary for the synthesis of stoichiometric PdH(D) as large amounts of hydrogen (deuterium) are absorbed by the sample. The DAC is mounted in an optical helium cryostat for operation between 1.5 and 300 K. The force on the diamonds can be changed at any temperature from the top of the cryostat. A continuous flow heat exchanger for operation with liquid helium is mounted in the body of the cell. This makes rapid temperature sweeps possible with relatively low helium consumption. The design of the press and cryostat are similar to that of Silvera and Wijngaarden [9] and will be described elsewhere [10]. For the determination of the pressure dependence of the superconducting T_c of stoichiometric PdH(D)) we had to solve two experimental problems. First the loading of the sample with hydrogen at pressures upto 30 kbar in a DAC and secondly the determination of T_c by means of in situ resistance measurements.

<u>In situ loading of a sample with hydrogen.</u> The loading of a sample with hydrogen is schematicaly shown in fig. 1. The metal sample is contained in the gasket hole with hydrogen as the pressure medium. The absorption of hydrogen into the metal takes place in two stages:

1) a hydrogen molecule from the pressure medium dissociates at the surface of the metal sample into hydrogen atoms.
2) the hydrogen atoms then diffuse into the metal if the temperature is high enough.

The absorption of hydrogen goes on until the chemical potential of hydrogen inside and outside the sample have become equal. As hydrogen is not only part of the sample but is also the presure medium

---

[1] Natuurkundig Laboratorium der Vrije Universiteit, de Boelelaan 1081, 1081 HV Amsterdam, the Netherlands
[2] Centrum voor Submicron Technologie, Technische Universiteit Delft, P.O Box 5046, 2600 GA Delft, the Netherlands

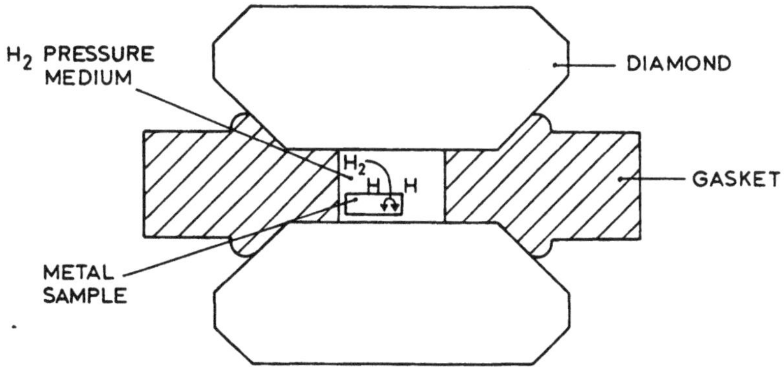

Fig. 1 Schematics of the absorpion of hydrogen in a metal sample in a diamond anvil cell.

we need it at a sufficient starting density to prevent the gasket hole from collapsing when applying pressure. This is the main reason why relatively large diamonds are used. There are two ways to obtain a sufficient starting density: load the DAC with hydrogen in a high pressure bomb under a few kbar hydrogen pressure or load the DAC at low temperatures with liquid hydrogen. As we are interested in the superconducting properties of PdH(D) we used naturally the second method to fill the gasket hole with hydrogen.

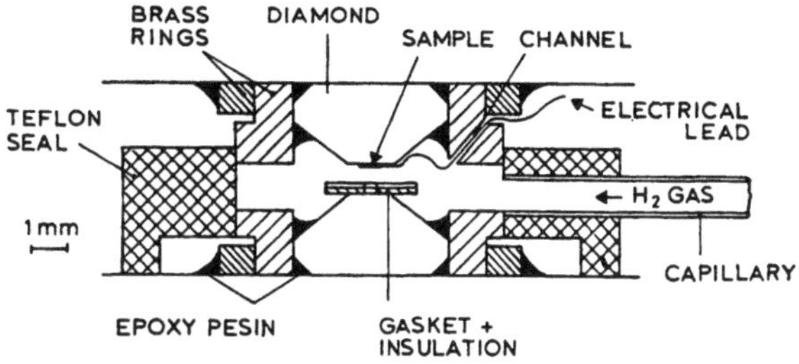

Fig. 2 Construction for the loading of the sample space with hydrogen and for the positioning of the electrical leads. For details see text.

The setup for condensing hydrogen in the gastket hole is shown in fig. 2. The diamonds are glued in a brass ring using an epoxy resin. A teflon ring surrounds the brass rings. In this way a small leaktight chamber is formed around the diamond tips. Gaseous hydrogen is introduced in the chamber through a capillary. When the temperature of the chamber is above the triple point temperature and the leakrate of the chamber is small enough for the pressure to rise above the triple point pressure (13.96 K, 0.072 bar for H_2 and 18.73 K, 0.171 bar for D_2) hydrogen (deuterium) condenses in the chamber sealed by the teflon ring. The condensation process as well as the sample can be viewed through the diamonds and the windows of the cryostat by means of a microscope coupled to a video system. After condensation a force is applied to the diamonds to seal the gasket hole and to pressurize the sample. Then the temperature is increased to allow for diffusion of hydrogen into the metal.

<u>Resistance measurements.</u> A major problem in making resistance measurements in a DAC are the contacts. For a sample of the form indicated in fig. 1, with a typical size of 100 μm, the making of contacts would be a formidable task. We therefore took another approach. We used evaporation techniques to deposit a thin film (0.1 μm) of the sample material (in this case Pd) on the culet of a diamond. Then we used optical lithographic techniques to pattern the film. Two examples of patterns used in our investigation are shown in figs. 3a and 3b. The electrical contact flaps extend all the way to the edge of the culet (see fig. 3c). In this way it is possible to connect wires to the sample near the edge of the culet and therefore to avoid to enter the sample space. The gasket hole used with these patterns is 300 μm in diameter.

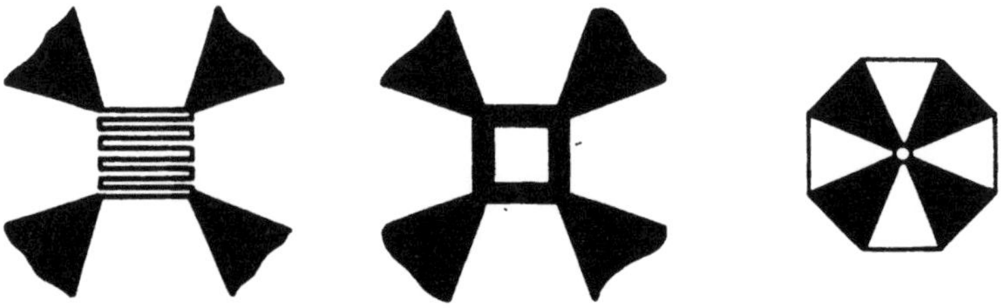

Fig. 3 Patterns used in our experiments on the Pd-H(D) system. a) Meander pattern with a trackwidth of typically 4 µm, b) the more robust square pattern and c) position of the contact flaps on the culet.

The electrical leads consist of two parts. First a Au wire of 25 µm diameter is fixed to the contact flaps using a thermo-bonding apparatus. The Au wire is connected to an insulated Cu wire (80 µm diameter) using a conducting glue. The Cu wire leaves the chamber through a channel drilled in the brass ring (see fig. 2). The channel is sealed with epoxy resin because a leaktight chamber is required for condensation of hydrogen.
For insulation between the gasket and the sample we use a polyimide foil (Kapton) with a thickness of 25 µm. The gasket is made of 100 µm thick brass foil. This is softer than stainless steel and can be used at lower pressures. The gasket is not preindented. With this combination of gasket and insulation we have been able to measure the superconducting T_c of PdH and PdD between 0 and 40 kbar.
As can be seen in fig. 2 we use a double ring system to define the position of the diamonds. The outer ring is fixed with epoxy resin. In this way we have the possibility to replace a diamond without destroying the alignment. This is a necessity for us because we cannot align the diamonds and preindent a gasket with the diamond on which the metal film has been deposited without destroying the sample.

Results

After loading with hydrogen a pressure of about 30 kbar is applied to seal off the gasket. We then start heating the cell at a rate of 2 K/min. The resistance of the sample during the heating is shown in fig. 4. The horizontal axis corresponds simultaneously to the time *t* from the start of the heating and to the temperature T(*t*), measured with a Platinum resistance thermometer.
At temperatures below 100 K we only see the temperature dependence of the resistance of the essentially pure Pd film as hydrogen diffusion into the sample is negligible. Just above 100 K we see a strong increase in resistance compared to that of the Pd film (dashed line). This is a result of the diffusion of hydrogen into the Pd which introduces disorder of occupied and unoccupied sites on the interstitial lattice [11,12]. Above a concentration [H]/[M]~0.7 disorder becomes smaller and leads to a reduction in the resistance. At stoichiometry ([H]/[M]~1.0) we have a perfect interstitial lattice in which all the interstitial sites of octahedral symmetry are occupied. It is interesting to note that the electrical resistivity of stoichiometric PdH is lower than that of pure Pd as a result of the d-band filling of the host metal upon loading with hydrogen. The lower resistance of the hydride after the peak in fig. 4 and the small upturn due to the thermal electron-phonon scattering give a good indication of the completion of the hydride formation.
Using the width of the peak in fig. 4 and the smallest dimension of the sample we can estimate the diffusion constant of hydrogen in Palladium at ~110 K. We find $D \sim 10^{-12} cm^2 s^{-1}$ which is in reasonable agreement with other experimental data [13,14]. For the Pd-D system we found similar results.
After the loading the sample with hydrogen we cool it down to measure the superconducting properties of the hydride as a function of pressure. The zero pressure transition temperatures are 8.8 K for PdH and 11.05 K for PdD. For the logarithmic pressure derivatives $d\ln T_c/dP$ we find -0.075 $kbar^{-1}$ for PdH and 0.049 $kbar^{-1}$ for PdD.

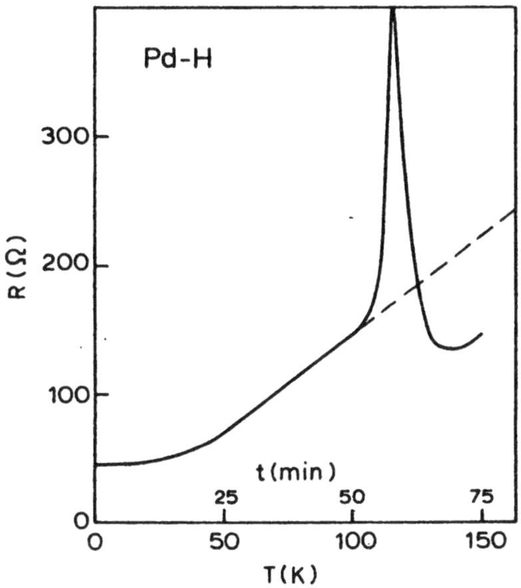

Fig. 4 Resistance of the metal film during the loading with hydrogen. For details see text.

These results are analyzed in ref. [8] in terms of the strong coupling theory of Allen and Dynes. From this analysis we conclude that $d\ln T_c/dP$ is mainly determined by the volume dependence of phonon frequencies.

Acknowledgement

This investigation was part of the research program of the Stichting voor Fundamenteel Onderzoek der Materie (FOM).

References

[1] T. Skoskiewicz: Phys. Stat. Sol. (a) **11**, K 123 (1972)
[2] D.A. Papaconstantopoulos, B.M. Klein, E.N. Economou and L.L. Boyer: Phys. Rev. B**17**, 141 (1978)
[3] for a summary see B. Stritzker and H. Wühl in: Hydrogen in metals 2 ch. 6, eds. G. Alefeld and J. Völkl; Topics in Applied Physics vol. 29; Springer Berlin 1978
[4] V.B. Ginodman and L.N. Zherikhina: Sov. J. Low Temp. Phys **6**, 278 (1980); (Fiz. Nizk. Temp. **6**, 582 (1980))
[5] R. Griessen and D.G. de Groot: Helv. Phys. Acta **55**, 699 (1982)
[6] A.P. Zhernov and Sh. L. Drekhsler: Sov. J. Low Temp. Phys. **11**, 495 (1985); (Fiz. Nizk. Temp. **11**, 899 (1985))
[7] J.M. Rowe, J.J. Rush, J.E. Schirber and J.M. Mintz: Phys. Rev. Lett. **57**, 2955 (1986)
[8] H. Hemmes, A. Driessen, R. Griessen and M. Gupta: to be published
[9] I.F. Silvera and R.J. Wijngaarden: Rev. Sci. Instrum. **56**, 121 (1986)
[10] H. Hemmes, A. Driessen and R. Griessen: to be published
[11] J.P. Burger, D.S. MacLachlan, R. Mailfert and B. Souffache: Solid State Commun. **17**, 281 (1975)
[12] B.M. Geerken and R. Griessen: J. Phys. F: Metal Phys. **13**, 963 (1983)
[13] F.M. Mazzolai and H. Züchner: Z. Phys. Chem. N.F. **124**, 59 (1981)
[14] A.H. Verbruggen, C.W. Hagen and R. Griessen: J. Phys. F: Metal Phys. **14**, 1431 (1984)

HIGH PRESSURE PHASE TRANSFORMATIONS IN Mn-CHALCOGENIDES

A. Kraft, B. Greuling 1)

1. Introduction

The phase relations in the Earth's lower mantle are subjects of considerable interest and controversy. The major phases below about 650 km have been proposed $(Mg,Fe)SiO_3$ with perovskite structure and (Mg,Fe)O with B1-structure (NaCl-type). Investigations in oxides and sulphides with MX-stoichiometry have resulted in phase transformations different to the expected B1-B2 transition. Interest has recently arisen by a paper of Navrotsky and Davies [3]. They suggested that the B8-structure (NiAs-type) is another possible candidate for (Mg,Fe)O at very high pressure. We took α-MnS and α-MnSe for model substances in order to investigate the B1-B8 transition.

2. Experimental Procedure

Samples of α-MnS and α-MnSe were prepared by the evacuated silica-glass tube technique at temperatures of 800°C. Powdered samples were investigated in a diamond anvil cell up to 20 GPa by high pressure X-ray diffraction. A gasketed diamond anvil cell has also been used for optical observations. The pressure was measured by the ruby-fluorescence technique.

The electrical conductivity measurements were carried out in a Bridgman opposed-anvil apparatus with an external heater. The anvils were made of WC/Co which permits pressures up to 8 GPa and temperatures from room temperature up to 300°C. The pressure was calibrated against the polymorphic phase transitions of bismuth at 2.55 and 7.7 GPa and of barium at 5.53 GPa. The temperature was measured by using a Ni/CrNi-thermocouple which is fixed at the anvils so that the deviation of the measured temperature from the actual temperature is negligible.

3. Structural transitions

The variation of the interplanar distances of α-MnS and α-MnSe as a function of pressure up to 20 GPa is shown in Figs. 1 and 2, respectively.

---

1) Central Institute for Physics of the Earth
Academy of Sciences of the GDR, Potsdam, 1561

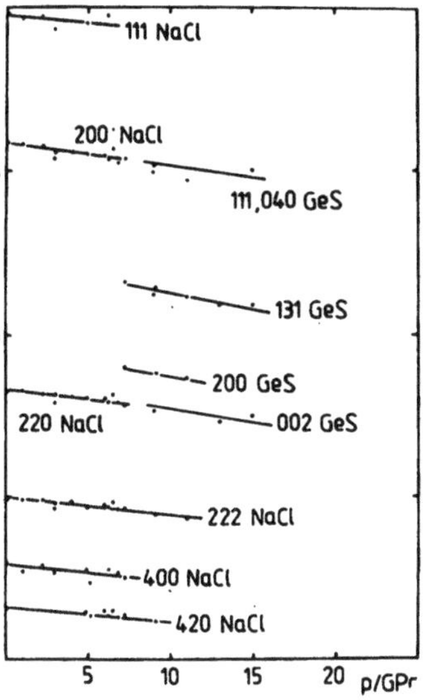

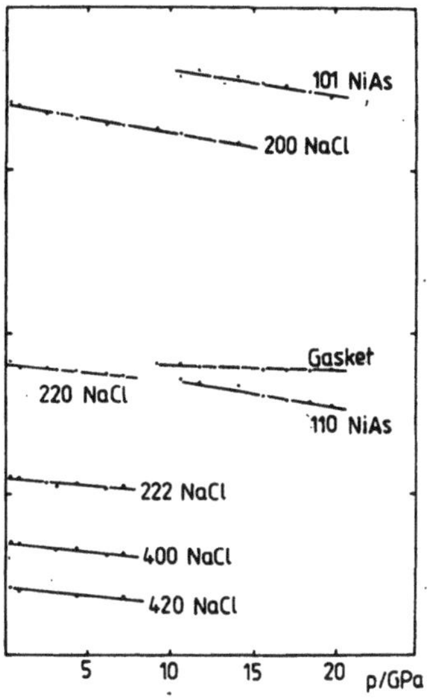

Fig. 1: Interplanar distances vs. pressure for α-MnS

Fig. 2: Interplanar distances vs. pressure for α-MnSe

We expected a B1-B8 phase transformation in both substances by reason of homology relations to MnTe. In α-MnSe we found this transition at about 10 GPa as indicated by Cemic and Neuhaus [1]. The (200)-peak of the B1-phase exists up to a pressure of 15 GPa. In α-MnS at 7.2 GPa new peaks appear in the pattern. We tried to index this pattern on the basis of cubic, tetragonal, hexagonal, and orthorhombic symmetry and to compare it with related structures. The best results could be obtained by assuming

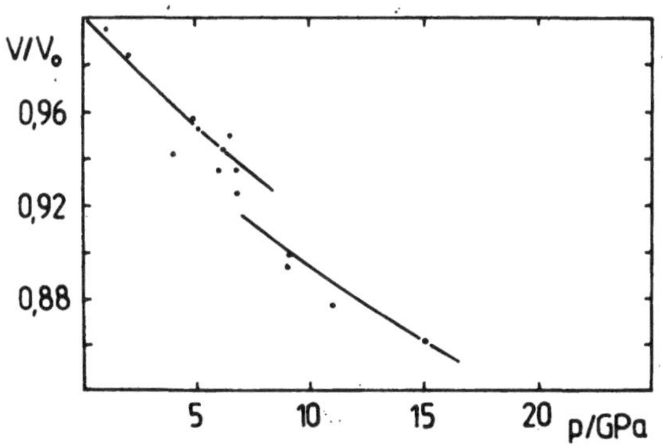

Fig. 3: Pressure dependence of the volume of α-MnS up to 15 GPa.

a transition to the B16-structure (GeS-type). This transition was also found in Pb-chalcogenides by Chattopadhyay et.al.[2]. The B1-type and the B16-type are closely related structures. They are explained geometrically by a translation of a/2. The transition in α-MnS at 7.2 GPa is connected with a volume difference of 1.6 % (Fig.3).
Optical observations during the phase transition show a colour change from yellow to red up to black in the pressure range up to 10 GPa.

4. Electrical measurements

The pressure dependence of the electrical conductivity of α-MnS and α-MnSe shows a nearly exponential decrease with increasing pressure. This is the behaviour of many semiconductors before the phase transition Also the temperature dependence of the conductivity at constant pressure shows semiconducting behaviour. In Fig.4 a log R vs.1/T plot in the case of α-MnS exhibits the expected linearity. From

$$R(T) = R_o \exp(E_A/2kT) \qquad (1)$$

one can estimate the activation energy E_A

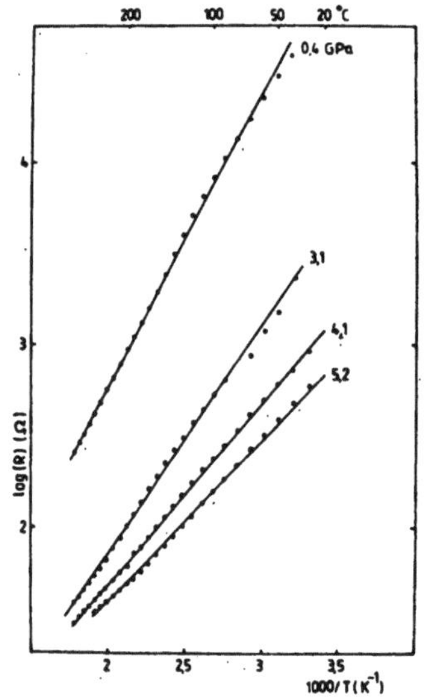

Fig.4: Temperature dependence of the resistance of α-MnS at various pressures

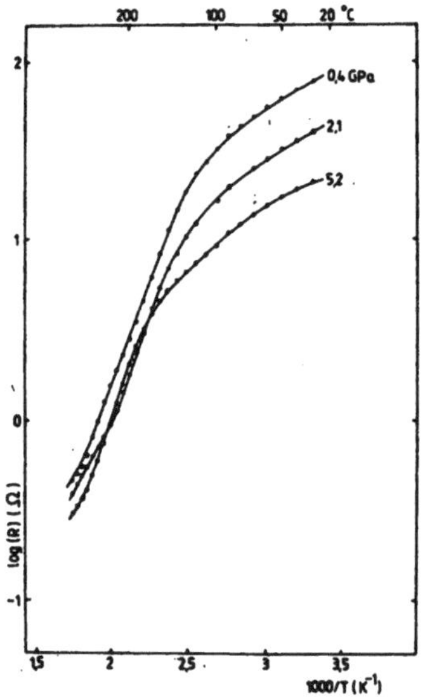

Fig.5: Temperature dependence of the resistance of α-MnSe at various pressures

Zero gap pressure is then extrapolated to be about 10 GPa which corresponds to the optical investigations of a semiconductor-metal transition This behaviour is shown in Fig.6. In the case of α-MnSe the log R vs.1/T plot shows two nearly linear branches i.e. two activation energies which do not significantly vary with pressure (Fig.5).

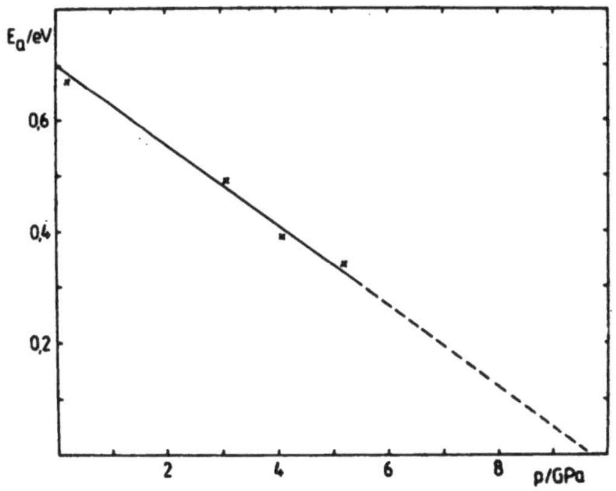

Fig.6: Activation energy vs. pressure for α-MnS

5. Conclusions

i) The high pressure behaviour of α-MnS and α-MnSe is different. α-MnSe undergoes a phase transition at about 10 GPa from B1- to B8-type. In the pressure range up to 15 GPa both phases coexist.

ii) In α-MnS a phase transition to the B16-type structure (GeS-type) takes place at 7.2 GPa which is connected with a volume change of 1.6 % and a colour change from yellow to red up to black.

iii) α-MnS and α-MnSe show semiconducting behaviour up to 8 GPa. in the case of α-MnS we assume a semiconductor-metal transition because of the linear decreasing gap energy and the optically observed colour change. In contrast the gap energy of α-MnSe does not significantly vary with pressure.

References

[1] Cemic, L. and A. Neuhaus: High Temp.-High Press. 6 (1974),203-215.

[2] Chattopadhyay, T., A. Werner, H. G. von Schnering and J. Pannetier: Rev. Phys. Appl. 19 (1984),807-813.

[3] Navrotsky, A. and P. K. Davies: J. Geophys. Res. 86 (1981),B5,3689-94

MÖSSBAUER SPECTROSCOPY OF ^{197}Au AT HIGH PRESSURES.
EXAMPLE: PHASE TRANSITION IN CsAu

S.S. Hafner 1), J.Stanek 2)

Abstract: The equipment for Mossbauer studies at high pressures and low temperatures is described. The Mossbauer spectra of ^{197}Au in CsAu at 4.2 K and at pressures up to 40 kbar show a phase transition from the CsCl-type structure to a structure of lower symmetry. The transition is indicated by abrupt decrease in the electron density on ^{197}Au, the occurrence of an electric field gradient at the Au position in the high pressure phase, and an increase of the Debye temperature.

Introduction

The intermetallic compound CsAu which crystallizes in the CsCl-type (B2) structure is semiconducting with an energy gap of 2.6 eV at room temperature. Relativistic band structure calculations /1,2/ which compare well with photoemission valence band data indicate a significant charge transfer from Cs to the Au site and suggest a metallization of CsAu at high pressure. Recent total energy calculations /3/ predict also a structural instability of CsAu at high pressure, i.e. when the interatomic discances decrease. The present study is a continuation of experiments /4/ carried out in order to obtain a structural insight into the predicted high pressure transformation.

Experimental

Mossbauer spectroscopy. The methods and results of ^{197}Au Mossbauer spectroscopy have been reviewed e.g. in /5/. Because of the relatively high energy of the recoil-free transition (77.4 keV) source and absorber must be held at liquid helium temperature. A source of 200 mCi ^{197}Pt with a half-life of 18 h in a Pt foil was used so that a cycle of measurements should be terminated within a few days. The nuclear spins of the ^{197}Au in ground and excited states are 3/2 and 1/2, respectively. The Mossbauer spectra are therefore, generally simple singlets if the site symmetry is cubic or quadrupole-split doublets if an electric field gradient is pesent at the nuclear site.

High-pressure apparatus. The apparatus for high-pressure Mossbauer studies at low temperatures, presented in Fig. 1, was similar to the one described in /6/. The absorber pellet had a diameter of 4 mm. Because of the high chemical activity of CsAu it was formed and sealed in a thin plastic bag under argon atmosphere. Subsequently, the pellet with the

---

1) Institute of Mineralogy, University of Marburg, 3550 Marburg, FRG
2) Institute of Physics, Jagiellonian University, 30-059 Cracow, Poland

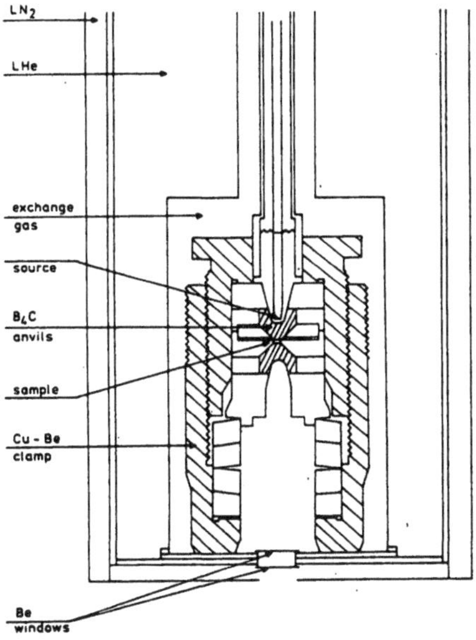

Fig.1. Apparatus for high pressure-low temperature Mossbauer measurements.

plastic bag was placed between two B_4C anvils. Pressure was applied at room temperature and maintained using a copper-berylium clamp while the system was cooled to liquid helium temperature. In contrast to the high pressure X-ray diffraction studies, the determination of the effective pressure in the Mossbauer experiments is a involved problem. The pressure was determined with a bismouth wire which exhibits sharp jumps in resistivity at 25.4 and 27 kbar at room temperature. After cooling to 4.2 K the pressure decreases by about 10% as determined from the measurements of the temperature of the metal to superconductor transition of a lead manometer /7/.

Results

The 4.2 K Mossbauer spectra of CsAu covered with a metallic gold foil are shown in Fig. 2. The 1 bar spectrum consists of two lines at -1.2 mm/s and +7 mm/s, arising from metallic gold and CsAu, respectively. The single line of CsAu splits at high pressure to a quadrupole doublet with a splitting of 1.6 mm/s, and shifts to lower energy: IS(1 bar) - IS(27 kbar) = 0.2 mm/s. Also a strong increase in the CsAu line intensity was observed at higher pressures.

From the spectra shown in Fig. 2, it was possible to estimate the ratios of recoil-free fractions of CsAu to metallic Au, and to determine the Debye temperature, Θ, of CsAu. It turned out that Θ_D(CsAu) = 72 K at 1

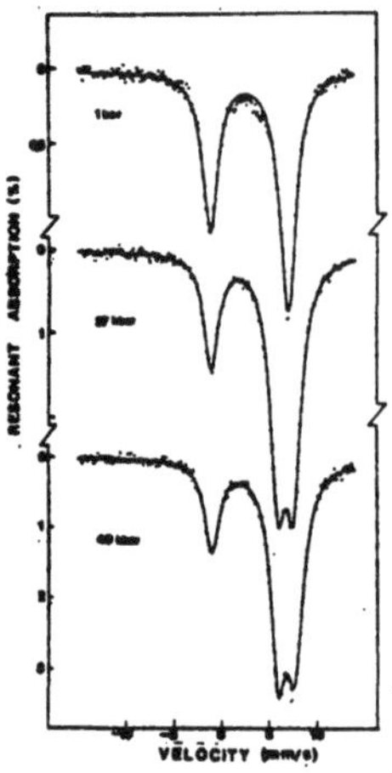

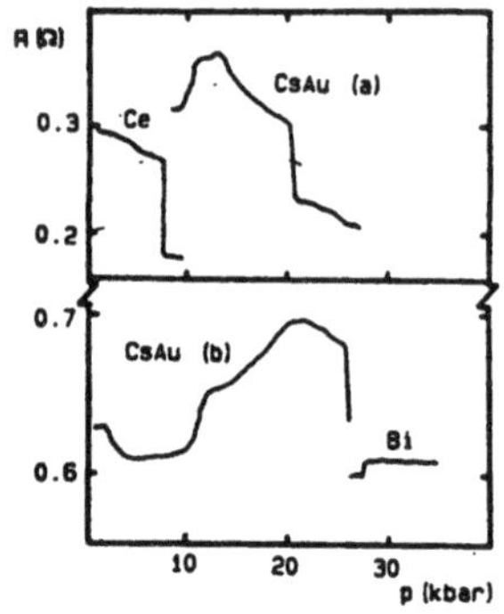

Fig. 2. Mossbauer spectra of CsAu (covered with Au-metal foil) measured at pressures as marked.

Fig. 3. CsAu resistivity vs. pressure. (a) measured with Ce calibration, (b) measured with Bi calibration.

bar increases to 96 K and 99 K at 27 and 40 kbar, respectively. While first experiments /4/ were indicative of a transition pressure near or below 25 kbar we now conclude that it may be as low as 10 kbar at 4.2 K. A precise determination would require more accurate calibration of the pressure between 1 and 20 kbar. The observed changes in hyperfine parameters and Debye temperature are reversible. Above 40 kbar the decomposition of CsAu, which appears to be sometimes even an explosive one, to metallic gold was observed. Also the pressure dependence of the resistivity of CsAu was measured at 295 K. The results are shown in Fig. 3. At about 25 kbar an abrupt decrease of resistivity was found.

Discussion

The observed strong and probably abrupt increase in Debye temperature, the decrease of the electron density at the Au nuclei and the pressure induced electric field gradient are indicative of a phase transition as predicted on the basis of theoretical calculations /3/. The exact value of the pressure at which these transformations of structure and chemical bonds occur was not determined exactly. However, some broadening of the CsAu resonant absorption line was seen already at about 10 kbar, i.e. at much lower pressure than concluded from the total energy calculation (40 kbar). This discrepancy may be explained by the fact that our measurements were performed at 4.2 K while the calculation referres to 295 K. It is also not quite clear if the high pressure phase is metallic or semiconducting. The abrupt decrease of the CsAu resistivity at about 25 kbar at 295 K may support the postulated metallization of CsAu /2/ at high pressure.

Conclusion

The observation of a new phase transition in CsAu at high pressure confirms the prediction of the theoretical calculations /2,3/. More substantial progress in the understanding of the appearing processes will be possible after determination of the crystal structure of the high pressure phase of CsAu.

References

1. H. Overhoff, J. Knecht, F. Fischer, F. Hensel. J. Phys. F. 8, 1609 (1978).
2. C. Koenig, N.E. Christensen, J. Kolar. Phys. Rev. B 29, 6481 (1984).
3. N.E. Christensen. Phys. Rev. B 32, 207 (1985).
4. J. Stanek, S.S. Hafner, F. Hensel. Phys. Rev. B32, 3129 (1985).
5. P. Gütlich, R. Link, A. Treutwein, in: Inorganic Chemistry Concepts. (Springer, Berlin, 1978).
6. J.S. Schilling, F.U. Klein, W.B. Holzapfel. Rev. Sci. Instrum. 45, 1353 (1974).
7. J. Stanek. J.Chem. Phys. 76, 2315 (1982).

DETERMINATION OF THE GRUENEISEN PARAMETER FROM THE PRESSURE DEPENDENCE OF THE THERMAL CONDUCTIVITY

By U. Seipold (1)

The Grueneisen parameter is one of the most important material parameter in high pressure physics. Therefore there exists a great interest for methods to determine its value and especially the values at high pressures and high temperatures. We tried different ways to determine γ from results for the pressure dependence of the thermal diffusivity or conductivity. It is clear that there is a principal restriction. The thermal conductivity is mainly determined by the acoustic phonons. Therefore, γ calculated by these methods will be the so-called acoustic Grueneisen parameter, this means it will be the mean value for the acoustic branches of the phonon spectrum.

By differentiating the Leibfried-Schloemann equation regarding the thermal conductivity there follows for the volume dependence [Slack 1979]

$$g \equiv -\frac{\partial \ln \lambda}{\partial \ln V} = 3\gamma + 2q - 1/3. \tag{1}$$

Applying the definitions of the pressure coefficient α and the isothermal bulk modulus B_T

$$\alpha = \frac{\partial \ln \lambda}{\partial p} \; ; \quad \frac{1}{B_T} = -1/V \frac{\partial V}{\partial p} \tag{2}$$

one gets

$$\gamma = 1/3 \; (\alpha B_T - 2q + 1/3). \tag{3}$$

The elimination of the second Grueneisen parameter $q = \frac{\partial \ln \gamma}{\partial \ln V}$ was tried in different ways:

<u>1. Continuum approximation</u>

According to Boehlin and Andersson [1974] in the continuum approximation holds

$$q = 2\gamma \tag{4}$$

Thus is follows

$$\gamma = 1/7 \; (\alpha \cdot B_T + 1/3). \tag{5}$$

---

(1) Central Institute for Physics of the Earth, Potsdam, Telegrafenberg

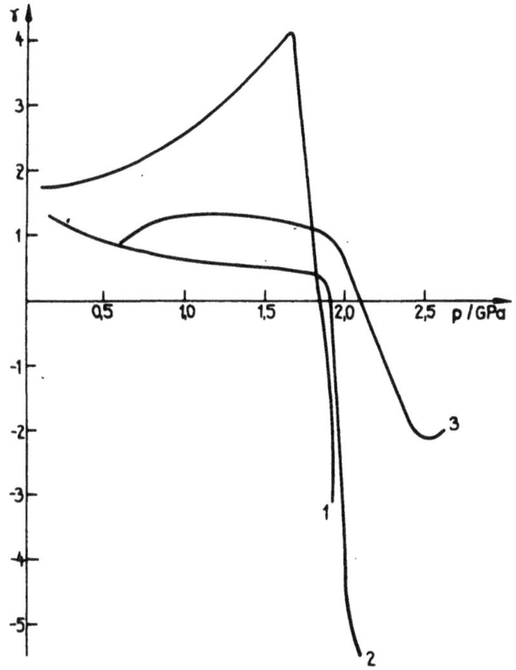

Application of the measuring values gained for the thermal conductivity of potassium bromide at pressures up to 2.2 GPa gives the results in figure 1. Values of the bulk modulus are adopted from Weir and Piermarini [1964].

Fig. 1 Pressure dependence of γ calculated from (5)
1,2 Results of the author
3 Values from Schloessin

2. Numerical solution of the differential equation for γ

Inserting

$$q = - \frac{B_T}{\gamma} \frac{\partial \gamma}{\partial p} \qquad (6)$$

in (3) delivers a differential equation for γ:

$$\frac{d\gamma}{dp} = \frac{\gamma}{2B_T} (3\gamma - \alpha \cdot B_T - 1/3). \qquad (7)$$

The values of $\alpha(p)$ and $B_T(p)$ are numerically obtained from the high pressure experiments. The solution of (7) for the initial value $\gamma_o = 1.5$ was calculated by means of the Euler-Cauchy-method. Results are shown in fig. 2.

3. Solution for linear pressure dependence

If one suppose a linear pressure dependence for the thermal conductivity and a pressure independent isothermal bulk modulus

$$\lambda = \lambda_o (1+\alpha_o p), \quad B_T(p) = \text{const.}, \qquad (8)$$

then the solution of the differential equation (7) can be determined as an analytical expression:

$$\gamma = a/b \;/\; [1 + \frac{a + b\gamma_o}{b\gamma_o} \exp(a\,p)] \qquad (9)$$

with the abbreviations

$$a = \frac{\alpha_0}{2} + \frac{1}{6B_T}, \quad b = 3/2\,B_T .$$

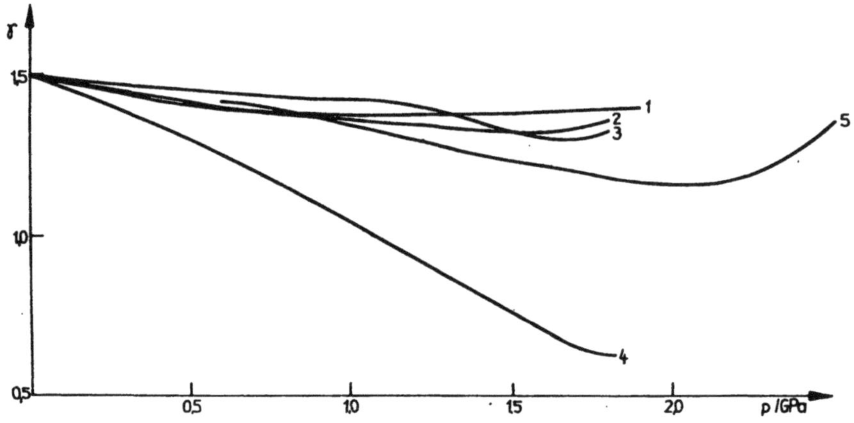

Fig. 2 Pressure dependence of γ calculated from (7)
1-4 Results of the author
5 Values from Schloessin

In tests with measuring values for KBr these curves only differ slightly from straight lines in the investigated pressure range up to 1.2 GPa (see figure 3).

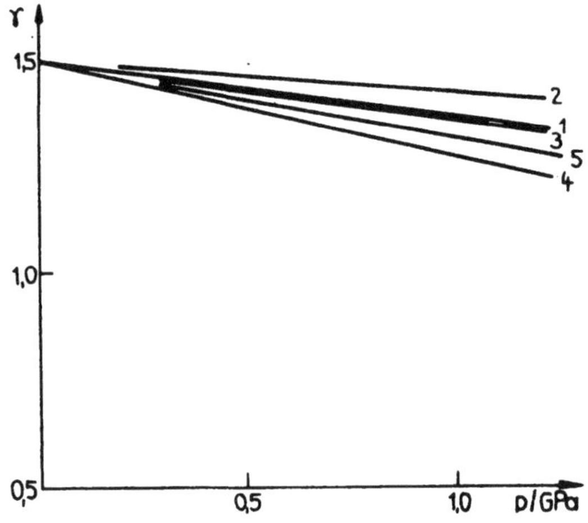

Fig. 3 Pressure dependence of γ calculated from (9)
1-3 Results of the author
4 Values from Andersson
5 Values from Schloessin

4. Temperature dependence of γ

The continuum approximation (5) had to be used for the calculation of the temperature dependence of γ. The pressure coefficient α of the thermal diffusivity was determined for KBr in the pressure range from 0.3 to 1.0 GPa at temperatures up to 800 °C. The calculations were carried out firstly for the case of a constant bulk modulus and otherwise using the value $\partial \ln \beta / \partial T = 4.8 \cdot 10^{-4}$ /K for the temperature dependence of the compressibility [Mathur and Singh 1968].

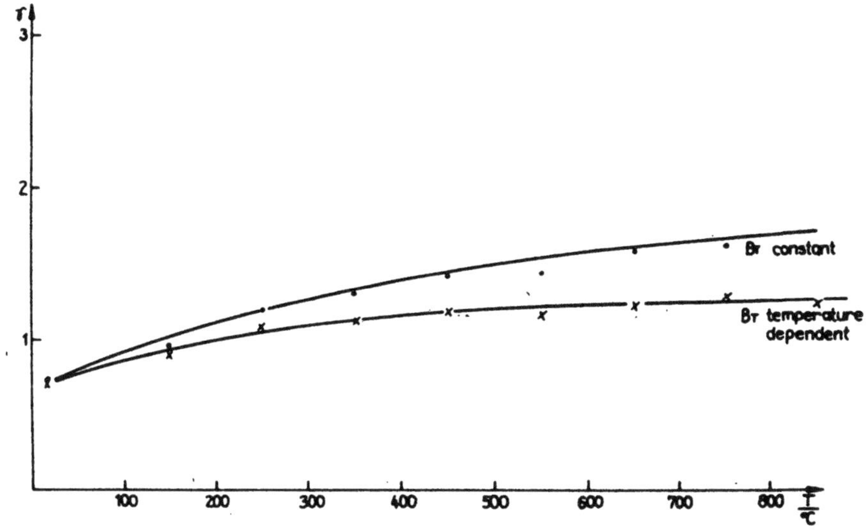

Fig.4 Temperature dependence of γ

5. Discussion

The curves γ(p) and γ(T) show the expected tendencies. However the downwards step of γ at the B1-B2-transition is controverse to the results of Boehler and Ramakrishnan [1980] using the adiabatic decompression method. This may be explained by the fact that we measured the acoustic Grueneisen parameter in contradiction to the thermal Grueneisen parameter by Boehler and Ramakrishnan.

References

Boehler, R., J. Ramakrishnan: Experimental results on the pressure dependence of the Grueneisen parameter J.Geophys.Res.85(1980),B12,6996-7002

Boehlin,L., P. Andersson: A comment on the thermal conductivity of insulating solids at high pressures Solid State Comm.14(1974),711-714

Mathur, V.K., S.P. Singh: Andersson Grueneisen parameter for NaCl type halides J.Phys.Chem.Solids 29(1968),959-962

Slack, G.A.: Thermal conductivity of non-metallic crystals Solid State Phys.34(1979)1-71

Weir, C.E., G.J. Piermarini: Lattice parameters and lattice energies of high pressure polymorphs of some alkali halides J.Res.Nat.Bur.Standards 687(1964)1,105-111

COHESION IN TRANSITION METALS UNDER HIGH PRESSURES

Uwe Walzer [1)]

A bonding theory for transition metals has been derived up to a computing program. The theory permits good to very good predictions for the total energy, static pressure, Hugoniot pressure, bulk modulus and its pressure derivative as a function of the relative volume. A combination of the pseudopotential theory with the muffin tin orbital theory serves for the calculation of the cohesive properties of d-state metals. The coupling of free-electron states and local, nonoverlapping d-states is also taken into account. The nearly free s-electrons contribute four terms to the energy: kinetic energy, exchange energy, a Madelung term, and its pseudopotential correction. On this basis, two new models are developed. In both models, it is assumed that the ratio K of nearest-neighbor separation to the atomic radius is independent of the pressure p. In the first model (MB) [15], the additional condition is that a redistribution of electrons between the two outer shells, which are not completely filled, is permissible. In the second model (MC) [14], the requirement is that for elements with a full d-shell the theoretical bulk modulus at vanishing pressure is exactly equal to the experimental bulk modulus at vanishing pressure and that the pressure exactly vanishes for the zero-pressure volume. In a comparison with various kinds of measured values, MC proves to be very successful for all d-state metals except Sc, Ti and Nb. In both models, the initial pressure derivative B_1 of the bulk modulus is not an input parameter, but can be predicted. The theories are applicable in a unified manner to all transition metals.

Below, only the final results for the two equations of state will be given. For the calculation of the static pressure p as a function of the volume Ω, we have, according to model B:

$$\begin{aligned} p = & -c_1 \frac{5}{3} b(a\Omega^{-1} + b)^{2/3} + c_1 \frac{2}{3} (a\Omega^{-1} + b)^{5/3} + c_2 \frac{4}{3} b(a\Omega^{-1} + b)^{1/3} - \\ & -c_2 \frac{1}{3} (a\Omega^{-1} + b)^{4/3} + 2b(a + b\Omega) (c_3\Omega^{-1/3} - c_4\Omega^{-1}) - \\ & - (a + b\Omega)^2 (c_3 \frac{1}{3} \Omega^{-4/3} - c_4\Omega^{-2}) - \\ & - c_5 [1 - \frac{c - b\Omega}{10}] (\frac{5}{3} c\Omega^{-8/3} - \frac{2}{3} b\Omega^{-5/3}) + \\ & + c_5 \frac{1}{10} b(c - b\Omega)\Omega^{-5/3} + c_6 (\frac{8}{3} c\Omega^{-11/3} - \frac{5}{3} b\Omega^{-8/3}) \end{aligned} \qquad (1)$$

with

1) Zentralinstitut für Physik der Erde, Institutsteil Jena, Burgweg 11, 6900 Jena

$$c_1 \equiv (2.210)\,\frac{\hbar^2}{2m}\,(\frac{4\pi}{3})^{2/3}\ ; \qquad c_4 \equiv 2\pi e^2 r_c^2\ ;$$

$$c_2 \equiv (0.916)\,\frac{e^2}{2}\,(\frac{4\pi}{3})^{1/3}\ ; \qquad c_5 \equiv (30.9) n^{1/2}\,\frac{\hbar^2}{2m}\,\frac{r_d^3}{K^5}\,(\frac{4\pi}{3})^{5/3}\ ;$$

$$c_3 \equiv \frac{e^2}{2}\,\alpha\,(\frac{4\pi}{3})^{1/3}\ ; \qquad c_6 \equiv (11.40) n\,\frac{\hbar^2}{m}\,\frac{r_d^6}{K^8}\,(\frac{4\pi}{3})^{8/3}\ . \qquad (2)$$

$\hbar$ is the Planck quantum of action, divided by 2π, e is the charge and m the mass of the electron, $\alpha = 1.8$ is fixed, r_c is the Ashcroft empty-core radius, r_d is the d-state radius of the free atom, n is the number of nearest neighbors of an atom. The three parameters a, b and c denote

$$a = Z_s^o - Z_s^{o'}\Omega_o\ , \quad b = Z_s^{o'}\ , \quad c = Z - Z_s^o + Z_s^{o'}\Omega_o \qquad (3)$$

where Z_s^o is the number of s-electrons at vanishing pressure, $Z_s^{o'}$ the volume derivative of the number of s-electrons at vanishing pressure and Ω_o the volume at zero-pressure. Higher derivatives were equated to zero. However, according to model C,

$$p = \frac{2}{3} k_1\Omega^{-5/3} - \frac{1}{3} k_2\Omega^{-4/3} + k_3\Omega^{-2} - \frac{5}{3} b_n k_4\Omega^{-8/3} + \frac{8}{3} c_n k_5\Omega^{-11/3} \qquad (4)$$

with

$$k_1 \equiv (2.210)\,\frac{\hbar^2}{2m}\,Z_s^{5/3}\,(\frac{4\pi}{3})^{2/3}\ ,$$

$$k_2 \equiv \frac{e^2}{2}\,[(0.916)\,Z_s^{4/3} + \alpha Z_s^2](\frac{4\pi}{3})^{1/3}\ ,$$

$$k_3 \equiv 2\pi e^2 r_c^2 Z_s^2\ ,$$

$$k_4 \equiv Z_d\,(1 - \frac{Z_d}{10})\,(30.9)\,n^{1/2}\,\frac{\hbar^2}{2m}\,\frac{r_d^3}{K^5}\,(\frac{4\pi}{3})^{5/3}\ ,$$

$$k_5 \equiv Z_d n\,(11.40)\,\frac{\hbar^2}{m}\,\frac{r_d^6}{K^8}\,(\frac{4\pi}{3})^{8/3} \qquad (5)$$

where Z_d is the number of d-electrons, Z_s that of the nearly free s-electrons. If the requirement is that, with equation (4), the pressure exactly vanishes for $\Omega = \Omega_o$ and that the theoretical bulk modulus is in precise agreement with the experimental bulk modulus, you obtain $b_n \equiv b_{MA}$ and $c_n \equiv c_{MA}$. If this obvious requirement is made only for elements with full d-shells, the equation of state for model C is obtained by using equation (4) with

$$b_n = (4Z_d)^{-1}[(5Z_d - 10)\,b_{MA} + 10 - Z_d] \qquad (6)$$

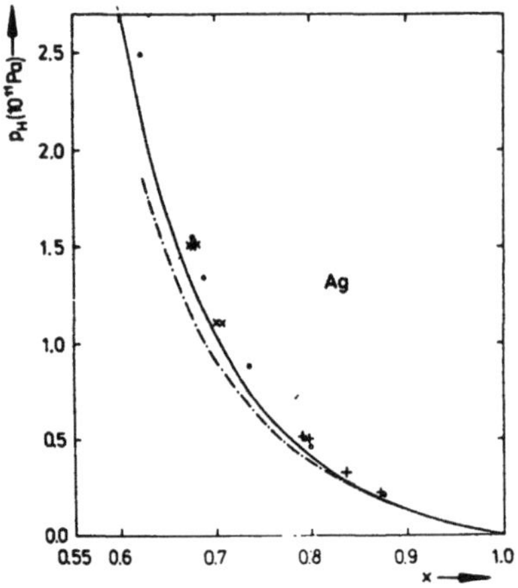

Fig. 1. Hugoniot pressure p_H versus relative volume x for silver. Theory MB is dash-dotted, theory MC continuous. The experimental values after [8] are solid circles, those after [9] upright crosses, those after [7] are oblique crosses and those after [1] empty circles.

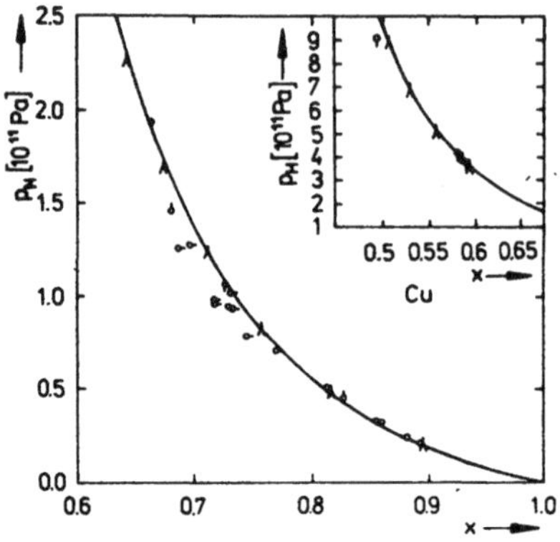

Fig. 3. Hugoniot pressure p_H versus relative volume for copper. The experimental values have been taken from [8]. The curve has been calculated according to the theory MC.

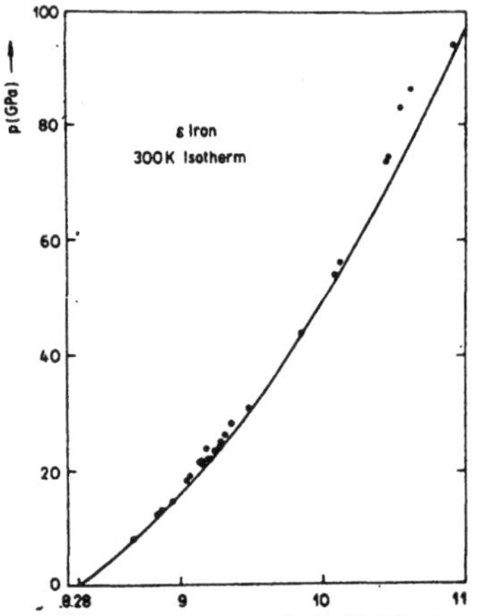

Fig. 2. Static high-pressure data p versus density for ε iron. The curve has been calculated according to theory MC. The solid circles represent experimental results from [5] and [6] plotted after [2].

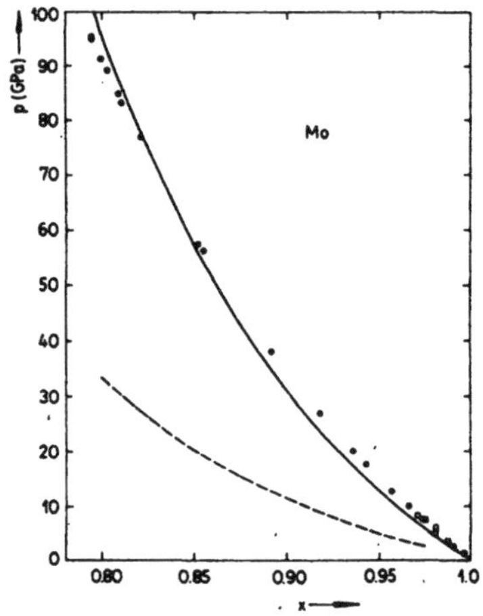

Fig. 4. Pressure p versus relative volume x. A comparision of the equations of state of MC (full curve) and of Wills and Harrison [16] (dashed curve) with experimental results for molybdenum. Full circles after [4], open squares after [3].

and

$$c_n = (4Z_d)^{-1} \left[(5Z_d - 10)\, c_{MA} + 10 - Z_d \right] . \tag{7}$$

The formulas have been derived in [14] and [15]. These papers are developments based on [11] and [12]. It is to be expected that the new theories [14] and [15] are better suited than [10, 13] for deriving the pressure dependence of melting temperature. In the following, a comparison with measured data is to demonstrate the applicability of the theories.

References

[1] L.V. Altshuler, K.K. Krupnikov, and M.I. Brazhnik, Soviet Phys. - J. exper. theor. Phys. 34 (1958) 614

[2] J.M. Brown and R.G. McQueen in: S. Akimoto, High-Pressure Research in Geophysics. Advances in Earth and Planetary Sciences 12 (1982) 611

[3] Li Chung Ming and M.H. Manghnani, J. Appl. Phys. 49 (1978) 208

[4] H.K. Mao, P.M. Bell, J.W. Shaner, and D.J. Steinberg, J. Appl. Phys. 49 (1978) 3276

[5] H.K. Mao, W.A. Bassett, and T. Takahashi, J. Appl. Phys. 38 (1967) 272

[6] H.K. Mao and P.M. Bell, J. Geophys. Res. 84 (1979) 4533

[7] R.G. McQueen and S.P. Marsh, J. Appl. Phys. 31 (1960) 1253

[8] M. Van Thiel, Compendium of Shock Wave Data Report, UCRL - 50108, Lawrence Livermore Laboratory, University of California 1977

[9] J.M. Walsh, M.H. Rice, R.G. McQueen, and F.L. Yarger, Phys. Rev. 108 (1957) 169

[10] U. Walzer, Exp. Techn. Phys. 31 (1983) 33

[11] U. Walzer, High Temp. - High Pressures 16 (1984) 335

[12] U. Walzer, Phys. Status Solidi B 125 (1984) 55

[13] U. Walzer, Exp. Techn. Phys. 32 (1984) 281

[14] U. Walzer, High Temp. - High Pressures 19 (1987) 161

[15] U. Walzer, Phys. Status Solidi B 140 (1987) 377

[16] J.M. Wills and W.A. Harrison, Phys. Rev. B 28 (1983) 4363

THERMODYNAMIC CONSIDERATIONS OF PHASE CHANGES IN THE EARTH'S MANTLE

Wolfgang Ullmann [1)]

First of all we must touch upon the dimensionless parameter Δ as a function of depth z [10]. Its computability by an Earth model such as PREM [4] is due to the relation

$$\Delta = \frac{\Phi}{\varrho g} \left(\frac{d\varrho}{dz}\right)^{-1} ,$$

where ϱ, g and Φ are the density, acceleration of gravity and seismic parameter at depth z, respectively. It can easily be shown that Δ amounts to the relative deviation of the effective compressibility of the material at depth z from that which could emerge under isentropic condition.

In view of a distinctive feature pointed out by Stacey [9] is should also be noted that $\Delta < 0$ is a necessary condition for convective transport of the material. Conversely, the material is convectively stable as long as $\Delta > 0$. In the latter case the region under consideration features phase changes (see also [3]).

Fig. 1 shows two graphs of Δ versus depth obtained by PREM and PEM [5]. The solid line is due to PREM, while the dashed line, drawn as a comparison only, is due to PEM. This chiefly goes to show that in either Earth model the upper mantle transition zone (TZ) is distinguished by outstandingly great absolute values of Δ.

If there exists a depth range in which the material is homogeneous with respect to phase as well as composition, then it can be shown that $\Delta + 1$ must be less than the ration C_P/C_V of the two principal specific heats C_P and C_V throughout that region, provided that the temperature is a strictly increasing function of depth as expected anyhow [11]. We particularly learn from Fig. 1 that Δ takes on values greater than 0.7 within the TZ. If such a postulate of homogeneity would be applicable to the TZ, then C_P/C_V must reach values greater than 1.7 down there.

However, since the material in the TZ can be assumed to behave like an isotropic (amorphous) solid, the appropriate values of C_P/C_V cannot exceed unity to such an extend without the occurrence of phase changes. Turning to the depth z suitably chosen we therefore consider a one-component system of two phases maintained at constant pressure P and temperature T. Ignoring any possible surface effects at the interface between the two phases at depth z, we can be sure that both temperature and pressure are uniform throughout the system. Further having conservation of the total mass of the system assumed to be closed we obtain

1) Dr. sc. W. Ullmann, Institut für Physik der Erde, Burgweg 11
6900 Jena, DDR

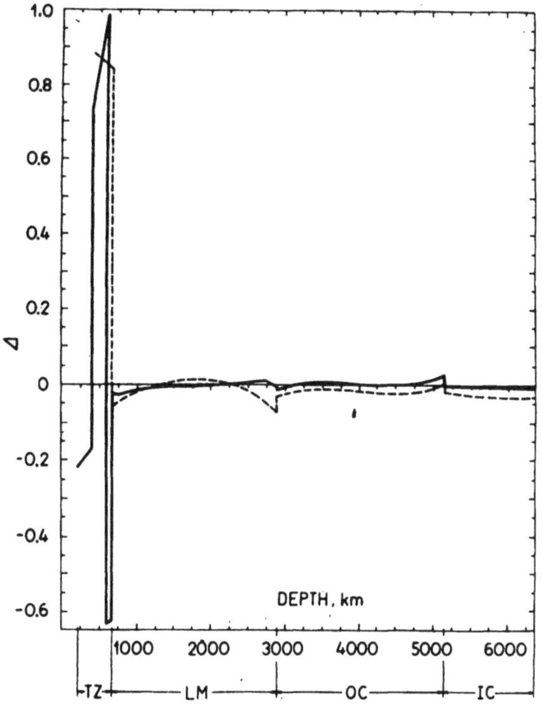

Fig. 1. The parameter Δ as a function of depth $z > 220$ km. The solid line is according to the PREM model. For comparison, the dashed line shows Δ after the PEM model.

equality between the specific Gibbs functions G_1 and G_2 of the two phases indicated by the subscripts 1 and 2, respectively.

As to the deriving of the condition for the mechanical and thermal equilibrium between the phases, the obvious choice is to minimize the specific Gibbs function G of the total system at depth z. Then G must have the same value in the two phases in mutual equilibrium. A first order phase change takes place at depth z if both

$$\frac{\partial}{\partial P} G_1(P,T) \neq \frac{\partial}{\partial P} G_2(P,T) \quad \text{and} \quad \frac{\partial}{\partial T} G_1(P,T) \neq \frac{\partial}{\partial T} G_2(P,T)$$

comes into question. In that case it is instructive to represent $G_i(P + \delta P, T + \delta T)$, $i = 1,2$, as a surface in three-dimensional space spanned by pressure, temperature and specific Gibbs function. The surfaces for the two different phases come into contact with each other in a curve along which

$$G_1(P + \delta P, T + \delta T) = G_2(P + \delta P, T + \delta T) , \tag{1}$$

as shown in Fig. 2. This equating is instrumental to the Clausius-Clapeyron equation.

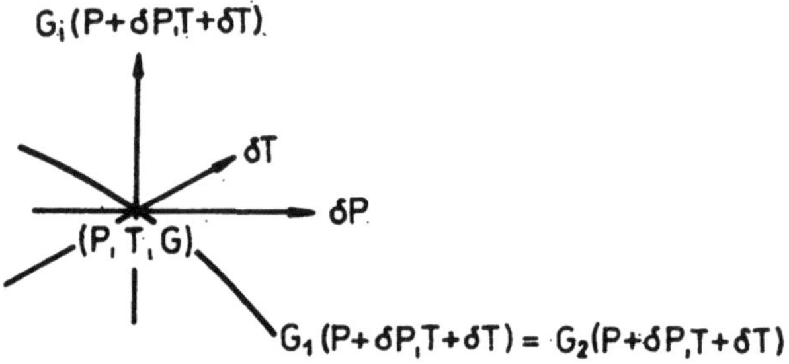

Fig. 2. The contact curve of the two surfaces given by the Gibbs functions $G_1(P + \delta P, T + \delta T)$ and $G_2(P + \delta P, T + \delta T)$, where P and T are kept constant.

In view of the relations

$$\left(\frac{\partial G_i}{\partial P}\right)_T = V_i \quad \text{and} \quad \left(\frac{\partial G_i}{\partial T}\right)_P = -S_i ,$$

where $V_i \equiv 1/\varrho_i$ and S_i are the specific volume and the specific entropy of the phase indicated by i, respectively, equation (1) leads to

$$\frac{\delta P}{\delta T} = \frac{S_2(z) - S_1(z)}{V_2(z) - V_1(z)} .$$

Replacing $V_i(z)$ and $S_i(z)$ by the appropriate one-sided limits

$$\lim_{\varepsilon \to 0} V(z \pm \varepsilon) \equiv V(z \pm 0) \quad \text{and} \quad \lim_{\varepsilon \to 0} S(z \pm \varepsilon) \equiv S(z \pm 0) ,$$

where ε is non-negative, we may write

$$\frac{\delta P}{\delta T} = \frac{S(z + 0) - S(z - 0)}{V(z + 0) - V(z - 0)} .$$

This formula gives the rate at which the pressure at depth z ought to be changed with temperature for the two phases to remain in stable equilibrium.

Since temperature is assumed to be strictly increasing with depth throughout the Earth's interior, it can be shown (see [12]) that a first order phase change at depth z leads to

$$S(z + 0) - S(z - 0) > 0 .$$

Further everything points to the fact that density as a function of depth is strictly increasing at every depth greater than 220 km (according to PREM) so that

$$V(z + 0) - V(z - 0) < 0$$

must be valid. From these inequalities having opposite senses we arrive at the conclusion that the first order phase change features a negative Clapeyron slope $\delta P/\delta T$ (cf. [12]).

Evidence of the Clapeyron slopes is of interest for the following reasons. If we consider the down-going material in mantle convection to be cooler than average (and the up-going material to be hotter) then the depth at which a particular phase change occurs in each of these is determined by the Clapeyron slope of the transition. By analysis of global gravity data McQueen and Stacey [6] found out that the only important phase changes occur within the TZ; the elevations and depressions of the two principal phases should be negatively correlated. In other words, the dominant phase changes occurring in the TZ involve Clapeyron slopes with unlike signs.

Taking up this idea we have no choice but suppose that not all phase changes in the TZ are of first order. In that case density and specific entropy vary continuously at the depth z where a phase change is assumed to occur. The idea suggests itself that the phase change under consideration is a second order transition according to Ehrenfest's classification scheme. From a detailed analysis it follows that a second order phase change taking place at depth z entails discontinuity of the two field functions generated by Φ and the Grüneisen parameter γ at that depth. We may then write

$$\frac{\delta P}{\delta \ln T} = \frac{\frac{\Delta(z+0)}{\gamma(z+0)} + \frac{\Delta(z-0)}{\gamma(z-0)}}{\frac{\Delta(z+0)+1}{\varkappa_S(z+0)} + \frac{\Delta(z-0)+1}{\varkappa_S(z-0)}}, \tag{2}$$

where $\varkappa_S = \varrho\Phi$ is the isentropic incompressibility.

Worth mentioning is that our theoretical result can be brought into line with recent experimental checks of the behaviour of γ at a phase change. For instance, in all cases that Boehler and Ramakrishnan [2] studied, whether the transition is structural or electronic, they found a sharp increase of γ at the phase change. Previous theoretical predictions using the Slater, Dugdale-MacDonald, and free volume formulas significantly underestimate this effect.

There is every indication that the relatively sharp seismic discontinuity of the first kind at about 400 km depth is due to a first order phase change. Ringwood [8] pointed out once again that in this case the transformation of olivine to $\beta(Mg, Fe)_2SiO_4$ is a possibility (see also [1]). Add to this that the corresponding Clapeyron slope is negative. That seem to be one of the two principal phase changes concluded by McQueen and Stacey. Following Ringwood's conception, we come to realize that the other is given by the further transformation of the β phase to $(Mg,Fe)_2SiO_4$ (spinel) which may cause a small seismic discontinuity of the first kind near 550 km depth. Hence, since Δ is positive everywhere in the depth interval between 400 km and 600 km (according to PREM), it follows from (2) that $\delta P/\delta T > 0$. If so, the assertion made by McQueen and Stacey should be born out.

References

[1] Bina, C.R., and B.J. Wood: Olivine-Spinel Transitions: Experimental and thermodynamic constraints and implications for the nature of the 400-km seismic discontinuity. J. Geophys. Res. 92 (1987) B6, 4853-4866.

[2] Boehler, R., and J. Ramakrishnan: Experimental results on the pressure dependence of the Grüneisen parameter: A review. J. Geophys. Res. 85 (1980), 6996-7002.

[3] Brown, J.M., and T.J. Shankland: Thermodynamic parameters in the Earth as determined from seismic profiles. Geophys. J. R. Astron. Soc. 66 (1981), 579-596.

[4] Dziewonski, A.M., and D.L. Anderson: Preliminary reference Earth model. Phys. Earth Planet. Inter. 25 (1981), 297-356.

[5] Dziewonski, A.M., Hales, A.L., and E.R. Lapwood: Parametrically simple Earth models consistent with geophysical data. Phys. Earth Planet. Inter. 10 (1975), 12-48.

[6] McQueen, H.W.S., and F.D. Stacey: Interpretation of low degree components of gravitational potential in terms of undulations of mantle phase boundaries. Tectonophysics 34 (1976), T1-T8.

[7] Navrotsky, A.: Lower mantle phase transitions may generally have negative pressure-temperature slopes. Geophys. Res. Lett. 7 (1980) 9, 709-711.

[8] Ringwood, A.E.: Phase transformations and differentiation in subduced lithosphere: implications for mantle dynamics, basalt petrogenesis, and crustal evolution. J. Geol. 90 (1982), 611-643.

[9] Stacey, F.D.: Application of thermodynamics to fundamental Earth physics. Geophys. Surv. 3 (1977), 175-204.

[10] Ullmann, W.: On the parametrization of inhomogeneities in the Earth. Gerlands Beitr. Geophys. 93 (1984) 1, 59-68.

[11] Ullmann, W.: Boundary conditions for the inhomogeneity parameter in the Earth: Implications for heat action within the upper mantle. Gerlands Beitr. Geophys. 93 (1984) 2, 142-148.

[12] Ullmann, W.: On thermal features within the transition zone. Phys. Earth Planet. Inter. 42 (1986), 154-164.

Phase Separation in Lithium Silicate Glasses under High Pressure-Temperature Conditions up to 2.0 Pa and Transformation Region

Dörte Stachel*, Wolfgang Götz*

Experiments

Two binary glasses consisting of 22.4 and 28.3 mole% Li_2O and 77.6 and 71.7 mole% SiO_2 resp. were melted in platinum crucibles by induction heating. After homogenization the clear melts were casted into brass tubes to from samples for pressure experiments. All samples were tempered at about 1.12 T_g (maximum of phase separation close before crystallization) to reach an equilibrium state as a starting point for the pressure experiments.

The experiments were performed in a piston-cylinder type apparatus shown in figure 1.

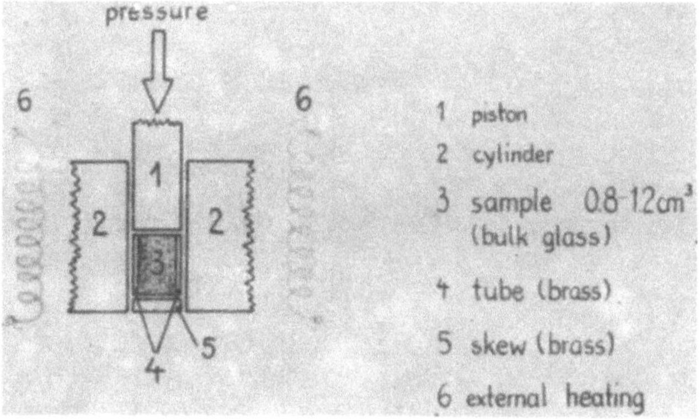

fig. 1

Consisting of special steel 55WCrV7 the device permits pressures up to 2 GPa and temperatures up to 500 °C.

After pressure treatment the compact glass samples were rapidly quenched to freeze their new structure.

Electron micrographs, X-ray scattering as well as density and refractive index measurements were carried out at room temperature and normal pressure and allowed to study only the permanent structure effects.

Results and Discussion

As known, crystalline $Li_2Si_2O_5$ is a pressure-temperature unstable substance /1/. Glasses in the system Li_2O-SiO_2 have a tendency to

*Friedrich-Schiller-University
Jena/GDR
Otto-Schott-Institute
Department of Chemistry

immiscibility. In the case of the used glass compositions the matrix glass phase is $Li_2Si_2O_5$-rich and the droplets consist of SiO_2 glass /2, 3/.
Figure 2 shows the change of the droplets diameters under pressure-temperature conditions.

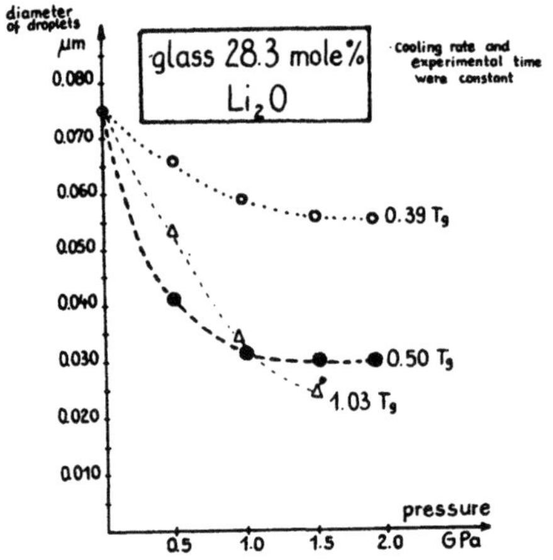

fig. 2

All temperatures are related to the transformation point. At higher pressures the diameters of the droplets decrease. The change below T_g is higher than should be expected for pure permanent network densification.
The pressure favours higher solubility of SiO_2 in the $Li_2Si_2O_5$ structure-like matrix glass and leads to its stabilization /4/. At high temperatures and low pressures the droplets grow. Pressures and temperatures exceeding this point of $Li_2Si_2O_5$ stability cause a decomposition of glassy $Li_2Si_2O_5$ matrix phase as shown, for instance, in figures 3 and 4.

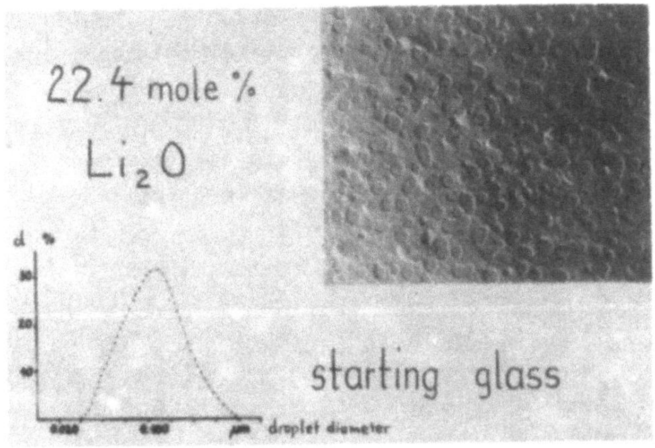

fig. 3

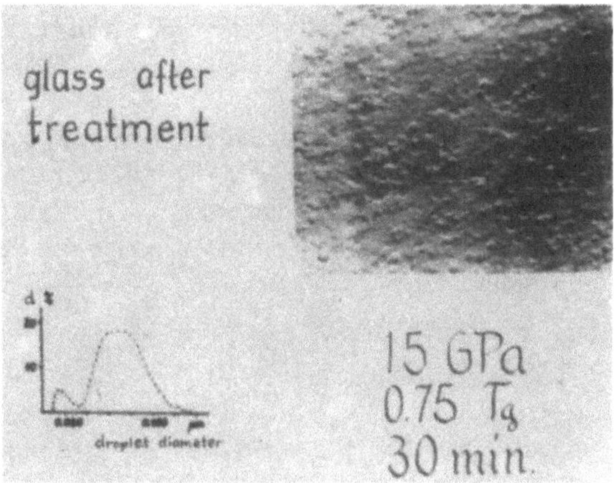

fig. 4

The distribution curve of the pressure-temperature tread glass has a little pre-maximum at smaller dimensions of the SiO_2 droplets and demonstrates a beginning secondary phase separation of the $Li_2Si_2O_5$ matrix glass into new (quite small) SiO_2 droplets and glassy Li_2SiO_3 matrix. Figure 5 shows the region including this decomposition of glassy $Li_2Si_2O_5$.

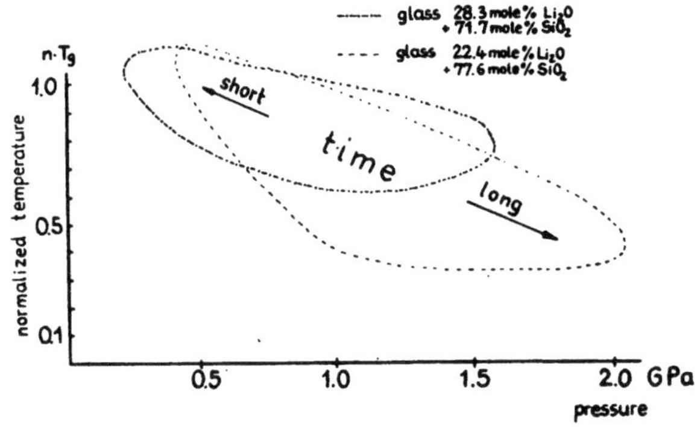

fig. 5

The limits of this range vary with glass composition and depend on the kinetics of the process.

Only pure $Li_2Si_2O_5$ decomposes completely. Excess SiO_2 concentration stabilizes the glass structure by a "dilution effect", as finally shown in figure 6.

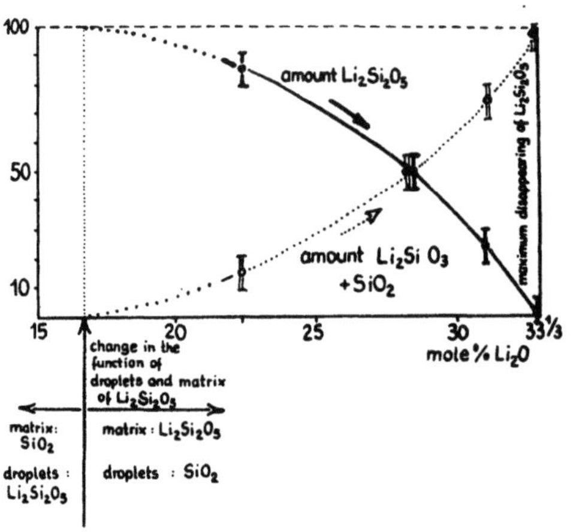

fig. 6

/1/ Tashiro, M.; S. Sakka & T. Yamamoto: Bull. Chem. Res. Kyoto Univ. 44 (1966), 5, 401 - 409

/2/ Vogel, W.: Ztschr. f. Chemie 3 (1963), Hefte 4, 5, 7, 8

/3/ Stachel, D.: Dissertation A, Jena 1975

/4/ Glasser, F. P.: Phys. Chem. Glasses. 8 (1967), 6, 224 - 232

IGH-PRESSURE INVESTIGATIONS OF LAYERED $CdInGaS_4$, $CdInAlS_4$, $TlGaSe_2$, ND $TlInS_{0.8}Se_{0.2}$ SINGLE CRYSTALS

K.R.Allakhverdiev[1], S.Guseinov[1], H.D.Hochheiemer[2], M.M.Shirinov[1], M.M.Tagiev[1]

All the investigated crystals have recently become the subject of extensive research. X-ray results on these crystals are controversy. Based on the data of Muller and Hahn [3] the authors [2] assigned the space group C2/C to the monoclinic ternary chalcogenides. In our previous report [4] concerning with pressure Raman experiments of $A^3B^3C_2{}^6$ crystals it was indicated that the best agreement with the group-theoretical predictions can be obtained if one suppose that a unit cell of these crystals contains four layers having the same space group as $TlSe(D_{4h}^{18})$ and the space group of the crystals being C_{2h}^6.

All compounds were grown from high-purity elements (at least 99.999%) in evacuated (10^{-5}Torr) quartz tubes by Bridgman ($TlGaS_2$, $TlGaSe_2$, $TlInS_{0.8}Se_{0.2}$) and chemical vapour transport ($CdInGaS_4$, $CdInAlS_4$) methods. The crystals obtained exhibited good optical quality and could easily be cleaved into plane-parallel plates perpendicular to (001) direction.

The phonon spectra of layered crystals contains low frequency IR or Raman-active phonon in which the layers vibrate as rigid units respectively to one another. For such a vibration the mode Gruneisen parameter

$$\Gamma_j \equiv \frac{1}{\beta}\gamma_i = \frac{1}{\beta}\frac{1}{\nu_j}\frac{d\nu_j}{dP} \qquad (1)$$

is one or two orders of magnitude higher in comparison with the same value for other so called intralayer vibrations, where β is the isothermal compressibility, ν_j is the frequency of j-th vibration. Raman-pressure measurements for layered crystals are very useful tool for identifieation of inter- and intralayer modes. These experiments are also useful for getting the information about the character of bonding between the atoms or group of atoms in crystals.

Raman scattering spectra of the investigated crystals were measured at room temperature under various pressure up to 50 kbar. The pressure were exerted on the specimens in an anvil diamond cell. Spectra were excited by the 676.4 nm and 647.1 nm lines of a Kr^+ laser, and were recorded using a Tarell-Ash as well as DFS-24 double monochromators and photon counting detection.

It was shown that $TlInS_2$ and $TlGaSe_2$ undergo a phase transition at about 5 kbar. The results for $TlGaSe_2$ are in good agreement with that

1)The Institute of Physics, Academy of Sciences of the Azerbaijan SSR, 33 Prospekt Narimanova, Baku, USSR

2)Max-Planck-Institut für FKF, Heisenbergstraße 1, Stuttgart, FRG.

with that published in [1, 4] . In the pressure interval about 5 kbar, where the phase transition occurs the pressure dependence of phonon frequencies shows discontinuities for some lines. No noticible change was observed for $TlInS_2$, $CdInGaS_4$, $TlGaSe_2$, $CdInAlS_4$ with pressure up to 50 kbar. All lines shifted to higher wave numbers with increasing pressure. The parameters $\gamma_j = (1/\nu_j)(d\nu_j/dP)_T$ were determined for all observed lines. The highest frequency shift under pressure was revealed by low-frequency vibrations which correspond to rigid-layer mode. The presence of at least one "rigid-layer" vibration for $CdInGaS_4$ and $CdInAlS_4$ crystals indicates that their elementary unit cell contains at least two layers. For $CdInGaS_4$ and $CdInAlS_4$ the value of γ_j for lowest frequency interlayer vibrations equal 38.2×10^{-3} kbar and 40.2×10^{-3} kbar, respectively. Such a high value could reflect a fact of a weak interlayer forces in these crystals in a comparison with the other one (for $CdInGaS_4$ this value at least 1.6 times higher than for GaSe).

Analysis of a set of value of γ_j parameters permitted to make some conclusion about the types of bonding between the atoms inside and between the layers. Atoms inside layers in the investigated crystals seemed to be bonded by ionic covalent forces. The bonding between the adjacent layers at the same time looks like to be of the van der Waals type.

In mixed $TlInS_{0.8}Se_{0.2}$, $TlInS_{0.7}Se_{0.3}$, $TlInS_{0.6}Se_{0.4}$, $TlInS_{0.5}Se_{0.5}$ and $TlInS_{0.4}Se_{0.6}$ reversible changes of colour in crystals with pressure from red to black were observed. For most of the investigated crystals Raman spectra disappeared at pressures about 18 kbar and higher. It seemed that the value of the critical pressure was linearly dependent on the composition.

Literature

[1] Allakhverdiev K.R., Babaev S.S., Bakhyshov N.A., Mamedov T.G., Peresada G.I., Shukyrov M.M., Salaev E.Yu.: The influence of hydrostatic pressure on the fundamental absorption edge of $TlInS_2$ and $TlInS_{1.8}Se_{0.2}$. Soviet Phys.Semiconducors. 18 (1984) 9, 1704-1706.

[2] Henkel W., Hochheimer H.D., Carlone C., Werner A., Ves S., Schnering H.G.: High-pressure Raman study of the ternary chalcogenides $TlGaS_2$, $TlGaSe_2$, $TlInS_2$ and $TlInSe_2$. Phys.Rev. B26 (1982) 6, 3211-3221.

[3] Muller V.D., Hahn H.: Zur Struktur des $TlGaSe_2$. Z .anorg.allg.Chem. 438 (1978), 258-272.

[4] Vinogradov E.A., Zhizhin G.N., Melnik N.N., Subbotin S.I., Panfilov V.V., Allakhverdiev K.R., Salaev E.Yu., Nani R.Kh.: Raman scattering and phase transformations of $TlGaSe_2$ and $TlGaS_2$ under pressure. Phys.Stat.Sol.(b) 95 (1979), 383-390.

HIGH PRESSURE STUDIES ON THE STRUCTURE OF ALKALINE-DISILICATE GLASSES

Goetz, W. and Hendrich, J. 1)

The glass-forming system $Li_2O - SiO_2$, $Na_2O - SiO_2$ and $K_2O - SiO_2$ were studied by maens of a piston-cylinder and an opposed-anvil device in a range up to 4 GPa and temperature to 400 °C.

The glass show different densification behaviour depending on its alkaline content. In the case of $Li_2O \cdot 2\,SiO_2$- and $Na_2O \cdot 2\,SiO_2$-glass there is a significant densification at temperature below T_g. But the metasilicate glasses show a higher densification only at temperatures near T_g . From these results was concluded, that the structure of the glasses of disilicate composition is formed by irregularly arranged elements of the layer structure of the crystalline disilicates. This causes a lower degree of space-filling and higher densification under pressure.

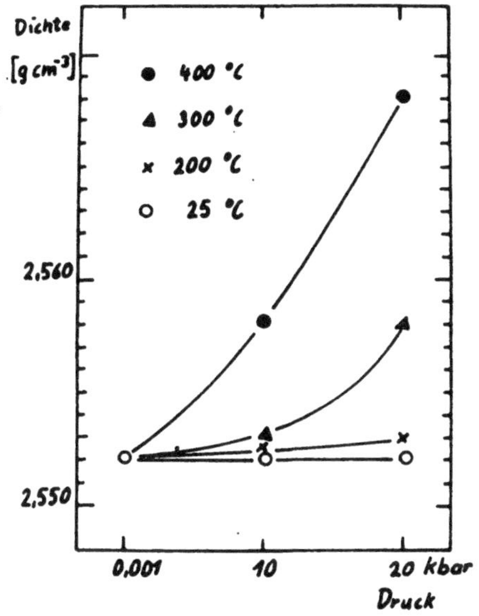

Fig. 1. The density of Na - disilicate in dependence of pressure

----

1) Friedrich-Schiller-University of Jena,
Otto-Schott-Institute, Department of Chemistry

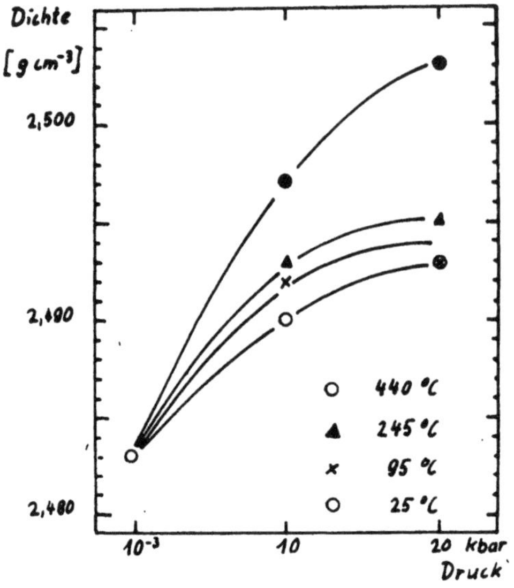

Fig. 2. The density of Na-metasilicate in dependence of pressure

One the contrary the glass-structure of metasilicate composition consists of relativly dense packed chain-like elements analogous to the crystalline metasilicates.

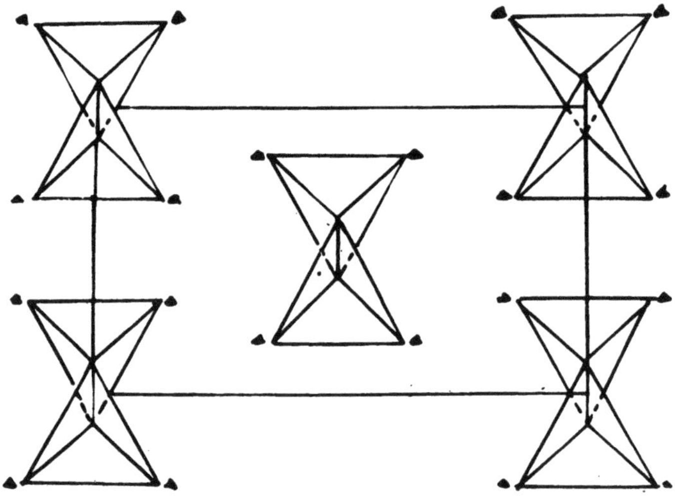

Fig. 3. The structure-model of Na-metasilicate

The structure model of the alkaline glasses could be supported by X-ray studies on the crystal phase precipitating from glass under pressure.

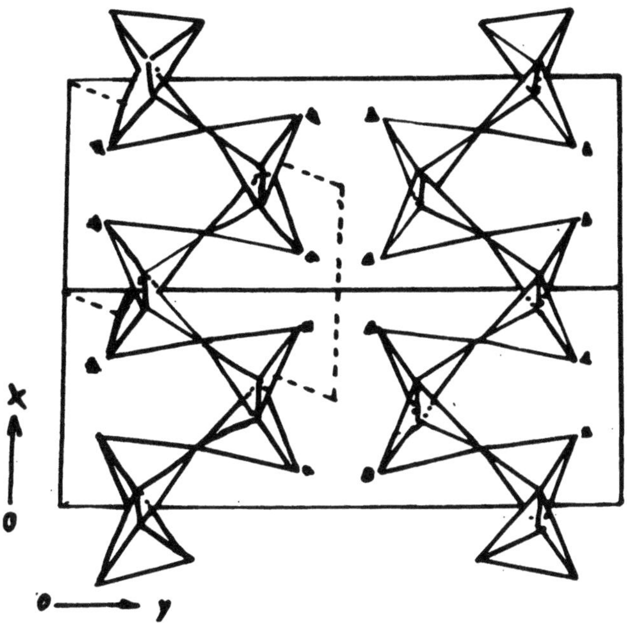

Fig.4. The structure-model of Na-disilicate

The stability of the disilicates was found to increase from lithium to potassium accordingly to the liquids-line in the phase diagrams.
For the first time could be shown that sodium disilicate (as also does Li-disilicate) decays under pressure into sodium metadisilicate and quartz. Potassium disilicate however was found to be stable in the studied pressure range up to 3 GPa.
The structure model describes above could recently be confirmed by Si^{19}-NMR(MAS)-studies.

References

/1/ Hendrich, J.: Dissertation, Friedrich-Schiller-Universitaet Jena, 1975
/2/ Soennichsen, K.: Diplomarbeit, Friedrich-Schiller-Universitaet; 1973
/3/ Tashiro, M.; Yamamoto, T.; Sakka, S.: Bull. Inst. Chem. Res. Kyoto Univ. 45; 318 (1967)

UNIAXIAL STRESS EFFECT ON THE ZERO-FIELD SPLITTING OF Cs_3CoCl_5 CRYSTAL

Yu Wan-Lun[1)]

Abstract: A relationship is established between the crystalline structure and the zero-field splitting parameter D of $3d^7$ ions in weak D_{2d} fields by using the superposition model. The relationship is employed to investigate the stress dependence of D of Cs_3CoCl_5 crystal and the theoretical results are reasonable.

The EPR spectra of Cs_3CoCl_5 crystal has been observed by Van Stapele et al [1] at normal pressure. Recently, Lukin and Teslya [2,3] have studied the stress dependence of the zero-field splitting parameter D with hydrostatic and uniaxial stresses. It was found that the value of D varies linearly with applied stresses and the variation coefficient $\partial D/\partial P$ was reported as $2.0 cm^{-1} GPa^{-1}$ for hydrostatic pressure and $6.6 cm^{-1} GPa^{-1}$ for stress along the C_4 axis at temperature 4K. This work makes an attempt to explain the observed effect.

In the crystal Cs_3CoCl_5, the Co^{2+} ion is surrounded by four Cl^{1-} ions and the local symmetry around Co^{2+} is D_{2d} [4].

Under the action of a tetrahedral D_{2d} field, the first excited cubic field level $^4T_2(F)$ of a $3d^7$ ion is split into $^4B_2$ and ^4E which hereafter will be denoted as $^4B_2(T_2)$ and $^4E(T_2)$, respectively.

The zero-field splitting parameter D of a $3d^7$ ion in a D_{2d} field can be calculated by taking the spin-orbit interaction as perturbation. The expression for the first and the second orders is as follows

$$D = -(4/9)\,\zeta^2\,(1/E_{b2} - 1/E_e) \tag{1}$$

where

$$E_{b2} = E(^4B_2(T_2)) - E(^4A_2), \quad E_e = E(^4E(T_2)) - E(^4A_2) \tag{2}$$

1) Institute of Solid State Physics, Sichuan Normal University, Chengdu 610066, People's Republic of China.

and ζ is the spin-orbit coupling constant. This expression is widely used in literature. The third-order perturbation formula can be derived in the same way. We have

$$D^{(3)} = (4/27)\,\zeta^3\,(1/E_e^2 - 1/E_eE_{b2}) \qquad (3)$$

Due to the action of the tetragonal components of the crystal field, the $^4E(T_2)$ state couples with $^4E(T_1)$ of ^4F and ^4P. To obtain the exact energy of $^4E(T_2)$ for the purpose of calculating the zero-field splitting parameter, one has to diagonalize a three-dimentional energy matrix. However, it is a good approximation to take the diagonal element of $^4E(T_2)$ in the weak field scheme as the energy of $^4E(T_2)$ for Cs_3CoCl_5, as in the weak crystal field the off-diagonal elements have values one order in magnitude less than the differences among the diagonal elements[5] .

Under this approximation we have, from the energy matrices in weak field scheme[5] ,

$$\begin{aligned} E_e &= -35b_2 + b_3 \\ E_{b2} &= 8b_3 \end{aligned} \qquad (4)$$

where b_2 and b_3 are crystal field parameters which can be expressed in the superposition model[6] as

$$\begin{aligned} b_2 &= (1/21)\bar{A}_4(R_0)\,(R_0/R)^{t_4}(35\cos^4\theta-30\cos^2\theta+3) \\ b_3 &= (5/3)\bar{A}_4(R_0)\,(R_0/R)^{t_4}\sin^4\theta \end{aligned} \qquad (5)$$

Here $\bar{A}_4$ is the intrinsic parameter and t_4 the power law exponent. In the point charge limitation, t_4 equals to 5 and $\bar{A}_4$ is given by

$$\bar{A}_4(R_0) = -\frac{e^2\langle r^4\rangle}{8\,R_0^5}\,(1 + \frac{5u}{eR_0}) \qquad (6)$$

u being the electric moment of Cl^{1-}. From the work of Zhao and Zhang 5 we obtain $\bar{A}_4$=535cm^{-1}.

Applied stresses will disturb the crystal structure, leading to changes in the interatomic distance R and in the bond angle θ. This will induce an additional crystal field which gives rise to changes in zero-field splitting parameter. For stress which is hydrostatic or along the C_4 axis, the local symmetry around the central metal ion Co^{2+} remains unchanged and the crystal field V at Co^{2+} is functions of R and θ only. Thus, in the first order, the stress-induced crystal field ΔV and the resultant zero-field splitting parameter ΔD can be expressed respectively as

$$\Delta V = (\frac{\partial V}{\partial R}\frac{dR}{dP} + \frac{\partial V}{\partial \theta}\frac{d\theta}{dP})\,dP \qquad (7)$$

$$\Delta D = \left(\frac{\partial D}{\partial R}\frac{dR}{dP} + \frac{\partial D}{\partial \theta}\frac{d\theta}{dP}\right) dP \tag{8}$$

The variations of R and Θ under stress can be known from the elastic constants. Detailed calculation gives

$$\frac{\partial D}{\partial R} = \frac{20}{27} \bar{A}_4(t_4/R_0)\, \zeta^2 \Big[\Big(\frac{1}{E_e^2} - \frac{8}{E_{b2}^2}\Big) \sin^4\theta_0 - \frac{1}{E_e^2}\,(35\cos^4\theta_0 - 30\cos^2\theta_0 + 3)\Big] \tag{9}$$

$$\frac{\partial D}{\partial \theta} = -\frac{40}{9}\bar{A}_4\, \zeta^2 \sin 2\theta_0 \Big[\Big(\frac{1}{E_e^2} - \frac{8}{E_{b2}^2}\Big)\frac{1}{3}\sin^2\theta_0 + \frac{5}{3E_e^2}\,(7\cos^2\theta_0 - 3)\Big] \tag{10}$$

in the second order, where R_0 and Θ_0 denote the bond distance and the bond angle at normal pressure, respectively. The third-order contribution can be calculated in the same way and is found to be negligible for Cs_3CoCl_5.

From the reported elastic constants 2 we calculate the stress variation coefficient $\partial D/\partial P$ of Cs_3CoCl_5. The results are listed in the table 1 for comparison with the experimental values 2 .

Table 1. The stress variation coefficient of D of Cs_3CoCl_5 crystal, in units of $cm^{-1}GPa^{-1}$.

	hydrostatic pressure			$P//C_4$		
t_4	5	6	7	5	6	7
300K	1.36	1.58	1.80	3.33	3.39	3.52
77K	1.28	1.46	1.50	2.77	2.88	2.94
obsd.[2]		2.0 ± 0.2			6.6 ± 0.3	

It is seen from the table that the theoretical values obtained for the variation coefficients are in good agreement with the experimental values for hydrostatic pressure but there is only a rough agreement for the stress along the C_4 axis. It is also observed that the $\partial D/\partial P$-value increases with increasing power law exponent t_4 and better values for hydrostatic pressure are arrived at $t_4=7$, although a little improvement can be seen for $P//C_4$. In addition, it is noted that the point charge limitation ($t_4=5$) gives reasonable results.

In summary, the crystal field theory has been shown to be successful in explaining the uniaxial stress effect on the zero-field splitting of Cs_3CoCl_5 crystal. It is expected that the present work can be developed to investigate the stress dependence of the Zeeman g-factor of this crystal.

---

[1] Van Stapele R.P., H.G.Beljens, P.F.Bongens and H.Zijilstra, J. Chem. Phys. 44 (1966) 3719.

[2] Lukin S.N. and O.P.Teslya, Phys.Stat.Sol(b) 132 (1985) 601.

[3] Lukin S.N. and O.P.Teslya, Fiz. Nizk. Temp. 10 (1984) 78.

[4] Figgis B.N., M.Gerloch and R.Mason, Proc. Roy. Soc. A279 (1964) 210.

[5] Zhao M. G. and Y.F.Zhang, IEEE: Trans. Mag. MAG19 (1983) 1972.

[6] Newman D.J., Adv. Phys. 20 (1971) 197.

MECHANICAL AND THERMAL INVESTIGATIONS OF PARAGONITE

Raab, S. 1) ; Ullrich, B. 2) ; Vollstaedt, H. 1)

I n t r o d u c t i o n

Paragonite is a plate-shaped dioctahedral layer silicate. It belongs to the glimmer-group. Contrary to the dioctahedral three layer silicates as pyrophyllite and talc, which show no intercalation of interlayer cations, a part of the Si^{+4}-ions are replaced by Al^{+3}-ions. As a result of the arising negative charge excess interlayer cations as K, Na, and Ca are inserted between the sheets. In the case of paragonite it is Na.

In the south of the GDR a small deposit exists where paragonite occurs in a compact, nearly monomineralic form. Up to now no commercial usage of paragonite has been known. Investigations were conducted to prove the application of paragonite as gasket material in high pressure devices from the type Bridgman or belt.

E x p e r i m e n t a l   m e t h o d s

Thermal experiments were performed in a derivatograph (system PAULIK and PAULIK) using either the DTA- (differential thermal analysis) TG- (thermal gravimetry) combination or the DTA- TD- (thermal dilatometry) combination. The sample material was crushed to a grain size of < 63 µm and studied up to 1000°C.

The mechanical behaviour of paragonite was examined in a Bridgman- anvil chamber using the critical thickness (h_c) methode after WAKATSUKI et al. [1]. An advantageous mixture of grain sizes was heattreated stepwise mingled with a small amount of an organic binder and pressed to tablets with 12 mm in diameter and variable thickness at 0,3 GPa. The thickness of these discs before and after the pressure test up to 1,5 GPa was measured within an accuracy of ± 0,005 mm.

---

1) AdW der DDR, Zentralinstitut fuer Physik der Erde, Telegrafenberg, Potsdam, DDR - 1561

2) Institut fuer mineralische Rohstoff- u. Lagerstaettenwirtschaft, Oskar- Roeder- Str. 10, Dresden, DDR - 8036

Results and Discussion

Mineralogical description and chemical composition

The mineralogical properties of the phyllosilicate paragonite are essentially determined by its position in the mica group. They include excellent cleavability according to {001}, flexibility and elasticity, a hardness of 2-3 and a density of some 2,85 g/cm^3.

The intercalation of Na into the interlayer lattices give rise to a higher plasticity of paragonite with respect to muscovite where potassium is the interlayer cation.

The paragonite rock from the Western Erzgebirge comprises more than 80% paragonite, small quantities of muscovite (up to 10%) as an independent mineral phase and traces of other minerals which, however, amount to a total of less than 2%.

All occuring varieties of the paragonite rock where investigated chemically (tab.1). The average chemical composition is in close proximity to the theoretical chemical composition.

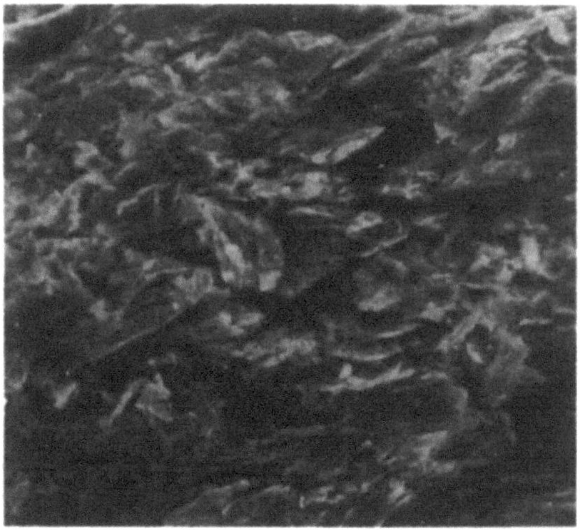

Fig.1 SEM-picture of paragonite rock from the Western Erzgebirge

The microscopic picture shows a fine-grained to dense paragonite pattern, partly of an irregular scaly structure and partly with a distinct texture of paragonite scales (Fig.1). The graine size lies between 10 and 50 µm.

Table 1

Average chemical composition of the paragonite rock from Western Erzgebirge compared to the theoretical chemical composition

	SiO_2	Al_2O_3	Fe_2O_3	MgO	CaO	K_2O	Na_2O	TiO_2	H_2O
mean	44,85	38,75	0,64	0,29	0,57	1,43	7,06	0,77	5,24
theor.	47,16	40,01					8,11		4,7

Thermal investigations

The sample is characterized by a small content of adsorbed water. The weight loss up to 400°C is less than 0,5% (Fig.2, TG-curve). The delivery of 4,5% hydroxyl water leads to an expansion beginning at 500°C and persisting up to the end of the dehydroxylation process at approximately

900°C (see TD-curve in Fig.2). Above 1000°C a conversion to high-temperature phases begins forming at first mullite and a glassy phase and from 1400°C onwards corrundum.
Therefore paragonite shows the typical thermal decomposition sequence for dioctahedral clay minerals:

natural sample --->
homogeneous medium product (dehydrated) --->
homogeneous medium product (dehydroxylated) --->
heterogeneous reaction products

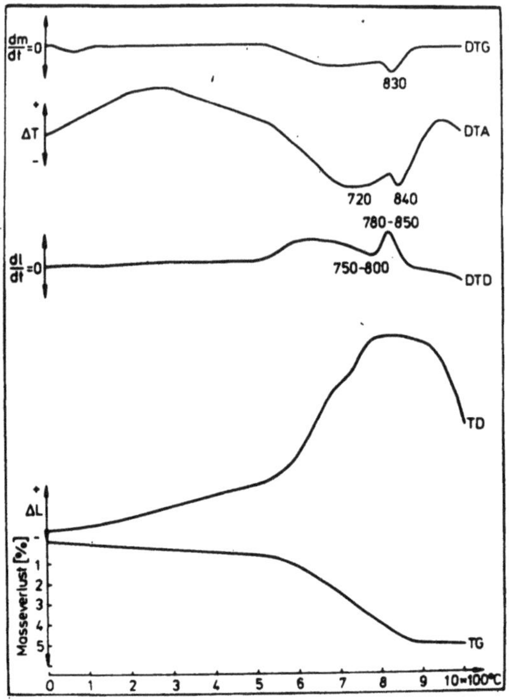

Fig.2 DTA-TG-DTG-curves and TD-DTD-curves of paragonite

<u>Mechanical investigations</u>
Paragonite was heattreated stepwise and then pressed to tablets. The endthickness h_E of the disc was measured after a pressure run in dependence upon the initial thickness h_A at various temperature levels.
For each temperature level there exists a critical thickness h_c, that means the highest endthickness value (Fig.3).

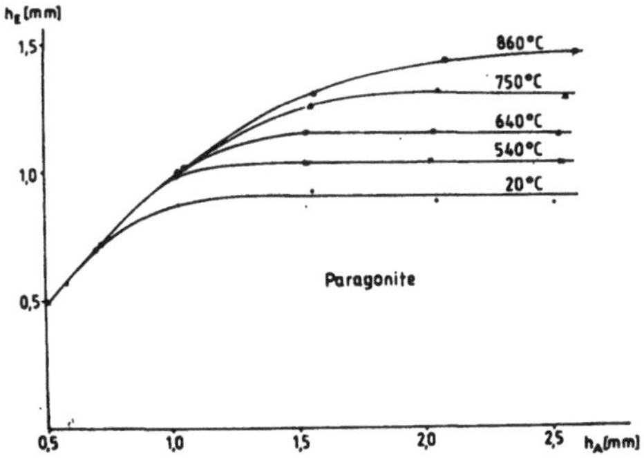

Fig.3 Endthickness h_E in dependence on initial thickness h_A at various temperature levels

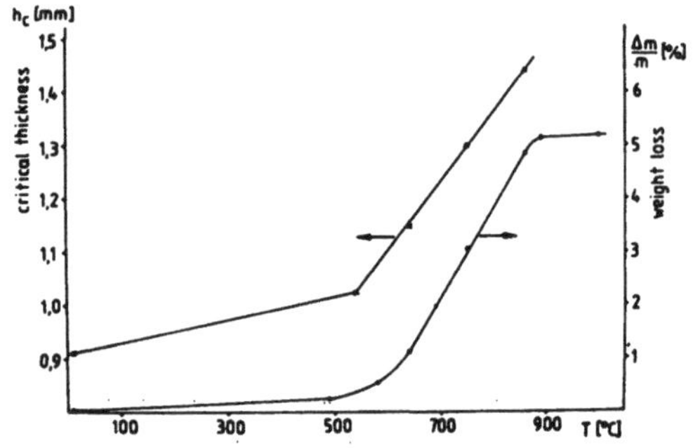

Fig.4 Critical thickness h_c and weight loss of paragonite in dependence upon temperature

Fig.4 shows the dependence of the critical thickness h_c and the weight loss on temperature. There is an obvious connection between the plastic flow properties represented by the h_c-value and the crystal water content.

The strong linear correlation between these two values is given in Fig.5. The straight line

$$y = -4,969\ 10^{-3}\ x + 1,514$$

fit the data with a correlation factor $r = 0,995$.

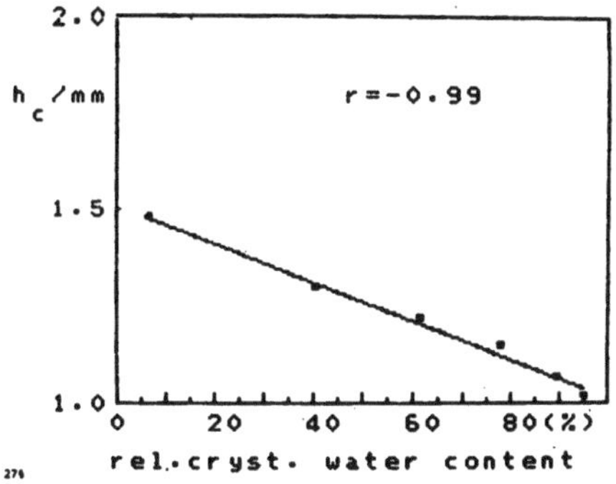

Fig.5 Correlation between the critical thickness h_c and the relative crystal water content of paragonite

Literature

[1] Wakatsuki, M., K. Ichinose, T.Aoki ; Jpn.J.Appl.Phys. 11(1972) No.4

P-V-T AND PHASE DIAGRAM OF IRON TO 420 KBAR AND 2300 K.

R. Boehler 1), M. Nicol 2), S. Johnson 2), M. Nielsen 3), I.M. Besson 4), G. Weill 4), J.P. Itié 5)

Abstract. Wire heating in diamond anvil cells provides high temperature stability and uniform temperature distribution for in situ measurements of pressure, temperature, resistance and X-ray diffraction patterns of iron. We have measured the solid-solid (hcp-fcc) transition and the melting curve of iron from resistance measurements and emission spectroscopy, and the density of fcc-iron using synchrotron radiation. The hcp-fcc ($\varepsilon - \gamma$) transition is in very good agreement with previous estimates to lower pressures. The melting temperatures are considerably lower than those obtained from laser heating and yield a melting temperature for the inner-outer core boundary that is on the low side of the many theoretical estimates. The extrapolation of our PVT data to outer core conditions yields a density of iron, that is about 15 % higher than the seismic model density in PREM.

Introduction. Cosmochemical and geophysical arguments suggest that the major element in the core of the Earth is iron. The precision of seismic models and density measurements at high pressure such as shock compression measurements, however, is not high enough to determine the precise chemicalcomposition of the Earth's core, which for example, puts tight constraints on accretion models of our planet. While seismic methods and models are being improved they have to be checked by laboratory experiments on the crystallography, density, and melting behavior of the core material, e.g. the melting temperature of iron at P-T conditions of the solid-liquid core boundary can be used to estimate the temperature in the earth at that depth and the measurement of the density puts a constraint on the chemical composition of the core.
In this study new methods to measure phase diagrams, melting temperatures and densities of iron (and metals in general) are introducted. The P-T range substantially exceeds that of previous methods and the accuracy for measuring T and V are greatly improved by optical emission spectroscopy and by angle dispersive technique using synchrotron radiation. These data represent the first density measurements of fcc-iron; do to the

----

1) Max-Planck-Institut fuer Chemie, Mainz, FRG
2) Department of Chemistry, University of California, Los Angeles, CA,USA
3) Research Establishment, Risø, Roskilde, Denmark
4) Université Curie, Paris, France
5) Université Paris-Nord; Villetaneuse, France

limited beam time at the Synchrotron, however, the amount of PVT data is too small to solve a full high-temperature equation of state of iron.

Method: A sandwiched gasket arrangement shown in figure 1 is used in the diamond anvil cell. The iron-wire (0.025 mm diameter), embedded in alumina, is resistively heated, where the gasked halves are used as the electrodes. The pressure were measured by the ruby method using a ruby line shift of 0.353 Å/kbar (Chopelas and Boehler, 1984). Temperatures are measured spectroscopically with a diode-array detector over a 300 nm spectral range. This method is described in more detail by Boehler(1986).

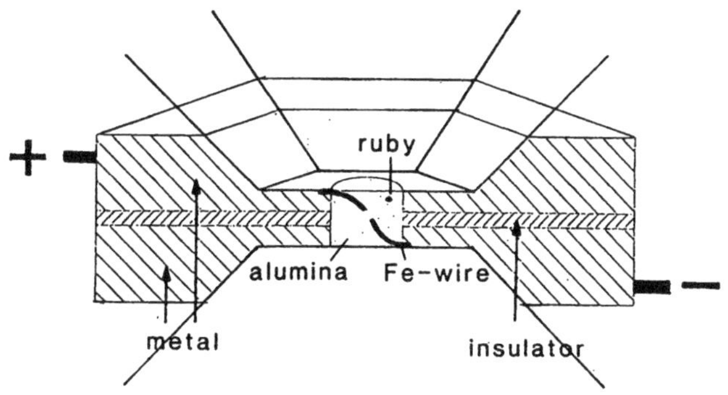

Fig. 1. Schematic of internally heated diamond anvil cell.

Due to the high thermal conductivity of diamond, the electrically heated wire develops a stable hot spot near the center of the cell of about 25 μm in diameter. The synchrotron beam which is collimated to about 30 x 30 μm and monochromated to 700 pm is focussed on the hot spot of the wire, shown schematically in figure 2.
Scattered X-rays were detected with a position sensitive flow proportional counter filled with Xe: 10% CH_4 located 1.1 m from the sample. Collection times were about 10 minutes. A typical diffraction pattern is shown in figure 3.

Phase diagram of iron. The ε-γ (hcp-fcc) transition and the melting curve of iron have been studied by this method. The results are published elsewhere (Boehler 1986). The ε - γ transition is in good agreement with lower pressure data and thermodynamical constraints on the α - ε - γ triple point. The melting curve, however, because of its low slope leading to a melting temperature of less than 2500 K at 500 kbar has been subject to debate because it differs significantly from laser heated experiments

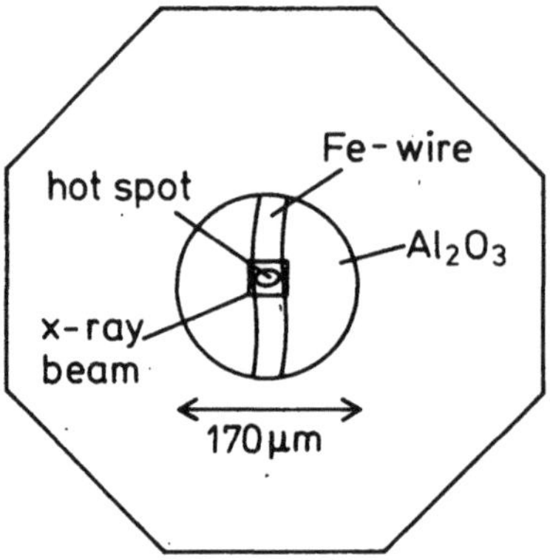

Fig. 2
Approximate dimensions of the high pressure cell, hot spot and synchrotron beam diameter.

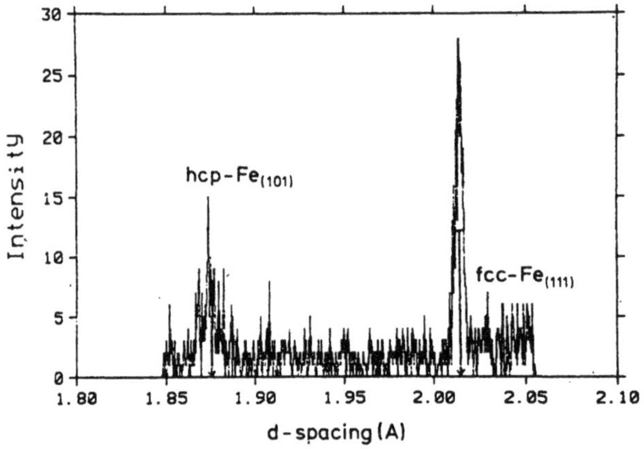

Fig. 3
Typical diffraction pattern of iron at high pressure and high temperature. fcc-iron is only present in the center of the hot spot and therefore yields sharper diffraction than ε-iron which is subject to large temrature gradients.

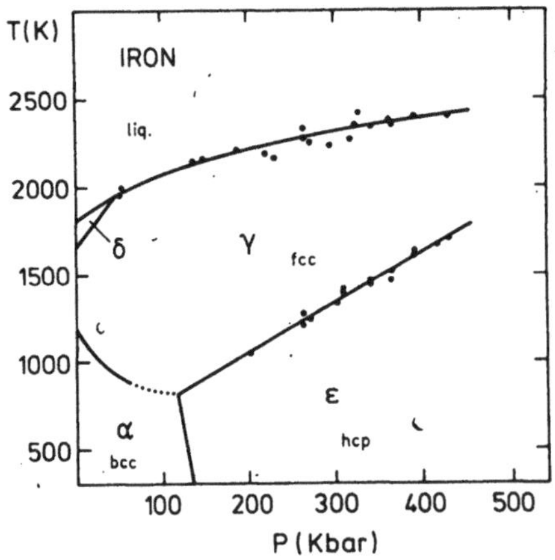

Fig. 4
Phase diagram of iron. Boehler (1986).

Williams et al., 1987). Laser heating experiments, however, yield very large data scatter (error bars over 1000 K) due to laser power fluctuation and very steep temperature gradients. Due to the large size of our (very stable) hot spot the consideration of the temperature gradients has has proven not to affect average temperature measurements. The trend, nevertheless, in that laser heating produces higher melting temperatures than resistive heating. One reason for this could be the significantly different temperature distribition in the sample that is heated from the inside in one case and from the outside in the outer case. Only a combination of the two methods will solve this problem.

<u>V(P,T) of fcc-iron.</u> The volumes of fcc-iron shown in table 1 are obtained from the shift of the (111) diffraction line. Pressure were obtained from the ruby line shifts. Independent estimates of pressures in the iron wire were made by comparing the positions of the (002) and (101) features of the X-ray diffraction patterns of ε-iron at room temperature with data reported by Jephcoat et al. (1986). In general, the positions of the relatively weak (002) features corresponded have been noted by Jephcoat et al. These deviations and the low intensities of the (002) diffraction features imply that the ε-iron is perferentially oriented by an axial pressure gradient in the alumina with the (001) axis aligned along the axial, high stress direction.

Table 1:
P - V - T of fcc-iron

P (kbar)	V (cm^3/mol)	T (K)
0	7.36	1200
112	6.81	1210
206	6.50	1140
216	6.43	1204
245	6.41	1680
255	6.40	1690
255	6.42	1737
264	6.34	1770
337	6.07	1628
337	6.07	1688
337	6.13	2093
337	6.10	2100
337	6.09	1761
337	6.07	1982
400	5.90	1622
422	6.05	2147
422	6.01	2185
422	6.09	2340

The temperatures of the γ-iron samples were determined in two ways. During the X-ray measurements, the current-voltage characteristics of the sample and the electrical power used to heat sample were measured. Boehler (1986) has shown that, at a given pressure, the temperature of the hot spot of the wire can be estimated with an accuracy of about $\pm$ 20 K by comparing the heating power with the needed to initiate (below 11.6 GPa) the α-γ or (above 11.6 GPa) the ε-γ transition at: T/K = 478 + 2.86 x (P/kbar). Optical (blackbody) determinations of the temperatures of the hot spots were made only during a few experiments with γ-iron because of geometrical constraints of the beam line and diffractometer arrangement. The temperature determined by fitting black-body functions to the emission spectra range within 100 K of the temperature estimated from measurements of the electrical power used to heat the sample.

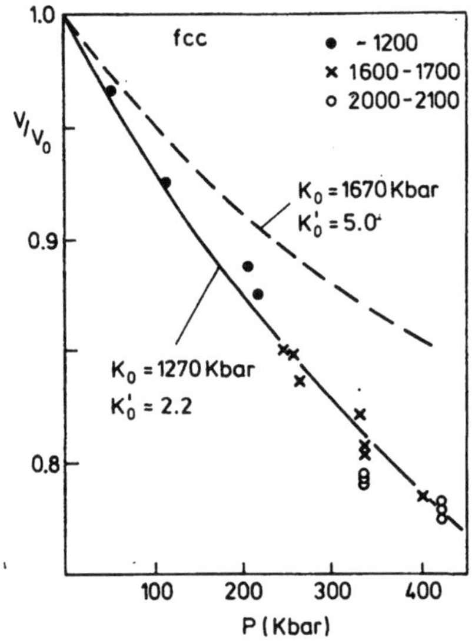

Fig. 5. Compression of fcc-iron to 422 kbar and 2340 K. The zero pressure volumes were calculated from one atmosphere thermal expansion data.

In figure 5 the compression of fcc-iron is plotted against pressure. The compression was not measured along isotherms and is therefore represented by 3 different temperature ranges. V_0 for each temperature was calcutated using handbook values of the thermal expansion of fcc-iron at one atmosphere. The solid line (least squares fit) represents an average bulk modulus of K_0 = 1270 Kbar and its pressure derivative K_0' = 2.2 in the temperature range 1200 - 2100 K. The dashed line is a suggested compression curve of fcc-iron (Anderson, 1987) at room temperature using K_0 = 1670 Kbar an K_0' = 5.0, values based on the α- and ε-iron compressions. These values and our data yield temperature derivatives of the bulk modulus of $dK_0/dT = 0.3 - 0.4$ Kbar K^{-1} and $dK_0'/dT = 0.0015 - 0.002\ K^{-1}$. We have calculated the density of iron at conditions of the core-mantle boundary using a Murnaghan equation and our average bulk modulus data. For solid iron this yields a value of $\rho = 11.5\ g/cm^3$ which is shown in figure 6.

This is in very good agreement with data obtained by Brown and McQueen (1986) from shock wave experiments. Thermodynamical calculations by Anderson (1986) and PREM are also shown. The vertical bars on our curve represents a temperature range 3500 - 4500 K which we consider the uncertainty of temperature estimates at the core mantle boundary. The arrows in figure 6 indicate the range of densities in the outer core derived from seismic models.

It is clear that our data base is too small to give a full thermal equation of state of iron suitable for the extrapolation to inner core conditions. The data nevertheless are the first on -iron and the first in this P-T range. The accuracy, due to the nature of the synchrotron radiation, cell geometry, and temperature stability, is high enough to yield meaningfull results when moderately extrapolated to outer core conditions.

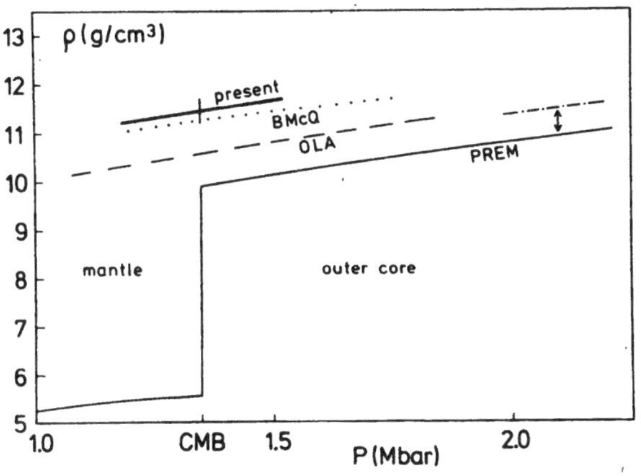

Fig. 6 Density of fcc-iron at P-T conditions of the outer core. PREM = Preliminary reference Earth model (Dziewonski and Anderson, 1981). The range of other density models is indicated by the arrow. Dashed line - O.L. Anderson (1986) thermodynamical calculation; Dotted Line - Brown and McQueen (1986) densities along the Fe-Hugoniot; the temperature at 1,35 Kbar is about 4000 K. Solid Line - present results, extrapolated from data shown in figure 5 using Murnaghan equation. The bar corresponds to a temperature range of 3500 - 4500 K.

References.

/1/ Anderson, O.L., Properties of iron at the Earth's core conditions, Geophy. J. R. Astr. Soc. (1986).

/2/ Boehler, R., The phase diagram of iron to 320 kbar, Geophys. Res. Lett., 13, 1153-1156 (1986).

/3/ Brown, J.M., and R.G.McQueen, Phase transitions, Grueneisen parameter and elasticity for shocked iron between 100 GPa and 400 GPa, J. Geophys. Res., 91, 7485-7494 (1986).

/4/ Chopelas, A., and R. Boehler, MgO: V^{2+} pressure scale for diamond anvil cells, Mat. Res. Symp. Proc., 22, 275-278, (1984).

/5/ Dziewonski, A.M., and D.L. Anderson, Preliminary reference Earth model, Phys. Earth Planet. Int. 25, 297-356 (1981).

/6/ Jephcoat, A.P., H.K. Mao, and P.M. Bell, The static compression of iron to 78 GPa with rare gas solids as pressure-transmitting media, J.Geophys. Res. 91, 4677-4684, (1986).

/7/ Williams, Q., R. Jeanloz, J. Bass, B. Svensen, T.J. Ahrens, and J. Verne, The melting curve of iron to 250 Gigapascals: A constraint on the temperature at the Earth's center, Science, 236, 181-182 (1987).

Acknowledgements. We are grateful for the warm hospitality and facilities to us by Professor C. Kunz, Dr. G. Materlik and the straff of the HASYLAB and for assistance in preparing programs, equipment, and samples provided by M.L. Johnson, P. Stanitzeck, Y. Talmi, and C.S. Zha. We are pleased to acknowledge that financial support and equipment for these experiments was derived from NSF Grants DMR 83-18812 and EAR 85-07755, an INSU Grant under ATD Contract 1627, and grants from the LLNL-IGPP and the UCLA Resarch Committee.

A MODIFICATION OF THE SIMON - GLATZEL - EQUATION

Rein, C. 1); Demus, D. 2)

The usefulness of the well-known Simon - Glatzel - equation [1]

$$\frac{p}{b} = \left(\frac{T}{T_0}\right)^m - 1 \tag{1}$$

for the description of melting and other phase transition lines is confirmed for many examples. However, the structure of this equation does not allow a description of a turning back of transition lines, that is, this equation is not able to describe a change of a positive to a negative slope with a temperature maximum on the transition line.
Usually the Simon - Glatzel - equation is derived from the Clausius - Clapeyron - equation by integration of

$$T dp/dT = \Delta H/\Delta V = m \cdot (p+b) \tag{2}$$

(m, b are constants).
But the same result we will get with the assumptions:

a, $\Delta H = k \cdot m(p+b)$ (3a)
$\Delta V = k = \text{const.}$ (3b)

or b, $\Delta H = k = \text{const.}$ (4a)
$\Delta V = k \cdot a/(p+b)$ (4b)

($a = 1/m$)
(It is trivial to set k = 1).
The case a is inconsistent with the experimental experience. On the other hand case b values the influence of ΔH and ΔV on the shape of the transition line in an appropriate way [2] and reflects the dependence of ΔV on the pressure with the correct trend.
The continuous change of ΔV is a result of the continuous change of the volumes of both phases participating in the transition. At high pressures the difference ΔV can remain positive or it may become zero or negative. All three cases (fig. 1) can be described by

$$\Delta V = a/(p+b) + c \tag{5}$$

(c is a constant).

----

1) Zentralinst. fuer Physik der Erde, AdW, Potsdam, Telegrafenberg, GDR
2) Sektion Chemie, Martin - Luther - Univ., Halle, Muehlpforte 1, GDR

Combined with eq.(4a) and the Clausius - Clapeyron - equation we get (after integration) a modified Simon - Glatzel - equation

$$T = T_0 \cdot (1 + p/b)^a \cdot e^{cp} \qquad (6)$$

For $c = 0$ eq.(6) is identical with the Simon - Glatzel - equation (1).

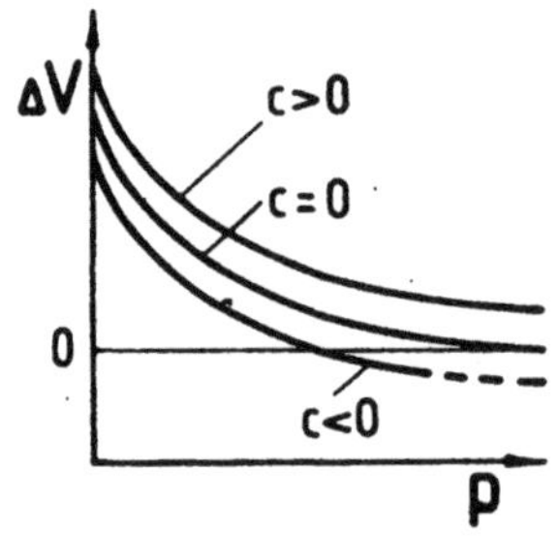

Fig.1 Schematic representation of of ΔV as a function of p dependent on c in eq.5

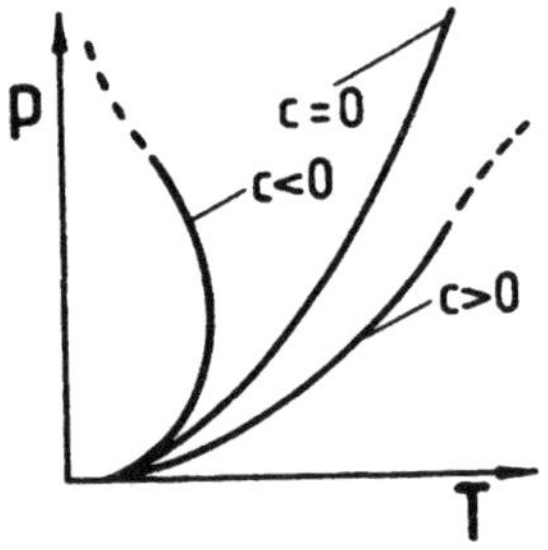

Fig.2 The shape of the transition lines dependent on c in eq.(6)

Eq.(6) takes into account the dominant role of volume changes for the shape of the phase transition lines and allows the description of a temperature maximum. If this point is passed, the influence of ΔV decreases and eq.(5) is not any more applicable on the supposition that changes of ΔH are of minor importance (eq.4)(dashed parts of curves in fig.2).
To investigate the validity of eq.6 we have carried out fits of reported melting data of several metals. Additionally we fitted own data of the nematic - isotropic transition (orientational melting) of the liquid crystal n - octyloxy - cyanobiphenyl (8OCB).

Table 1: Results of fits. a,b,c are constants of eq.(6). S is the standard deviation, n is the number of data points and p_m is the highest pressure, which was used for calculations.

Subst.	a	b/MPa	$10^5 \cdot c/MPa^{-1}$	S/K	n	p_m/GPa	ref.
Li	0.1747	2010.3	-1.4664	0.01	9	1.3122	[3]
Na	0.3441	1320.2	-2.3019	0.06	14	2.1720	[4]
K	0.2739	512.9	-2.5378	0.11	13	1.2000	[1]*
Rb	0.3064	428.7	-4.2624	0.01	9	1.2995	[3]
Rb	0.3066	434.2	-3.7506	0.01	7	0.3000	[5]*
Cs	0.4022	444.7	-1.2116	0.02	5	0.6000	[6]
Cs	0.4143	417.4	-1.8276	0.06	8	0.3500	[5]*
8OCB	0.3915	407.5	+1.8462	0.19	11	0.8265	-

Table 2: Calculated parameters at zero pressure by use of the fitting results of tab.1. K is the curvature $K = \frac{d^2T/dp^2}{(1+(dT/dp)^2)^{3/2}}$ of the melting line. p(Tmax) is the calculated pressure of the temperature maximum far in the extrapolated range.

Subst.	T	dp/dT	K	Tdp/dT	p(Tmax)	ref.
	K	MPa/K	(GPa,K)	GPa	GPa	
Li	453.76	30.50	-17	13.84	9.9	[3]
Na	370.96	11.34	-51	4.21	13.6	[4]
K	335.6	5.85	-251	1.96	10.3	[1]*
Rb	312.1	4.77	-355	1.49	6.8	[3]
Rb	311.9	4.79	-344	1.49	7.7	[5]*
Cs	301.5	4.22	-393	1.27	2.9	[6]
Cs	302.9	4.08	-477	1.23	1.8	[5]*
8OCB	353.2	2.89	-417	1.02	-	-

* Used p - data were not transformed to SI-units ([1]: atm, [5]: kg/cm^2).

The low standard deviations demonstrate the satisfactory interpolation properties of eq.(6). On the other hand, the extrapolation results agree roughly with experimental trends (p(Tmax): Cs ≈ 2 GPa [6]; Rb > 8 GPa [7]; K > 8 GPa [7]; Na ? [7]; Li > 10 GPa [7]). No tendency for a temperature maximum is observable in the case of the liquid crystal 8OCB. This shows the high stability of the liquid crystalline order. Furthermore we can state, that the demand of the Simon-Glatzel-equation, the linear dependence of $\Delta H/\Delta V$ on pressure, cannot be maintained.

References

[1] F.E.Simon, G.Glatzel: Z.Anorg.allg.Chem. 178 (1929) 30

[2] P.W.Bridgman: Proc.Amer.Acad.Arts Sci. 74 (1942) 399

[3] I.N.Makarenko, A.M.Nikolaenko, S.M.Stishov: Proc.6.AIRAPT conf., 1977,K.D.Timmerhaus, M.S.Barber,edit.,Plenum Press, N.Y.(1977) 347

[4] V.A.Ivanov, I.N.Makarenko, A.M.Nikolaenko, S.M.Stishov: Phys.Lett. 47A (1974) 75

[5] P.W.Bridgman: Proc.Amer.Acad.Arts Sci. 60 (1925) 385

[6] I.N.Makarenko, V.A. Ivanov, S.M.Stishov: JETP Lett. 18 (1973) 187

[7] H.D Luedemann,G.C.Kennedy: J.Geophys.Res. 73 (1968) 2795

HIGH PRESSURE INVESTIGATIONS OF LATERAL SUBSTITUTED LIQUID CRYSTALLINE COMPOUNDS

Rein,C. 1); Demus,D. 2)

I n t r o d u c t i o n

Because of the positive slope of the nematic - isotropic - transition - line [1], there is a stabilizing effect of the pressure on the nematic state. In connection with the existence of liquid crystalline compounds diverging considerably from the "ideal" spherocylinder - like shape [2], we investigated whether and in which extent voluminous lateral substituents are interfering to the nematic state, especially at high packing fractions.

S u b s t a n c e s

Table 1: List of compounds studied. T_d is the depression of the transition temperature T_c owing to the thermal decomposition at the end of all measurements. p_m is the highest applied pressure. n is the number of measuring points.

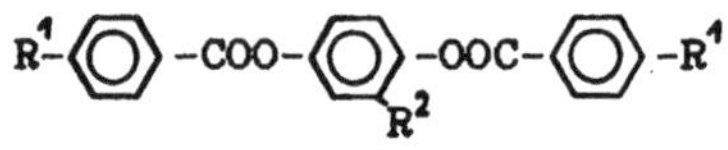

abbr.	R^1	R^2	n	p_m/MPa	T_d/K
80	$n-C_8H_{17}O-$	H -	4	48.5	0.0
1		CH_3-	8	292.1	0.2
2		C_2H_5-	8	409.6	0.4
3		$n-C_3H_7-$	10	522.8	0.0
4		$n-C_4H_9-$	9	512.0	0.0
8t		$t-C_4H_9-$	10	502.7	0.0
16		$n-C_{16}H_{33}-$	9	519.8	0.0
60	$n-C_6H_{13}-$	H -	3	24.6	0.0
6t		$t-C_4H_9-$	10	504.2	0.0

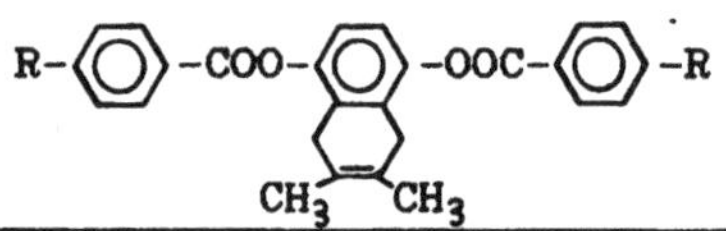

abbr.	R	n	p_m/MPa	T_d/K
N1	$n-C_6H_{13}-$	3	25.5	0.0
N2	$n-C_6H_{13}O-$	6	119.6	0.3

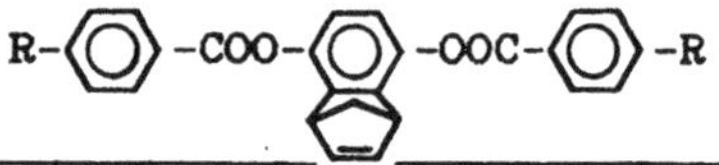

abbr.	R	n	p_m/MPa	T_d/K
NB	$n-C_6H_{13}O-$	11	409.6	0.9

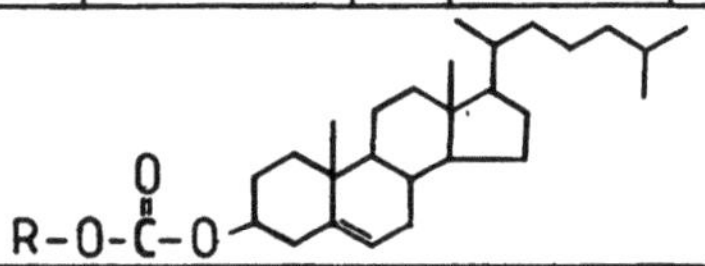

abbr.	R	n	p_m/MPa	T_d/K
CG	$-CH_2CH=C(CH_3)C_2H_4CH=C(CH_3)_2$	17	300.0	?

---

1) Zentralinst. fuer Physik der Erde, AdW, Potsdam, Telegrafenberg, GDR
2) Sektion Chemie, Martin - Luther - Univ., Halle, Muehlpforte 1, GDR

E x p e r i m e n t a l

The transition temperatures were determined by a high - pressure - DTA - apparatus. The sample was completely encapsulated in a manner, which allows an ideal hydrostatic transmission of pressure. Details will be reported elsewhere [3]. The measuring head was placed in a pressure chamber of a high - pressure gas compressor (manufacturing firm: unipress, Warsaw). For determination of pressure and temperature the resolution was 0.1 MPa respectively 0.05 K and the absolute accuracy up to 500 MPa and 550 K was better than 1 MPa respectively 1K.

R e s u l t s

The experimental p - T - points were fitted by the equation :

$$T = T_0 \cdot (1+p/b) \cdot e^{cp}, \qquad (1)$$

which is a modification [4] of the well-known Simon-Glatzel - equation. Because of their thermal instability, 80, 60 and N1 could be investigated at low pressures only. The few data were fitted to the equation:

$$T = T_0 + mp + np^2 \qquad (2)$$

Table 2: Results of fits. a,b,c are the constants of eq.(1), m,n of eq.(2). T_{c0} is T_c at p = 0.1 MPa. S is the standard derivation. V_B is the Bondi - volume of one molecule [5]. The length-to-breadth-ratio X is calculated from a cylinder with a length, determined from a calott - model, and a volume V_B.

abbr.	a	b	$10^5 \cdot c$	m	$10^4 \cdot n$	S	T	$(dp/dT)_0$	X
	–	MPa	1/MPa	K/MPa	K/MPa^2	K	K	MPa/K	–
80	-	-	-	0.5253	-2.09	-	466.2	1.90	11.75
1	0.336	275.9	-8.22			0.09	430.0	2.10	11.57
2	0.165	193.1	13.70			0.05	391.9	2.57	11.37
3	0.235	271.5	3.26			0.11	371.8	2.99	11.20
4	0.236	277.4	2.49			0.17	363.6	3.14	11.08
8t	0.213	254.7	4.14			0.03	357.9	3.18	11.08
16	0.247	312.3	2.19			0.11	341.9	3.60	9.63
60	-	-	-	0.5973	-3.17	-	441.6	1.68	9.40
6t	0.839	649.2	-31.12			0.17	308.2	3.31	8.79
N1	-	-	-	0.4531	-10.0	-	388.8	2.21	8.67
N2	0.692	409.0	-59.22			0.07	450.6	2.02	9.39
NB	0.161	190.0	12.57			0.15	385.4	2.66	9.63
CG	3.070	994.9	-180.30			0.10	325.0	2.40	7.59

Substance CG was not measured by us. For calculations we used reported data from Feyz and Kuss [6].

D i s c u s s i o n

The trends of T_{C0} and $(dp/dT)_0$ are in agreement with the theory [7-12]. At higher pressures the relative relations remain for the substances 2, 3, 4, 16, 8t, NB. However the compounds CG, N2, 1, 6t show a different behaviour, which is not yet seen immediately at the change of T_C and dp/dT - but at the curvatures of the transition lines (Fig.1). All these compounds with the highest curvatures at the highest pressures have a common characteristic: A bulky lateral loading (cyclic group in the cholesteric ester CG and in N2, t-C_4H_9 in 6t, CH_3 in 1 and N2 too).

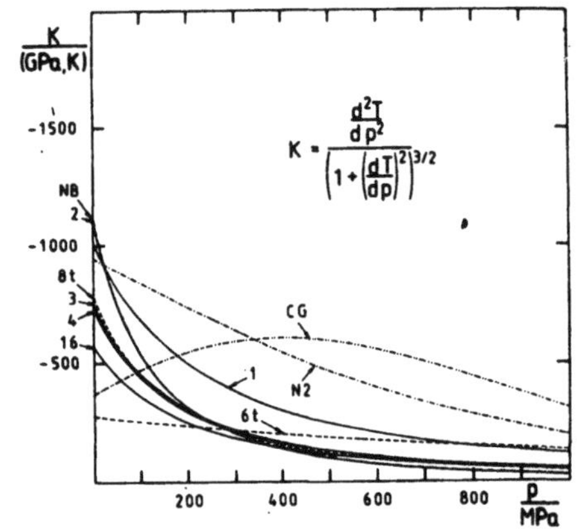

Fig.1 Curvatures of the transition lines

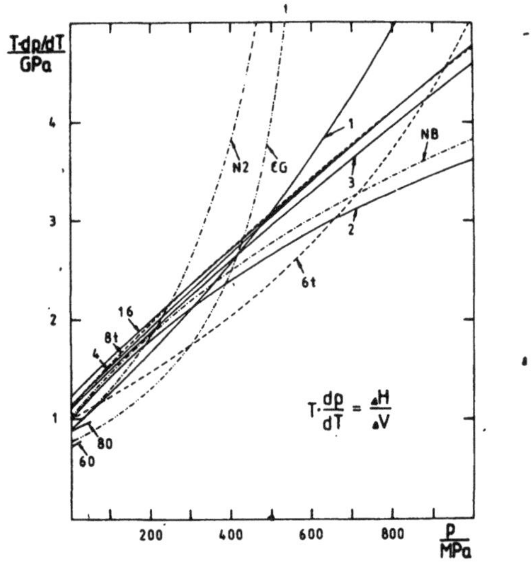

Fig.2 Relative slopes

Theories, which use hard - rod - potentials [7-12] or an inverse - power potential [13] yield more and more straight transition lines with increasing pressure. Because potential changes are not possible, the structure will be stabilized by pressure along the transition line. Therefore we believe in accordance to melting theories of Stillinger [14] and Weeks [15], that high values of the curvatures of the transition lines are due to strong changes of the intermolecular potentials connected with a great pressure dependence of the molecular

arrangement. This consideration is confirmed also by the pressure dependence of the relative slope (Fig.2), which is convex for compounds the transiton lines of which have high curvatures (CG, N2, 1, 6t). In our cases such changes of the molecular arrangement are explicable by the picture of increasing hindrance of the longitudinal translation of the molecules by the bulky lateral interference as the pressure rises. In this way the pressure may destabilize the nematic order.
For all other substances the lateral interference seems to be not powerful in this direction. Several effects are imaginable: It is too flexible (4, 16), it is screened by the "softness" (lubricant - effect) of the terminal chains (2, 3, 8t, NB) or it has a good stream-lined shape (NB?). Therefore in these cases the pressure will cause a stabilizing influence.

A c k n o w l e d g m e n t s

We are indebted to Dr. W. Weissflog, VEB Spezialchemie Leipzig, for providing the substances 8O,1,2,3,4,16,6O,6t,8t and to Dr. Ch. Mohr, Martin - Luther - Univ. Halle, for providing the substances N1,N2 and NB. We thank Dr. A. Hauser for valuable discussions.

R e f e r e n c e s

[1] S.Chandrasekhar, R.Shashidhar: Adv.Liq.Cryst. 4 (1974) 83
[2] W.Weissflog, D.Demus: Cryst.Res.Technol. 18 (1983) K21
[3] C.Rein, D.Demus: in preparation
[4] C.Rein, D.Demus: p. 143 in this volume
[5] A.Bondi: J.Phys.Chem. 68 (1964) 441
[6] M.Feyz, E.Kuss: Ber.Bunsenges.Phys.Chem. 78 (1974) 834
[7] M.A.Cotter: Phys.Rev. A10 (1974) 625; J.Chem.Phys. 66 (1977) 1098
[8] R.Alben: Mol.Cryst.Liq.Cryst. 13 (1971) 193
[9] W.M.Gelbart, B.A.Baron: J.Chem.Phys. 66 (1972) 207
[10] F.Dowell, D.E.Martire: J.Chem.Phys. 68 (1977) 1088, 68 (1977) 1094
[11] K.L.Savithramma, N.V.Madhusudana: Mol.Cryst.Liq.Cryst. 74(1981)243
[12] K.Singh, S.Singh: Mol.Cryst.Liq.Cryst. 108 (1984) 133
[13] J.A.Pople, F.E.Karasz: J.Phys.Chem.Solids 18(1961)28, 20(1961)294
[14] F.H.Stillinger: J.Chem.Phys. 65 (1976) 3968
[15] J.D.Weeks: Phys.Rev. 24 (1981) 1530

PHASE TRANSITIONS OF A_2MX_4 TYPE COMPOUNDS

F. Hanic 1)

Introduction

Information on the material properties deep within the Earth are of particular interest as models of the dynamic state of our planet. Much research is presently centered on the interconnection of the internal structure of the planet, its thermal and chemical state, with the dynamics and evolution of the mantle and crust.

The upermost mantle comprises mostly of olivines, which coexist with lesser amounts of pyroxenes and garnets to form a garnet peridotite. With the increasing depth, these minerals transform to more dense high pressure phases [1-14]. The high pressure transitions are presumably responsible for the observed discontinuities in seismic velocities at 390-400 km and 670-700 km depth due to the different phase assemblages existing on each side of these boundaries [1,2,15].

The new technique at ultrahigh pressures have increased the number and quality of data available on the physical properties of mineral systems and on the phase equilibria which occur within the mantle.

The aim of this work is to present a scheme of the high pressure - high temperature phase transformations based on a structure field approach [16-28], which may be useful for deciphering the nature of the mantle discontinuities and the structure and dynamics of the mantle. An attempt was made to find the fundamental parameters which reveal the most general relationships between the structure, chemical composition and physical properties of inorganic solids.

Four parameters are closely related to the physical and structural properties of solids. They were successfully applied in the sorting work over extensive bases of compounds and in the prognostication of the new phases and their properteis:

1. The <u>interatomic distances</u> d_{ij} within the coordination polyhedra constituting the crystal structure. Whereever it is possible, the average distance within the polyhedron is taken into consideration together with the symmetry of the polyhedron as well as with the symmetry of the investigated solid. This parameter is available from experiment from X-ray electron, neutron and synchrotron structure analysis. It is the only fundamental parameter directly related to the high pressure - high temperature phase transitions in a predictive way. It is also the basic parameter in determination of lattice energy, compressibility and bond

---

1) Institute of Inorganic Chemistry, Centre for Chemical Research, Slovak Academy of Sciences, Dúbravská cesta, 842 36 Bratislava, Czechoslovakia

order in the solids and it reflects the spin state of atoms. It can be simply determined from the set of self-consitent atomic radii which satisfy the condition $r_i + r_j = d_{ij}$ (experimental value) with an accuracy better than 1%.

2. The pseudopotential orbital radii r_s^i and r_p^i of the i-th atom representing crossing points of the nonlocal pseudopotential $V_{eff}(r)$ with = 0, 1, 2, ... for s, p, d. ... orbital electrons. They may be determined via atomic spectra of nontransition elements from the ionization potentials of atoms 25. They scale inversely with the orbital ionization potential and linearly with the radial modes of valence orbitals correlating therefore with electronegativities and atomic sizes. They are well sorting such structural parameters like the site preference and polytypism, using as sorting parameters $r_\sigma^A = r_s^A + r_p^A$ and $r_\pi^A = r_p^A - r_s^A$.

3. Energy b that measures the strength of the interactions between d electrons on neighboring atoms or ions of transition metals. This phenological parameter can be related qualitatively to the cation-cation separation d_{ij} and/or the strength of the covalent mixing with neighbòr anions. Specificly, it measures the transition energy of eletrons from locallized to collective state, or from semiconductive to metallic state, or from spontaneous ferromagnetism to Pauli paramagnetism, or from cation to anion transfer, etc. The fields of homogeneous physical properties in the phenomenological phase diagrams b-T (ferromagnetism, antiferromagnetism, Curie-Weiss and Pauli paramagnetism, spiral ferromagnetism, etc. are separated by critical valus of b for transitions and by critical temperatures (T_N = Neel temperature, T_C = Curie temperature, T_{CS} = critical temperature of onset of superconductivity).

4. The average number n_1 of electrons per valence orbital ($0 < n_1 \leq 2$)

d_i - space. Systematic investigation of the d_i - space, in which different structure types of solids (condensed oxides, sulphides, halogenides, etc.) of the same general formula, e.g. A_2MX_4, are plotted as a function of the interatomic distances d_i within the polyhedra constituting the structure, shows the occurence of a closed space (area) for each structure type. The boundary of each structure type space (area or field) can be definde by a minimum or a maximum value of the space filling coefficient $\phi = V/V_0$, where V_0 ist the space occupied by atoms in the unit cell and V is the unit cell volume. It means that the structure of solid is no more stable, when the atoms fill the unit cell space less than for 47% or more than 80% (the space filling rule).

In general, two types of regression equations can be derived in form of polynoms which define the unit cell volume as a function of d_i or r_i powers:

$$V/Z = (a_0 + a_1 d_1 + a_2 d_2 + \dots + a_i d_i)^3 \quad (1)$$

$$V/Z = b_0 + b_1 r_1^3 + b_2 r_2^3 + \dots + b_i r_i^3 \quad (2)$$

where Z is the number of the formula units in the unit cell. The relation

(2) can be used only in case, when ionic radii represent a selfconsistent set. A simultaneous application of eq. (1) and (2) improves the definition of the structure field boundaries.

Overlap of the structure fields

The structure fields of the different structure types are not strictly separated. Two or more structure fields can overlap (Fig. 1). Overlap of the structure fields allows us to predict possible transformations at high pressures and temperatures and the corresponding volume changes (Fig. 1). Within the overlapping regions, the more dense structures are realized at high pressures (the high pressure polymorphs) and the less dense structure types at high temperatures (the high temperature polymorphs). When the coordination number of atoms varies during transition, the interatomic distances change discretly. They increase at increasing coordination number (usually at high pressure transitions) or they decrease when the coordination number is reduced. Such changes can shift compounds from their low density structure field into the stability field of the high pressure polymorph.

The multiple overlap of structure fields allows us to predict possible high pressure transformations. However, such overlap is only one of several preconditions of the high pressure transition. Another determinative parameter is the smalles possible structural and coordinative change during transformation, which controls the specific transition pass. It means that reconstructive transformations connected with large volume changes usually pass through several transition states.

Decomposition reactions under the high pressures

This kind of transformation reactions is important for explanation of the high pressure changes in the Earth mantle contributing to the densification process at the observed discontinuities in seismic velocities, especially at 670-700 km depth of the mantle [29-34].

References

[1] JEANLOZ,R., THOMPSON,A.B., Rev. Geophys. Space Phys. 21 (1983) 51

[2] McKENZIE,D.P., Scient. Amer. 249 (1983) 67

[3] AKIMOTO,S.,AKAOGI,M.,KAWADA,K.,NISHIZAWA,O.,Mineralogic Distribution of Iron in the Upper Half of the Transition Zone in the Earth Mantle, in the Geophysics of the Pacific Ocean Basin and its Margin, Geophys. Monogr. Ser. 19 (G.H.Sutton, M.H.Manghnani, R.Moberly, ed.), AGU Wash. 1976

[4] AKAOGI,M., AKIMOTO,S., Phys. Earth Planet. Inter. 19, (1979) 31

[5] LIU,L., Phase Transformations and Constitution of the Deep Mantle, in the Earth: Its Origin, Structure and Evolution (M.W. McElhinny, ed.) Academic, New York, 1979.

[6] YAGI,T., BELL,P.M., MAO,H.K., Carnegie Inst. Year Book, Wash. 78 (1979),614-618

[7] AKAOGI,M., AKIMOTO,S., Phys. Earth Planet. Inter. 15 (1977) 90.

[8] AKIMOTO,S., YAGI,T., INOUE,K., in High -Pressure Research (M.H. Manghnani, S.Akimoto, ed.),Academic, New York, 1977, pp. 585-692.

[9] AKIMOTO,S., SYONO,Y., Phys. Earth Planet. Inter. 3 (1970) 186.

[10] BASSETT,W.A., MING,L., Phys. Planet. Inter., 6(1972) 154.

[11] MING,L.C., BASSETT,W.A., Science, 187 (1975) 66.

[12] MING,L.C., BASSETT,W.A., Earth Planet. Sci. Lett., 25 (1975) 68

[13] LIU,L., Phys. Earth Planet. Inter. 11 (1976) 289.

[14] SUITO,K., Phase Relations of Pure Mg_2SiO_4 up to 200 kb, in High - Pressure Research (M.H. Manghnani, S.Akimoto, ed.), Academic, New York 1977, pp. 255-266.

[15] FITCH,T.J., in Maurice Erving Ser., Vol. 1/III (M.Talwani, W.C.Pitmann, ed.), AGU, Washington 1977, pp. 123-136.

[16] MULLER,O., ROY,R., The Major Ternary Structural Families, Vol. 4 (R.Roy, ed.), University Park - Springer, Berlin 1974.

[17] MOOSER,E., PEARSON,W.B., Acta Crystallogr., 12 (1959) 1015.

[18] JOHN,J.ST., BLOCH,A.N., Phys. Rev. Lett., 33 (1974) 1095.

[19] PHILLIPS,J.S., CHELIKOWSKY,J.R., Phys. Rev.B.17 (1978) 2453.

[20] BLOCH,A.N., SCHATTEMAN,G.C., in Structure and Bonding in Crystals (A.Navrotsky, M.O'Keeffe, ed.), Academic Press, New York, 1981.

[21] ZUNGER,A., Phys. Rev. Lett., 44 (1980) 582.

[22] ZUNGER,A., Phys. Rev. B, 22 (1980) 5839.

[23] MACHLIN,E.S., CHOW,T.P., PHILLIPS,J.S., Phys. Rev. Lett., 38 (1977)

[24] BURDETT,J.K., PRICE,G.D., PRICE,S.L., Phys. Rev. B 24 (1981) 2903.

[25] PRICE,G.D., PRICE,S.L., BURDETT,J.K., Phys.Chem.Minerals,8 (1982) 69

[26] BURDETT,J.K., PRICE,G.D., PRICE,S.L., J.Am.Chem.Soc.,104 (1982) 92

[27] HANIC,F., HANLOVIČ,M., BURDOVÁ,K., MAJLING,J.,J.Cryst.Spectr.Res.,12 (1982) 99

[28] HANIC,F., Čs.Čas.Fyz.A, 31 (1981) 370

[29] YAGI,T.,MAO, H.K., BELL,P.M., Carnegie Inst. Wash.Year-Book,77(1978) 837

[30] YAGI,T., BELL,P.M., MAO,H.K., Carnegie Inst.Wash.Year-Book,78,(1979) 614

[31] OHTANI,E., J. Phys. Earth, 27 (1979) 189

[32] NAVROTSKY,A., PINTCHOVSKI,F.,AKIMOTO,S., Phys. Earth Planet.Inter.19 (1979) 275

[33] ITO,E., YAMADA,H., in High-Pressure Research in Geophysics, Centre Acad. Pub.(S.Akimoto, M.H.Manghnani, ed.),Japan, Tokyo, 1981

[34] JEANLOZ,R., AHRENS,T.J., Geophys. J.R.Astr.Soc.,62 (1980) 505

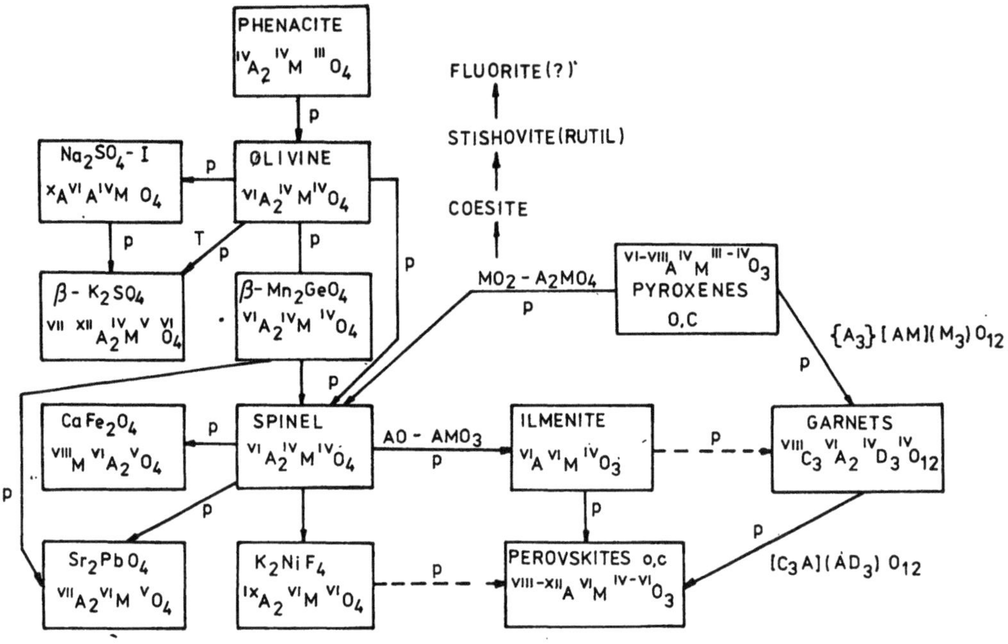

Fig. 1

Structure type fields of A_2MO_4 compounds evaluated from eq. for space-filling factor $\emptyset = V_0/V$, where V_0 is the space occupied by atoms in the unit cell and V is the unit cell volume calculated from regression equation of the type: $V/Z = (a_0 + a_1 d_{AO} + a_2 d_{MO})^3 \approx a'_0 + a'_1 r_A^3 + a'_2 r_M^3$ (nm)

Structure type	Z	Space group	Symmetry	a_0	a_1	a_2	a'_0	a'_1	a'_2
Phenakite	18	$R\bar{3}$	Rhomboedric	0,037	1,40	0,74	0,059	136	55
Olivine	4	Pnma	Orthorhombic	0,017	1,46	0,57	0,055	39,2	118
Spinel	8	Fd3m	Cubic	0	4/3	0,77	0,0528	46,5	42
β-Mn_2GeO_4	8	Imma	Orthorhombic	0,011	1,46	0,55	0,049	46	76
K_2NiF_4	2	I4/mmm	Tetragonal	$V/Z=4d_{AX}\, d^2_{MX}(\sqrt{3}+1)$			0,050	12,6	71
$CaFe_2O_4$	4	Pnam	Orthorhombic	-0,032	0,46	1,66	0,0517	51	6,55
β-K_2SO_4	4	Pnam	Orthorhombic	0,063	0,90	0,98	0,0603	13,3	236
Sr_2PbO_4	2	Pbam	Orthorhombic	-0,026	1,33	0,72	0,0506	25	30

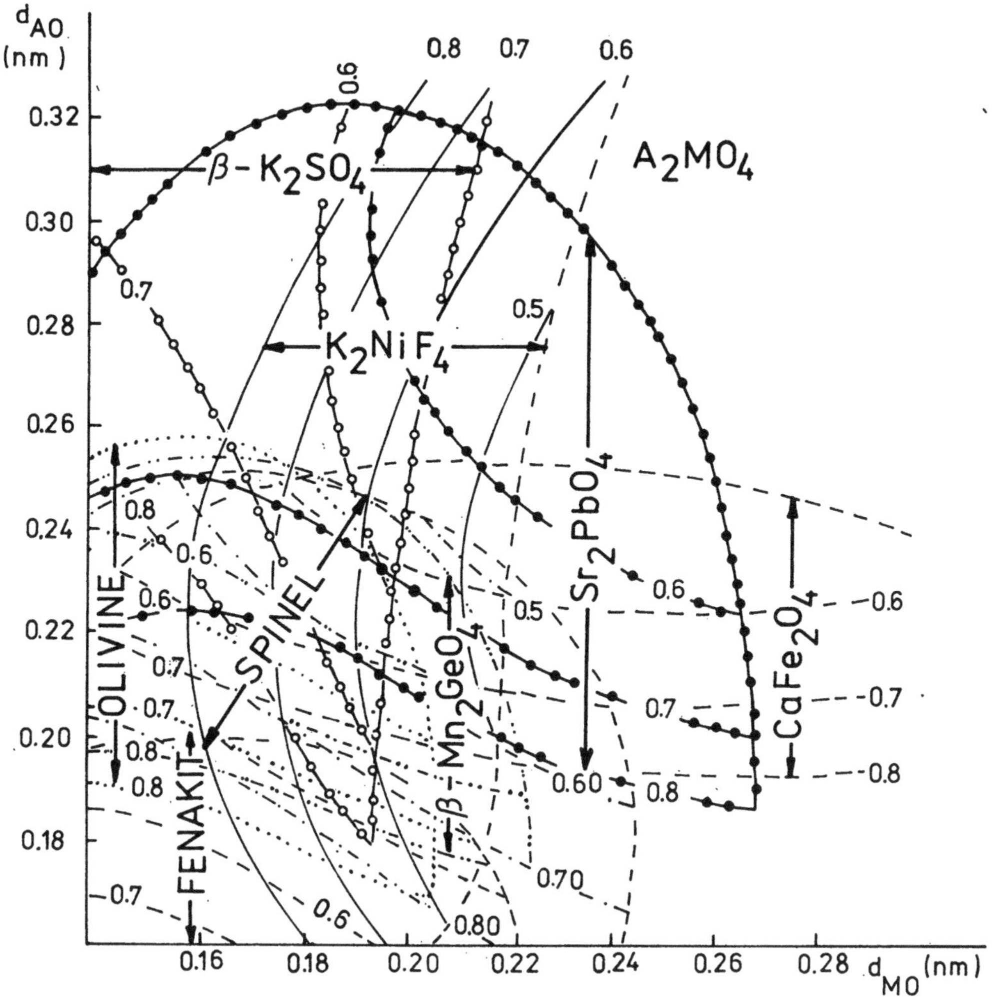

Fig. 2

Scheme of the high pressure transformations of solids related to the high pressure transitions in the Earth mantle. The scheme was constructed from the overlap of corresponding structure fields. The densest phases which can occur at the seismic velocity discontinuity at 670-700 km depth are: perovskite type, K_2NiF_4 type, $CaFe_2O_4$ type, Sr_2PbO_4 type, 3-K_2SO_4 type, CaF_2 type, rutil type structures. Several of these final phases were experimentally proved.

This work was closely developped under the cooperation with the Central Institute for Physics of the Earth, i.e. Prof.Stiller, Prof. Vollstaedt, Dr. Kraft, Dr. Franck.

ELASTIC PROPERTIES OF $SrTiO_3$ PEROVSKITE UNDER HIGH PRESSURE

M. FISCHER, B. BONELLO
Département de Recherches Physiques (C.N.R.S. U.A. 71)
Université Pierre et Marie Curie, Tour 22, E 4.
4, place Jussieu, 75252 Paris Cedex 05, FRANCE

A. POLIAN
Physique des Milieux Condensés (C.N.R.S. U.A. 782)
Université Pierre et Marie Curie, Tour 13, E 4
4, place Jussieu, 75252 Paris Cedex 05, FRANCE

INTRODUCTION

The purpose of our studies is the determination of the elastic properties under high pressure of compounds having the ideal structure of the perovskite. This determination is of high interest for the study of the properties of constituants of the Earth's lower mantle. |1|

$SrTiO_3$ perovskite is considered as a model for the study of ferroelastic phase transition. Under normal conditions its structure is cubic and at T_c which is about 105 K, $SrTiO_3$ undergoes a transition to a tetragonal phase (Fig. 1).

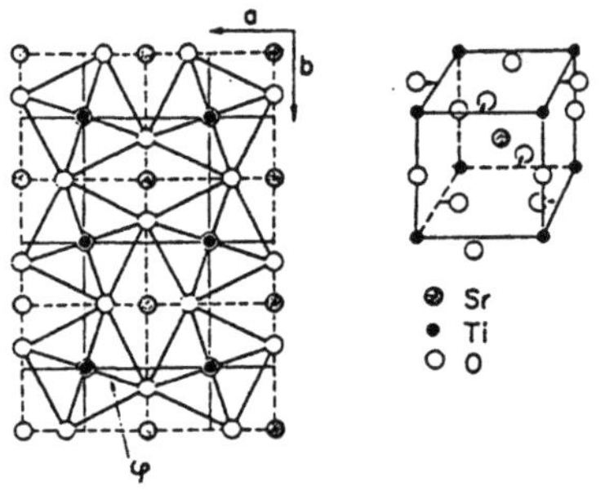

Figure. 1. A projection down a cube axis of the distorted phase of $SrTiO_3$. |2|

This structural phase transition produces no change in the cell volume, thermal expansion coefficient or dielectric constant but is accompanied by anomalies in the elastic constants |3-5|. This is a consequence of the coupling between the rotational motion of the oxygen octahedra and the elastic strain.

In terms of the soft mode theory, this transition is explained by the decreasing of the frequency ω_s of the R_{25} zone boundary Brillouin mode towards zero as $T \to T_c$.

The hydrostatic pressure dependence of T_c is known up to 3 GPa. Below 0.85 GPa, Tc rises linearly with increasing pressure. This increase of Tc is in agreement with the well known behaviour under pressure of a zone boundary soft mode for a ferroelastic phase transition |6|. Above 0.85 GPa, T_c increases non linearly with increasing pressure |7| (Fig. 2).

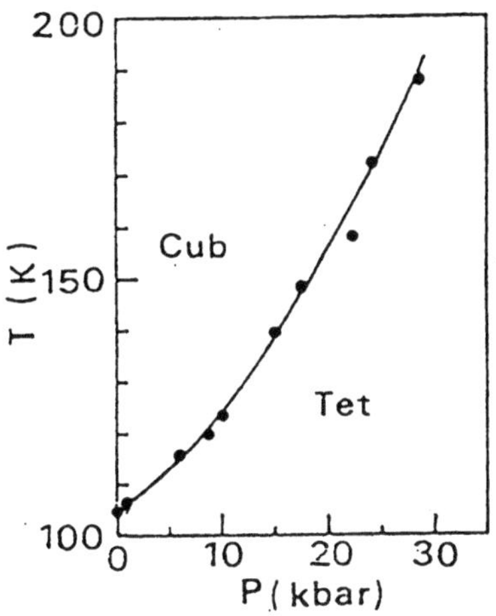

Figure. 2. Pressure dependence of T_c |7|.

We report in this paper the first measurements of the elastic properties of $SrTiO_3$ by Brillouin scattering between 0 and 36 GPa in a diamond anvill cell.

EXPERIMENTAL METHOD

We have used a Block Piermarini type diamond anvil cell which allows experiments in the backscattering geometry only. Typically the experimental volume is a cylinder of 200 μm in diameter and about 30 μm thick. The pressure transmitting medium is the 4:1 methanol ethanol mixture and the pressure is measured using the non-linear ruby scale |8|.

We have studied the Brillouin scattering of very small samples (20 μm thick). The Brillouin peaks of such samples are very weak. So we have used a piezoelectrically scanned Fabry-Pérot five pass interferometer.

The free spectral range are :

$$F.S.R = 5.3868\ cm^{-1} \quad \text{and} \quad F.S.R = 2.7273\ cm^{-1}$$

The wavelength of the exciting laser light is 514.5 nm.

In these experiments the phonon direction of propagation is parallel to the [111] direction of the cube.

RESULTS

In these experiments we observed simultaneously the Brillouin lines of the crystal and those of the ethanol methanol mixture. Around 10 GPa these two kinds of lines are superposed and the Brillouin shift measurement in the crystal is impossible.

At high pressure, the Brillouin scattering is weak and the Brillouin peaks are obtained after 200 scans at least.

The quantitative results which we obtained for the [111] direction are the following :

- The transverse acoustical mode has been measured up to 25 GPa. The variation of the Brillouin shift $\Delta\sigma$ with pressure is not continuous (Fig. 3). Around 6 GPa, its slope becomes smaller. A negative jump of $\Delta\sigma$ occurs between 5.5 and 7.5 GPa. The jump value is about 0.05 cm^{-1}. These two observations are consistent with a pressure phase transition. The accurate value P_c of the pressure of transition has not been determined because the points are too much scattered in the pressure range 5.5-7.5 GPa.

The observed transition is the cubic to tetragonal one. Indeed, the extrapolation of the pressure dependence of T_c towards high pressure (Fig. 2) is consistent with a transition around 5.5 GPa.

Moreover, the cubic-tetragonal phase transition observed by Brillouin scattering at room pressure and low temperature produces a discontinuity in $\Delta\sigma$ of the same order of magnitude with the same sign |9|

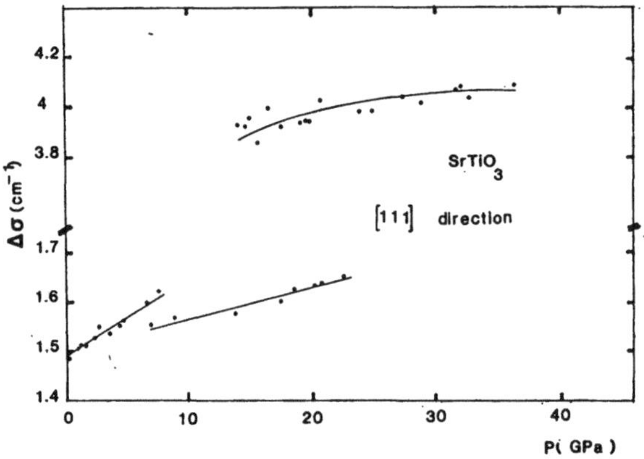

Figure. 3. Brillouin shift versus pressure along the [111] direction of the cubic phase.

The sound velocity is deduced from our results using the relation :

$$\Delta\sigma = \frac{2nv}{\lambda c}$$

where v is the sound velocity, n the refractive index, λ the wavelength in vaccum of the exciting laser light and c the light velocity. The pressure dependence of the refractive index is deduced from the Lorentz-Lorenz relation and the Birch Murnagham equation of state.

We used the following values of the bulk modulus B_o and its pressure derivative B'_o at room pressure |10|:

$$B_o = 173.3 \text{ GPa}$$

$$B'_o = 5.81$$

Using these results we calculate the variation of the transverse elastic modulus

$$\rho V_T^2 = \frac{C_{11} + C_{44} - C_{12}}{3}$$

Between 0 and 7.5 GPa ρV_T^2 increases linearly with pressure with a slope :

$$\frac{d\rho V_T^2}{dP} \sim 1.8$$

In the high pressure phase, we used the same extrapolation for the refractive index. So the slope of the elastic modulus becomes smaller and is equal to 0.2.

- Above 9 GPa, when the pressure is increased, new lines appear gradually (Fig. 3). The high frequency Brillouin shift corresponds certainely to a longitudinal or quasi longitudinal mode. The extrapolation at high pressure of the refractive index gives an elastic modulus which decreases when the pressure is increasing. This extrapolation is not suitable to the determination of the elastic moduli in the high pressure phase.

In conclusion, we have shown that the cubic to tetragonal phase transition occurs at about 6GPa in $SrTiO_3$. We determined the pressure dependence of a transverse acoustical phonon up to 25 GPa and a quasi longitudinal phonon from 13 to 36 GPa.

The Brillouin efficiency in this compound is quite small, 200 scans or more are often necessary to obtain acceptable spectra. Nevertheless, the Brillouin scattering technic in a diamond anvil cell is a key for the study of elastic properties of the perovskite structure under high pressure. We intend to study the Brillouin scattering along other crystallographic directions in order to obtain the whole set of elastic constants under pressure.

REFERENCES :

|1| Jeanloz, R., and A.B. Thompson : Rev. Geophys. Space Phys., 21 (1983), 51

|2| Bruce, A.D., and R.A. Cowley : Structural Phase transitions, Taylor and Francis LTD (1981)

|3| Lytle, F.W.: J. Appl. Phys., 35 (1964), 2212

|4| Samara,G.A.,: Phys. Rev., 151 (1966), 378

|5| Rehwald, W.,: Solid State Com. 8 (1970), 1483

|6| Samara, G.A., T. Sakudo, and K. Yoshimitsu : Phys. Rev. Letters, 35 (1975), 1767

|7| Okai, B., and J. Yoshimoto : J. Phys. Soc. Japan, 39 (1975), 162

|8| Mao, H.K., P.M. Bell, J.W. Shaner, and D.J. Steinberg : J. Appl. Phys. 49 (1978), 3276

|9| Kaiser, W., and R. Zurek : Phys. Letters, 23 (1966), 668

|10| Beattie, A.G., and G.A. Samara : J. Appl. Phys. 42 (1970), 2376

DETERMINATION OF THE STATE EQUATION OF $KCr(SO_4)_2 \cdot 12H_2O$ FROM ZERO-FIELD SPLITTING UNDER HIGH PRESSURE

Shang Bo and Zhao Min-Guang[1)]

Abstract: The state equation of $KCr(SO_4)_2 \cdot 12H_2O$ has been determined on the basis of the relationship between zero-field splitting and crystal structure. The pressure dependence of the zero-field splitting in the range of $-50 \leqslant T \leqslant 30^{\circ}C$ and $0 \leqslant P \leqslant 12$kbar has been calculated. The theoretical results agree well with the experimental findings.

Differing from the current way to determine the state equation, we determined the state equation of $KCr(SO_4)_2 \cdot 12H_2O$ on the basis of the relationship between the zero-field splitting and the crystal structure. This relationship is established by using the d-orbital[1] and the perturbation formula of D.

In the $KCr(SO_4)_2 \cdot 12H_2O$, the Cr^{3+} is surrounded by six O^{2-} ions. The field is predominantly octahedral, with a small trigonal field component. The bond distance and the bond angle are R=1.9Å and $\theta=54.8^{\circ}$[2]. Fig.1 shows the stereostructure of the Cr^{3+}-$6O^{2-}$ complex.

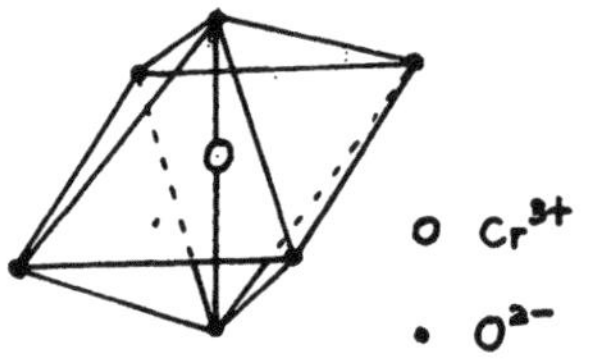

Fig. 1. Stereostructure of the Cr-6O complex.

With the consideration of the combined effects of trigonal field and the spin-orbit coupling we employ the third-order perturbation theory[3] and the parameterisation double zeta d-orbital[1] which is given by

$$R_d(r) = 0.591474(9.3232^7/6!)^{\frac{1}{2}}r^2\exp(-4.6616r) + 0.6014980(3.3052^7/6!)^{\frac{1}{2}}r^2\exp(-1.6526r) \quad (1)$$

Furthermore, following the average covalency approximation 4,5 we have

$$\langle r^2\rangle=N^2\langle r^2\rangle_0,\ \langle r^4\rangle=N^2\langle r^4\rangle_0,\ B=N^4B_0,\ C=N^4C_0,\ \zeta=N^2\zeta_0 \quad (2)$$

with

$$\langle r^2\rangle_0=2.4843\text{a.u.},\ \langle r^4\rangle_0=16.4276\text{a.u.},\ B_0=920.48\text{cm}^{-1},\ C_0=3330.71\text{cm}^{-1},\ \zeta_0=240\text{cm}^{-1} \quad (3)$$

1) Institute of Solid State Physics, Sichuan Normal University, P.R.China.

N being the average covalency reduction factor.

With the use of Eq.(1)-Eq.(3) and the formulae of g-factor and magnetic susceptibility[6] and by taking N=0.935, we obtain the results at normal pressure and at room temperature, as listed in table 1.

Table 1. The bands of $^4T_2(F)$ and $^4T_1(P)$ and magnetic values.

	Calculated	Observed[7,8,9]
$E(^4T_2)$, cm^{-1}	17600±100	17600
$E(^4T_1)$, cm^{-1}	24673±100	24700
g	1.974±0.01	1.98
χ, 10^{-3}e.g.s.	6.10±0.06	6.34
D, cm^{-1}	0.070±0.001	0.073

The pressure dependence of the zero-field splitting was measured by Ahantanu Sinha[10] in the range of $0 \leq P \leq 12$kbar and $-50 \leq T \leq 30^oC$. The value of D varies approximately linearly. By using the experimental

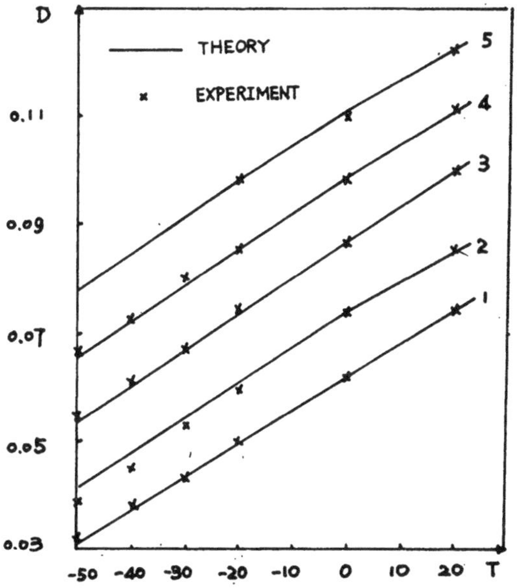

Fig.2. The comparison between theory and experiment. Line 1, 0kbar; line 2,3.1kbar line3, 6.2kbar; line 4, 9.3kbar; line 5, 12.4kbar.

findings, the formula of D and the average covalency reduction factor N which has been given previously we obtain the state equation of $KCr(SO_4)_2 \cdot 12H_2O$ as follows

$$V= (1-1.799\times10^{-2}P)(1 + 1.971\times10^{-4}T + 2.28\times10^{-6}T^2) + 0.142 \qquad (4)$$

$$\Theta = (1+0.438\times10^{-2}P)(1 + 6.3\times10^{-4}T + 0.11\times10^{-6}T^2) + 53.8 \qquad (5)$$

where V is volume in $Å^3$, Θ the bond angle in deg., P pressure in kbar and T temperature in ^oC.

Reversely from the state equation (4) we established theoretically the relationship between pressure, temperature and zero-field splitting and calculated the pressure and temperature dependences of D. The comparison between the theory and the experiment is shown in Fig.2. From the Fig.2 we can see that the theoretical results are in good agreement with the experimental findings. Furthermore, the behaviours of the zero-field splitting of $KCr(SO_4)_2\cdot12H_2O$ in the range of $P > 12$kbar and $-50 > T > 30^oC$ can be predicted. The prediction will be tested by further experiments.

[1] Zhao Min-Guang et al, Phys. Rev. B27 (1983) 1516.

[2] Allen C. Larson and Don T. Cromer, Acta Cryst. 22 (1967) 793.

[3] Macfarlane R.M., J.Chem.Phys. 39 (1963) 3118; 47 (1981) 2066.

[4] Zhao Min-Guang et al, Phys.Rev. B28 (1983) 6481.

[5] Zhao Min-Guang, Gui-Ru Bai and Huei-Cheng Jin, J. Phys. C 15 (1982) 5959.

[6] Griffith J.S.: The Theory of Transition-Metal Ions, Cambridge University Press, London 1961.

[7] Coggon P. et al, Chem. Soc. (A) (1970) 3269.

[8] Owen G.H. et al, J. Chem. 26 (1957) 1686.

[9] Bowers K.D. and J. Owen, Rep. Progr. Phys. 18 (1955) 304.

[10] Shantanu Sinha et al, Chem. Phys. Lett. 88 (1982) 115.

A NEW METHODOLOGY IN THE DETERMINATION OF THE STATE EQUATION OF CoX_2 FROM HIGH PRESSURE SPECTRA SHIFTS

Zhao Min-Guang and Shang Bo[1)]

Abstract: The high pressure V-P state equation and the α-P correlation of CoX_2 (X=F, Cl, Br, I) crystals are determined from the pressure shifts of the optical spectra of $CoCl_2$ and $CoBr_2$ based on the $\mu k \alpha$ model. By using the state equation and the α-P correlation the pressure behaviours of the optical spectra of CoF_2 and CoI_2 crystals are predicted.

The effect of pressure has been measured by some authours on the absorption spectra of $CoCl_2$ and $CoBr_2$. The pressure shifts in every case correspond to increasing in crystal field strength. To account for the shifts in a quantitative manner, it is necessary to presume that the Racah electrostatic parameter B and the Trees correction parameter a decrease with increasing pressure. This can be interpreted as an increase in covalency at pressure. Recently Zhao[1,2] developed a new methodology to predict the state equation of NiX_2 (X=F, Cl, Br, I) from the high pressure spectra shifts based on both the $\mu k \alpha$ correlation model and the parameterisation double zeta d-orbital model. In this methodology, once the correlations are established, it is not needed to introduce the fitting parameters for us to investigate the optical and the magnetic properties of M^{n+}-6 X^- (X=F, Cl, Br, I) clusters with various crystal structure. In a general sense, this kind of method can be referred as a " like ab initio method". In the present report, this methodology is generalized to determine the state equation of CoX_2 (X=F, Cl, Br, I) from the high pressure spectra shifts. After analysing a numer of experimental data and making a large amount of numberical calculations, we find that the correlation of N-α and μ-α for Co^{2+}-$6X^-$ cluster can be expressed as

$$1+\frac{5\mu'}{eR} = 0.5307\,\alpha_0^{1.1} + 0.833 \qquad (1)$$

$$N^{-4} = 0.1092\,\alpha^{0.8} + 1.0663 \qquad (2)$$

1) Institute of Solid State Physics, Sichuan Normal University, Chengdu 610066, People's Republic of China.

In addition, following the reference[1] the α-P dependence is given by

$$\alpha(P)/\alpha(P=0) = 1 + 0.002983P - 0.0000102P^2 \tag{3}$$

In the expressions, μ is the electric dipole moment of X^-, e the electronic charge, R the metal-ligand bonding length, α the electron polarizability of X^- and N the average covalency reduction factor 3 ; the α is in unit of $10^{-24}cm^3$ and P in kbar. So that a series of the optical and the magnetic properties of Co^{2+}-$6X^-$ cluster with various crystalline symmetry can be calculated by using the parameterisation double zeta d-orbital of Co^{2+} ion which is given by[4]

$$R_d = 108.66122r^2\exp(-5.4r) + 0.95856r^2\exp(-1.5r) \tag{4}$$

Furthermore, following the average covalency approximation 3,4 we have

$$\langle r^n\rangle = N^2\langle r^n\rangle_0, \quad \zeta_d = N^2\zeta_d^0, \quad B=N^4B_0, \quad C=N^4C_0, \quad a=N^4a_0 \tag{5}$$

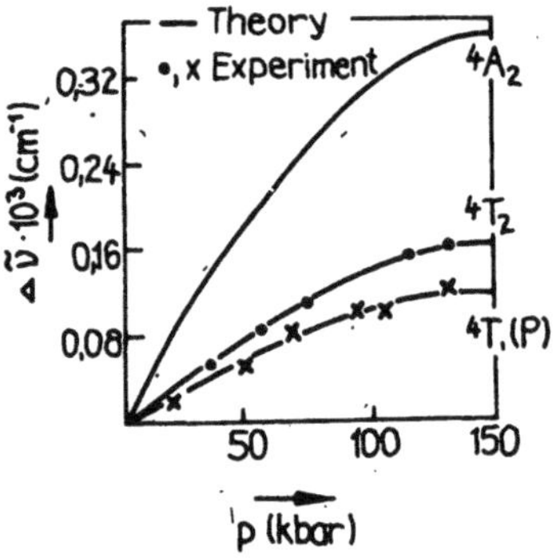

Fig.1. Pressure spectra shifts of $CoCl_2$ crystal.

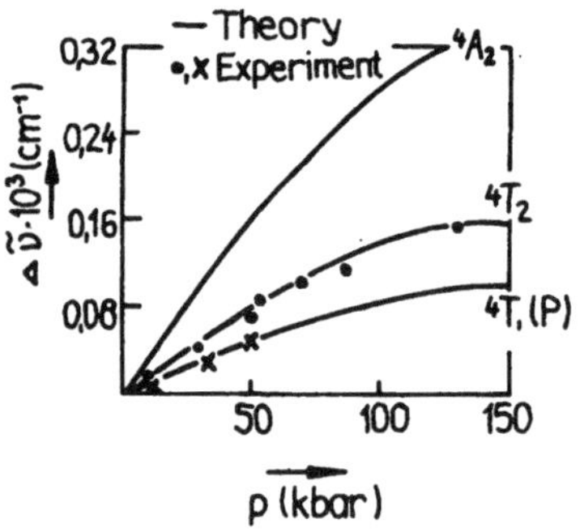

Fig.2. Pressure spectra shifts of $CoBr_2$ crystal.

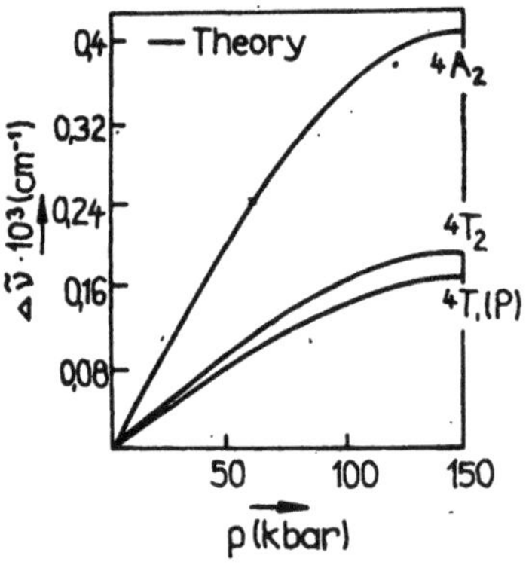

Fig.3. Predicted pressure spectra shifts of CoF_2 crystal.

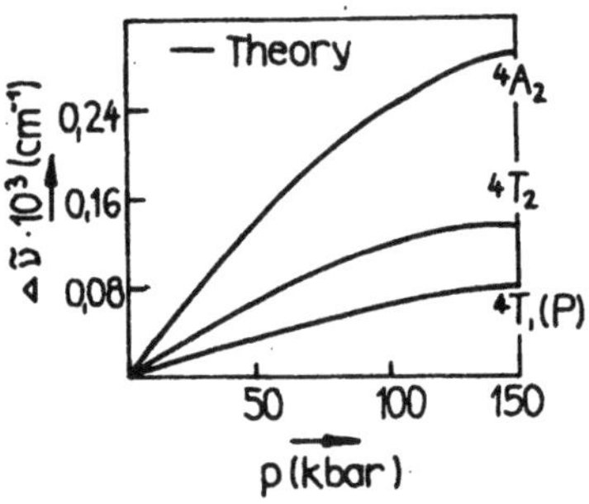

Fig.4. Predicted pressure spectra shifts of CoI_2 crystal.

where $\langle r^n \rangle_0$, ζ_d^0, B_0, C_0 and a_0 denote the "free-ion" values which can be calculated from the orbital (4). The results are as follows:

$$\langle r^2 \rangle_0 = 2.357 a.u., \langle r^4 \rangle_0 = 19.4538 a.u., \zeta_d^0 = -535 cm^{-1},$$

$$B_0 = 1063.2 cm^{-1}, C_0 = 3879 cm^{-1} \quad (6)$$

The Trees correction parameter a_0 is determined to be $70cm^{-1}$ from the free-ion spectrum. Analysing the experimental data of high pressure spectra shifts with the aid of Eq.(1)-Eq.(6) leads to the state equation V-P:

$$V(P)/V(P=0) = 1 - 0.00177P + 0.0000059P^2 \quad (7)$$

The comparison between the theory and the experiments[5] are shown in the Fig.1 and Fig.2. From the state equation (7) we predict the high pressure shifts of the optical spectra of CoF_2 and CoI_2 crystals as displayed in Fig.3 and Fig.4. The deduced state equation and the predicted pressure dependence of the optical spectra of CoF_2 and CoI_2 await experimental tests.

[1] Zhao Min-Guang, Du Mao-Lu and Sheng Guo-Yin, 24th EHPRG Conf. 1986.

[2] Zhao Min-Guang and Du Mao-Lu, J. Phys. C 20 (1987) 2931.

[3] Zhao Min-Guang et al, Phys. Rev. B27 (1983) 1516.

[4] Zhao Min-Guang and Zhang Yong-Fong, IEEE: Trans.Mag. M19 (1983) 1972.

[5] Zahner J.C. and H.G.Drickamer, J. Chem. Phys. 35 (1961) 1483.

BIRCH'S LAW AND THE PROPERTIES OF HIGH TEMPERATURE FLUID METALS[1)]

J. W. Shaner[2)]

Abstract

By comparing acoustic velocities in fluid metals over a very wide range of densities we have established Birch's Law as an approximate representation over the entire liquid range. For a given liquid metal the acoustic velocity is close to linear in density, with a slope determined by the atomic weight. The measurements include isobaric expansion to less than half normal density, ultrasonics on molten metals at 1 atmosphere, and shock-melted metals to greater than twice normal density. We also find unusual behavior of the Grüneisen gamma, which can be explained in terms of simple fluid models.

1. INTRODUCTION

In an attempt to identify what materials lie deep in the earth from seismic velocities, Birch proposed an empirical law that bulk and elastic wave velocities should scale linearly with density and inversely as the square root of the atomic weight [Birch 1968]. This law was based on elastic constant measurements on rocks to 10 kbar [Birch 1960], representing a density change of only 10%, and some early shock wave data from which rarefaction velocities were calculated or measured [Birch 1961]. Since Birch's original work, several people have tried to explain the linear density dependence as an approximation over a limited density range to more general solid state models [Shankland 1973]. For example, for a normal mode of a simple solid the acoustic wave dispersion relationship is $\omega = ck$, where c is the wave velocity. Since the k vector scales as the reciprocal of the interparticle spacing, $\partial\ell nk/\partial\ell n\rho = 1/3$. Using the mode gamma definition of Grüneisen's gamma, $\partial\ell n\omega/\partial\ell n\rho = \gamma$, one has

$$\frac{\partial\ell nc}{\partial\ell n\rho} = \gamma - 1/3 \quad , \tag{1}$$

which integrated becomes

$$c = A\rho^{\lambda} \quad , \tag{2}$$

where $\lambda = \gamma - 1/3$. Since γ is generally in the range 1 - 2 for most simple solids, λ should be around 1.

Over the past few years we have used the optical analyzer technique to measure the pressure at which shocked materials lose their ability to support longitudinal elastic waves [McQueen et al. 1982]. At higher shock pressures, a release wave moves into a fluid-like material at the bulk wave velocity. Above the shock melting pressure we

1) This work performed under the auspices of the U.S. Department of Energy.
2) Los Alamos National Laboratory Los Alamos, New Mexico, 87545, USA.

then have measurements of the acoustic velocity of the fluid. New data for liquid lead, obtained by this technique, are presented by Boness et al. [1987].

Another method of measuring thermodynamic properties of fluid metals is the isobaric expansion experiment (IEX). In this experiment, a metal wire is heated in a gas pressure vessel quickly enough so normal hydrodynamic instabilities play no role in the heating dynamics, but slowly enough so that isobaric conditions apply, and the electrodynamic skin effect is insignificant [Gathers et al. 1976]. The function of the gas pressure is to stabilize a homogeneous fluid phase over the widest density range possible. After heating the wire to 5000-10,000 K in several tens of microseconds, we often find a stable liquid column for 10 microseconds, during which time we can perform detailed optical pyrometry or an acoustic velocity measurement.

The acoustic velocity measurement consists of a stress pulse induced by a Q-switched ruby laser pulse of ~0.1 J in 20 nsec, focussed to a spot roughly 0.1 mm on the central portion of the fluid column. This pulse propagates at the bulk sound velocity across the diameter of the wire and initiates a stress wave in the gas at a point diametrically opposite from the laser focal spot. The gas stress wave is detected by Schlieren photography with a streak camera. The velocity is simply the diameter divided by the time interval between the exciting laser pulse and the initiation of the gas stress wave. This technique has allowed us to make measurements of acoustic velocity in lead, for example, expanded to almost one third normal density [Hixson et al. 1986].

The combination of acoustic velocity and a track on the equation of state (EOS) surface in the form of a (P,V,T) isobar, or a Hugoniot, allow one to determine two orthogonal derivatives of the EOS. Swenson et al. [1986] show how to obtain the Grüneisen gamma,

$$\gamma = V \left.\frac{\partial P}{\partial E}\right)_V \quad , \tag{3}$$

from the Hugoniot and acoustic velocity. This derivative is, up to a simple metric, orthogonal to the isothermal bulk modulus. With isobaric expansion data, one can obtain the Grüneisen gamma from

$$\gamma = \left.\frac{\partial V}{\partial H}\right)_P c^2 \quad , \tag{4}$$

where H is the enthalpy, the electrical energy input at constant pressure, and c is the acoustic velocity. Temperature measurements along the Hugoniot permit an experimental determination of the isothermal bulk modulus by straightforward but tedious arithmetic [Swenson et al. 1986]. Along the isobar, the isothermal bulk modulus, B_T, can be determined from

$$C_p = C_V + \alpha^2 VTB_T \quad ,$$

and

$$c^2 = VB_T \; C_P/C_V \quad , \tag{5}$$

where α is the volume thermal expansion coefficient and $C_{P,V}$ are the heat capacities. Everything is measured experimentally in Eq. (5) except C_V and B_T.

2. EXPERIMENTAL DATA

We present acoustic velocity data for expanded liquid lead and tantalum in Fig. 1.

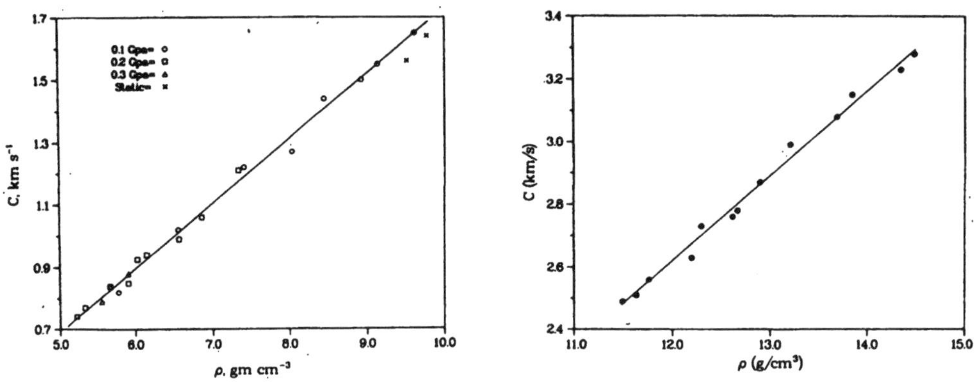

(a) (b)

Figure 1. Acoustic velocity in heated, expanded liquid metals: a) lead [Hixson et al. 1986a]; b) tantalum [Hixson et al. 1986b].

The lead data at high temperatures and expansions show no pressure dependence. Since the P,V isobars separate at high temperatures we can say that the acoustic velocity is only weakly temperature dependent, at least at high expansions. These isobaric expansion data show the normally observed linear dependence of acoustic velocity on density, but to much higher expansions than are usually obtained for metals [Shaner 1979].

The combination of data presently available for acoustic velocity in metals in expansion, shock compression, and at the one-atmosphere melting point is shown in Fig. 2, including the recent data of Boness et al. [1987]. The N_2 data are from Monte Carlo calculations [LeSar]. The change of slope in the Pb data from expansion to compression is well outside experimental error. It is important to realize that in the shock compression regions the density derivative of the temperature is positive, while it is negative in the expansion region, so any temperature dependence might tend to skew the data with respect to a straight line fit. In principle one should plot isotherms of c vs. density, but under the extreme energy density conditions of these experiments, well defined thermodynamic data are limited.

In Fig. 3 we show the other part of Birch's law for fluid metals - that the sound velocity should scale as the reciprocal square root of the atomic mass. This relationship works well for metals in the same column of the periodic table, which therefore have similar core structures. The offset of one metal group from another has no such simple interpretation yet.

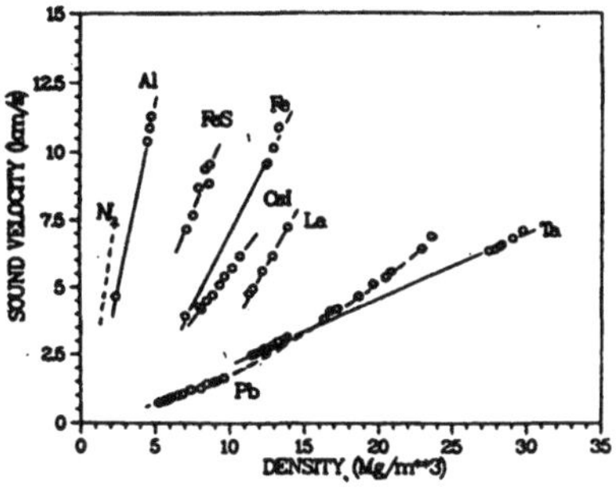

Figure 2. Density dependence of sound velocity for liquid metals.

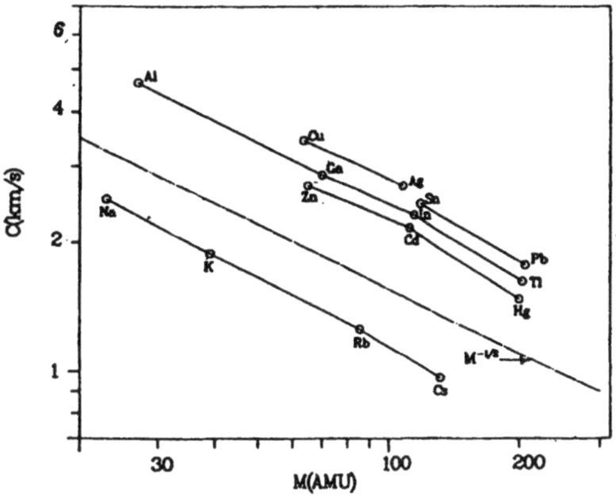

Figure 3. Acoustic velocity of liquid metals as a function of atomic mass. The thermodynamic state is the 1 atmosphere melting point in all cases [Webber and Stephens 1968].

Grüneisen's gamma shows unexpected behavior for Pb when plotted against density alone, as is shown in Fig. 4. The data of Boness et al. [1987] show $\rho\gamma$ to be weakly density dependent. On the other hand, the IEX data show γ/ρ to be roughly constant. Therefore, the slope of $\gamma(\rho)$ changes sign around normal density in this representation.

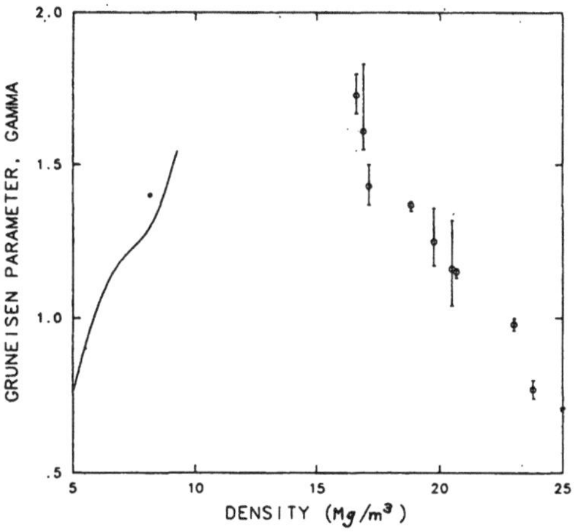

Figure 4. Density dependence of Grüneisen's Gamma for liquid lead.

3. DISCUSSION

The data of Fig. 2 show that the Birch's law suggestion that c is linear in density works reasonably well, even over a density range of three to five, for metals in the liquid range. The average slopes decrease monotonically with atomic mass. At low densities the linearity must break down as one approaches the ideal gas limit, which is independent of density.

We should comment that the display of c and γ vs. density alone is somewhat misleading, since the temperature varies dramatically along these curves. For example, one can prove rigorously that the density derivative of c along the Hugoniot must be greater than or equal to the same derivative along the isobar, which leads to some of the curvature observed when combining isobaric with shock data. If isotherms were plotted, we might expect an even better representation of Birch's law.

For a linear shock velocity vs. particle velocity, $U_s = C + SU_p$, the pressure derivative of the isentropic bulk modulus is given by $B' = 4S-1$. Thus, for a simple two parameter equation of state, B_s should scale like density to the 4S-1 power, or

$$c = (B_s/\rho)^{1/2} \sim \rho^{(2S-1)} \quad . \tag{6}$$

Since most Hugoniots have S between 1 and 1.5, one expects c to scale between linear and quadratic in density, at least in the compression range. Our data show that this scaling extends to significant expansions as well. The curious density dependence of gamma can be explained simply by a temperature dependent γ in the fluid phase. In fact, using the soft sphere model, which has been applied to the IEX data [Gathers et al. 1974], one obtains an explicitly temperature dependent γ, and the observed shape of the $\gamma(\rho)$ curve [Shaner]. The explicit temperature dependence of γ must be included for any accurate calculation of the adiabatic gradient, in a convecting fluid, for example. Recognition of an explicitly temperature dependent γ in the liquid is also of particular importance in calculating isothermal compression curves from Hugoniot data extending well into the liquid.

REFERENCES

Birch, F., J. Geophys. Res. 65, 1083 (1960).

Birch, F., Geophys. J. Royal Astron. Soc. 4, 295 (1961).

Birch, F., Phys. Earth and Planet. Interiors 1, 141 (1968).

Boness, D. A., J. M. Brown, and J. W. Shaner, Proceedings of the 1987 APS Topical Conference on Shock Waves in Condensed Matter, Monterey, CA, July 20-23, 1987.

Gathers, G. R., J. W. Shaner, and D. A. Young, Phys. Rev. Lett. 33, 70 (1974).

Gathers, G. R., J. W. Shaner, and R. L. Brier, Rev. Sci. Inst. 47, 471 (1976).

Hixson, R. S., M. A. Winkler, and J. W. Shaner, Physica 139 & 140B, 893 (1986a).

Hixson, R. S., M. A. Winkler, and J. W. Shaner, High-Temp. High-Press. 18, 635 (1986b).

LeSar, R., private communication.

McQueen, R. G., J. W. Hopson, and J. N. Fritz, Rev. Sci. Inst. 53, 245 (1982).

Shaner, J. W., in Thermal Expansion 6, I. D. Peggs, ed. [Plenum,Press (1979) 69].

Shaner, J. W., to be published.

Shankland, T. J., J. Geophys. Res. 77 3750 (1972), D. H. Chong, Science 177, 261 (1972), and O. L. Anderson, J. Geophys. Res. 78, 4901 (1973).

Swenson, C. A., J. W. Shaner, and J. M. Brown, Phys. Rev. B34, 7924 (1986).

Webber, G.M.B. and R.W.B. Stephens in "Physical Acoustics," vol. IV B, W. Mason, ed. Academic Press, NY, 1968.

THE p-Q TYPE DIAGRAM AS A MODEL OF FLUID FORMATION PROCESS.

K. Kravchuk 1)

Water and other volatile components are invariably present during evolution of magmatic melts, which goes on in the depths of the Earth at high temperatures and pressures. Volatile components are not only contained in aluminosilicate melts (in dissolved form) but are also capable of yielding independent phases - fluids, hydrothermal solutions - which, in turn, can solve and transport silicate materials.
The first question that arises in the description of magmatic source evolution deep in the Earth's core is the problem of the phase state of a substance as a function of its chemical composition, temperature, and pressure. In other words, it is necessary to ascertain the conditions which make magmatic melt homogeneous, the P-T-X parameters and the way melt is crystallized, the procedure which governs distribution of main chemical components therein, the state parameters under which the fluid phase separates and hydrothermal solution is yielded, and the substances (and their quantity) which can be dissolved in these non-crystalline solutions depending on temperature and pressure in the system.
This basic information is traditionally - and quite successfully - obtained via investigation of state diagrams for simple systems with low component indices, which are regarded as models for complex natural systems /1,2/. The results of such investigation show that rock-forming silicates and aluminosilicates react with water to form so-called p-Q type systems, which are characterized by critical phenomena among immiscible non-crystalline solutions in the presence of a solid phase. The use of such diagrams as models calls for such cooling stages of the natural magmatic source as the stage of magmatic distillation, the fluid stage, during which there cannot be more than one non-crystalline solution - fluid - in the system regardless of pressure (pneumatolytic rock formation occurs at this stage), the hydrothermal stage which is characterized by greater solubility of rock-forming minerals on lower temperature. This means that the evolution of the magmatic source containing a volatile component is viewed as stages of predominat existence of melt, fluid and hydrothermal solution in the form of non-crystalline phases; the stages are separated from each other with certain P-T-X borders within which non-crystalline phases are transferred into each other due to interphase reactions.

----

1) Kurnakov Institute of General and Inorganic Chemistry,
USSR Academy of Sciences, Moscow, USSR

Added complexity and examination of systems of the following types: two rock-forming aluminosilicates - water, water - carbon dioxide - rock - forming aluminosilicate, aluminosilicates - mixture of water and carbon dioxide, confirms the abovedescribed principal features /3/.
However, the data obtained on phase equilibria in systems containing components of varying volatility in the last few decades have shown novel properties of p-Q type systems, which must be included into the description of the generalized p-Q model in order to extend the range of phenomena conditioned by this model /4/.
There's hardly any need for such clear-cut definitions of aggregate states on non-crystalline solutions at high temperatures and pressures, and their categorical division into "liquid" (solution or melt), "gas" (vapour) and "fluid", which seems to be universally assepted. The alterations of the properties of non-crystalline solutions on changing P-T-X parameters are both considerable and multiform. Gases, liquids and fluids may be in mutual equilibrium, appear and disappear in interphase reactions. On the other hand, these phases can gradually transfer into other phases after changing their physico-chemical properties on alteration of state parameters. At times it is impossible to say with any amount of certainty, what the non-crystalline phase in question is: a melt, an aqueous solution or a fluid. The drive to cut the borders between aggregate states on non-crystalline solutions which is reflected in the abundance of diverse terms does more bad than good, since it presupposes the existence of certain "barriers" between them and masks their important ability to transfer into each other without a phase reaction.
As an example, it is possible to cite a transformation of a fluid into a hydrothermal solution which was described and experimentally confirmed in /5/ for the $NaAlSi_3O_8$-$Na_2Si_2O_5$-H_2O system. Retaining all properties of the accepted p-Q type model, the system exhibits a gradual transformation of the water-containing silicate melt into a high-concentration water fluid and then - into a hydrothermal solution. It is noteworthy that the temperature factor of silicate solubility remains positive throughout the entire process. One of the conditions for this process is the comparatively large, but feasible, pressure of 5-7 kbar at 700-800°C and 1-2 kbar at 300-400°C.
Another example is even more convincing. The silicate melt is transformed directly to hydrothermal solution due to gradual rise of water solubility in the melt on declining temperature. The "melt-solution" phase can reside in equilibrium with another non-crystalline phase whose composition and properties are also constantly changing. In various stages it could be defined as gas, fluid, and vapour. This transformation is theoretically argumentated in /6/, and experimentally confirmed in /7,8/ for the SiO_2 - $Na_2Si_2O_5$ - H_2O system. Model /6/ does not contradict the

universally accepted scheme for the magmatic source evolution /1,2/, but merely supplements it by describing its new properties and referring to P-T-X thresholds which make these properties active. The model is based on the existence of liquid immiscibility equilibria /9/ in all p-Q type systems (notably, the water - rock-forming silicate systems). Added components can lead to a situation when the water-containing melt will not fully crystallize even at the most moderate temperatures, which seem outrigt improbable in connection with silicate melts (up to 200°C in the example cited). In such systems, the appearance of the bulk of the fluid is not related to the retrograde boiling stage. At each stage of evolution the mass and composition of the fluid depend only on the overall composition of the system and solubility of volatile components in the silicate melt. Another feature of the model described in /6/ is the presence of not one source that generates hydrothermal solution (in accordance with the traditional p-Q scheme - cooling fluid), but two: the fluid and a low-temperature solution-melt. The hydrothermal solutions, which are radically different in composition and properties, can reside in mutual equilibria in a sufficiently wide range of P-T-X conditions. As opposed to the first example, there are practically no pressure limits for the presently disclosed transformation, the only restriction being the need for a pressure of at least 0.5 kbar throughout the temperature interval above 400-500°C (otherwise, water will evaporate from the melt).

The examination of the evolution followed by immiscible fluids in complex systems is also of interest. Here, two systems were analyzed: H_2O - CO_2 - NaCl /4/ and H_2O - CO_2 - NaCl - granite /10/.

In the widest range of state parameters, these systems contain two immiscible liquid phases. Moderate parameters provide for an aqueous NaCl solution and practically pure CO_2. Higher P-T para-meters trigger changes in mutual solubility of CO_2 and H_2O, salt and water. Continuous alteration of component concentration in the phases results in that on high parameters the immiscible phases are really different chemical substances - the mixture of CO_2 and H_2O in equilibrium with a water-containing salt melt. Using this experimental material as a model, and relying on data on H_2O, CO_2 and NaCl content in natural granites, the authors of /10/ came to the conclusion that two immiscible fluid phases, whose composition and properies continue to change, can exist in the natural granite source originally, and throughout its evolution.

Knowledge of the properties of fluids and hydrothermal solutions cannot be complete without the investigation of these phases' ability for solving a number of petrologically important minerals. Opinions on this subject are usually substantiated with experimental data on mineral solubility under given P-T-X conditions.

The following comment is called for at this point: as shown by experimental results of mutual salt influence on water solubility on varying P-T conditions /9/, the conclusions which apply to elementary systems cannot be transferred to more complex ones since even qualitative errors may result. For instance, solubility of NaF in water at average conditions is just over 4% by weight. With increasing temperature it declines to infinitesimal values at 374°C (the critical point for water). The pattern of NaF solubility is retained upon dissolution of NaF in KCl and NaCl solutions of relatively low concentration (up to 15-20% by weight). However, high chloride concentrations make solubility of NaF grow continuously with rising temperature. At 50% by weight of chlorides in the "solvent", for example, NaF solublity follows this pattern: 250°C - 0.5% by weight, 400°C - 2% by weight, 500°C - 3.5% by weight.

In conclusion, it seems expedient to point out once more (as has already been done in /4/ and /11/), that the above features of complex systems do not contradict the p-Q type model, although in the universally accepted version /1,2/ the model precludes melt-solution transformation and heterogenation of non-crystalline phases within the limits imposed by the so-called "fluid area". The above phenomena can certainly be detected only in individual cases, but then they cannot be left without attention altogether. It is therefore the goal of this work to promote wide discussion of physico-chemical studies of systems with components of varying volatility in order to correct, extend and enrich knowledge of p-Q type model characteristics.

REFERENCES

/1/ P. Niggley. Magma and its Products. Part I. Physico-Chemical Provisions (in Russian). Moscow-Leningrad, Gosgeokitizdat Publ., 435 p., 1947

/2/ F.G. Smith. Physical Geochemistry (in Russian), Moscow, Nedra Publ., 457 p., 1968.

/3/ A.A. Kadik, E.B. Lebedev, N.I. Khitarov. Water in Magmatic melts (in Russian), Moscow, Nedra Publ., 267 p., 1972.

/4/ M.Yu. Korotaev, K.G. Kravchuk, Heterophase Nature of Hydrothermal Solutions in Endogenous Mineral Formation (in Russian). IEM AN SSSR, Chernogolovka, 62 p., 1985.

/5/ D.A. Mustart. Phase Relations in the Peralcaline Portin of the System Na_2O - Al_2O - SiO_2 - H_2O. Diss. D.Ph. Thesis. Stanford Univ., Stanford, Calif., USA, 1972.

/6/ V.M. Valiashko, M.G. Valiashko. On the possibility to isolate the water phase from silicate melt in the light of equilibrium analysis in systems with components of varying volatility. In: Ocherki sovremennoi geokhimii i analiticheskoi khimii. Moscow, Nauka Publ., p. 142, 1972.

/7/ K.G. Kravchuk, V.M. Valiashko. Analysis of Phase Equilibria in the SiO_2 - Na_2O - H_2O System (in Russian). In: Soviet National Conference on Experimental an Technical Mineralogy an Petrography. Theses of reports. Kiev, Naukova Dumka Publ., 1978.

/8/ K.G. Kravchuk, V.M. Valiashko. Investigation of Phase Equilibria in the SiO_2 - Na_2O - H_2O Systems at High Temperatures and Pressures (in Russian), VINITI, RG KHIMIA, 2B887-dep., 1979.

/9/ M.I. Ravich. Water-salt Systems at High Temperatures and Pressures (in Russian), Moscow, Nauka Publ., 150 pages, 1974

/10/V.A. Zharikov, Yu.V. Alekhin, K.I. Shmulovich. Properties and Evolution of High-temperature Aqueous Systems (in Russian) In: 6th simpozium MAGRM. Tezisy Dokladov. Tbilisi, p. 278, 1982

/11/K.G. Kravchuk. Model Conceptions of Phase Equilibria in Hydrothermal Systems. In: Proc. VI IAGOD Symposium. Stuttgart, Germany, p. 375 - 378, 1984.

ON THE INFLUENCE OF WATER ON HIGH-PRESSURE MELTING OF SILICATES AND PLANETARY EVOLUTION

S. Franck and I. Orgzall 1)

A simple model for the thermal evolution of terrestrial planets shall be presented here shortly starting from an initially homogeneous planet already differentiated into core, that is in a fluid state and mantle. An important fact that strongly influences the evolutionary scenario is the mantle solidus. The usually accepted model is that of Green and Ringwood [4] for pyrolitic composition. It is well established and adopted in evolutionary calculations for all terrestrial planets. But looking at the temperature distribution in the preplanetary nebula and an appropriate condensation scheme (for instance Grossman and Larimer [5]) reveals that the conditions were different for each planet. Especially Earth and Mars differ in the amount of condensed water or hydrated minerals. While this amount for the Earth is relatively low so that possible influences remain insignificant, at Mars' distance from the Sun large amounts should have condensed to cause remarkable alterations in the evolutionary path. This effect is due to the changed melting properties of both systems: while the slope of the melting curve for dry systems is always positive, it is negative in water-bearing systems in the low-pressure region, that means decreasing melting temperature with increasing pressure. In the high-pressure region the slope changes its sign and the system behaves normally but resulting from the temperature decrease in the low-pressure range there remains a large difference in the melting temperatures of both systems.

Modelling the Martian thermal evolution thus requires the knowledge of the melting behaviour of a silicate-water system. While the high-pressure melting regime for dry systems is known up to highest pressures, experimental data for water-bearing systems cover only the range to about 4GPa caused by increasing experimental difficulties. So an extrapolation based on reasonable physical arguments has to extend this range to higher pressures so that it is applicable for the model formalism. This extrapolation is founded on experimental data fitted to a polynomial in the low-pressure region and on the assumption that the influence of water is nearly constant in the high-pressure range. The result is shown in Fig.1. The most striking feature is the large difference between the two solidi, that occurs first at low pressures and that persists up to higher pressures characterizing planetary interiors (a detailed analysis is given in Franck and Orgzall [3]).

Another important fact to be mentioned is the influence of water on the

1) Central Institute for Physics of the Earth
Academy of Sciences of the GDR, Potsdam, 1561

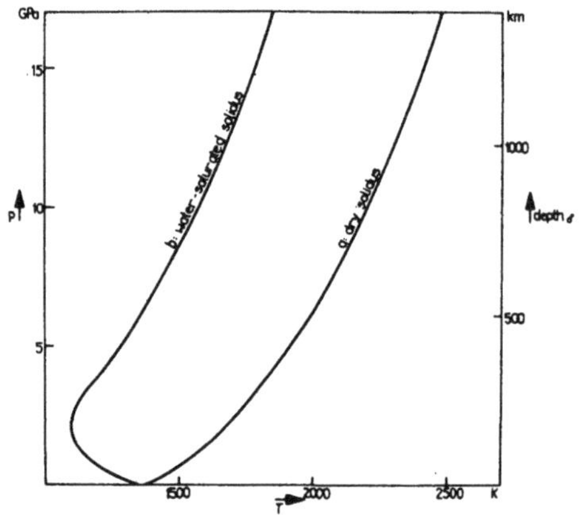

Fig. 1
Extrapolated model solidi of dry and water-saturated systems.

rheologic properties of the silicate. The viscosity of the melt is reduced by a factor of about 20 to 10^5 caused by depolymerization as experimentally observed for different systems and that of the solid decreases due to grain-boundary weakening; in this case only small amounts of water are sufficient to produce a measurable effect.
Thermal history models for the planets have been previously worked out by a variety of authors. In the following the parameterized convection formalism for whole-mantle convection of Schubert et al. [8], Stevenson et al. [10] or similar papers of other authors is adopted in general lines. Starting from the energy conservation equations of core and mantle regarding the effects of solidification of the inner core and resulting gravitational differentiation in the liquid outer core connected with the release of latent heat and gravitational energy, respectively, and additional influences of radioactive heating in the mantle, assuming adiabatic core temperature and introducing an appropriate melting curve for the core alloy as well as some results of convection and boundary layer theories a coupled system of ordinary differential equations is derived. The solution of this system supposing suitable initial conditions provides with the thermal history of the planet (the mathematical formulation of the problem may be found in Franck and Orgzall [2]).
For the Earth the usual dry solidus is adopted for the mantle material because of the negligible influence of water (in the subsequent crustal and upper mantle evolution this effect may be of certain interest, as for instance for subducting plates, but the model described here reveals only the basic features of whole-planet evolution). An initially hot planet is assumed whose temperature corresponds to the solidus in the mantle, due to accretional heating during accumulation and the release of gravitational energy in the course of core formation. In the beginning it cools rapidly by vigorous convection in the mantle. This process is slowed down after inner-core freeze-out starts. But it occurs only late in the planetary history, so that at present only about 4.5% of the

whole core volume are solidified and the process still continues. This gives the opportunity that a planetary dynamo may operate in the liquid shell for instance driven by compositional convection as proposed by Loper and Roberts [6]. Other obtained data like heat fluxes, viscosities, or Rayleigh numbers are in good agreement with previously published values of different authors. Smaller differences may be ascribed to a great lack of reliable experimental input parameters or relatively high uncertainties in these data.

For Mars the situation is quite different. Two models were tested: one with a familiar dry solidus and another with the extrapolated water-saturated solidus according to Fig.1. The core composition and therefore the melting curve is relatively unknown, so that an eutectic composition was chosen for having the lowest melting point. Because of a great difference of core and mantle melting temperatures in the first case the evolution time of 4.5 billion years was not sufficient that mantle convection could remove a substantial part of the interior energy so that no solid inner core could freeze out. In this model the whole core remains fluid. As many authors point out, thermal convection is not efficient enough to drive a planetary dynamo (for instance Stacey [9]), so it follows that this model would explain the lack of a Martian magnetic field (despite the opposite opinion of Dolginov [1] who concludes the existence of such field but his conclusions are somewhat in doubt, see Russell [7]). But the other parameters are not in satisfactory agreement with measured or estimated values of other authors, so that it seems that this model that does not take into account the appropriate condensation conditions, has to be rejected.

The second model appliing the water-saturated solidus gives a strongly changed picture. The difference between the melting temperatures at the core-mantle interface is small. So due to whole-mantle convection with increased efficiency because of the above mentioned viscosity reducing effect of water enough energy is removed and inner-core formation starts early. The growth proceeds up to the present and nearly the whole core is solidified. Only a small molten shell with thickness of about 100km may exist at the top of the core what prevents operations of a large-scale planetary dynamo. Assuming non-eutectic composition as starting point of these considerations the core melting temperature would be higher and possibly the whole core could be solid (as already Young and Schubert [11] pointed out regarding mantle viscosity; their critical value could be reached by introducing water as weakening medium). In this case a gravitationally powered dynamo could have operated during some evolutionary steps. Due to the solidification of the inner core and the existence of a melting interval of the core alloy the composition of the fluid outer core approaches the eutectic and the dynamo gets weaker and stops (despite possible thermal power sources) when the eutectic is reached. Other parameters are in better agreement with Martian data so that this model seems to be appropriate to describe the evolution of this planet.

So the evolution of Mars may be characterized shortly regarding the condensation conditions in the preplanetary nebula and the resulting influence of water. Mars grows by the impact of small water- and hydrosilicate-rich planetesimals. The temperature rises due to the release of gravitational and radioactive energy and earlier (as compared to other models) core formation starts. Energy release raises the temperature so that whole-planet melting occurs. Vigorous convection cools the interior efficiently. At the same time surface features form (features of ancient extensive volcanism suggest early crustal differentiation). Soon inner-core formation starts and a dynamo could have operated. With increasing time the inner core grew and today nearly the whole core is solidified. The dynamo possibly has stopped as mentioned above (or is very weak and does not contain a strong dipolar component). Eventually there may exist some features of remanent magnetism but these would be very weak. So the lack of a Martian magnetic field becomes obvious in the light of the assumption of realistic accretion conditions. Further theoretical investigations about the evolution of the magnetic field and space experiments -for instance the Soviet Phobos missions- will help to understand the process of the Martian evolution.

References

[1] Dolginov, Sh.Sh.: Geophys. Res. Lett. 5(1978),89-92; ibid. 5(1978), 93-95; Earth, Moon and Planets 37(1987),17-52.

[2] Franck, S. and I. Orgzall: Earth, Moon and Planets 37(1987),287-313.

[3] Franck, S. and I. Orgzall: Gerlands Beitr. Geophysik 96(1987), 251-259.

[4] Green, D.H. and A.E. Ringwood: Phys. Earth Planet. Inter. 3(1970), 359-371.

[5] Grossman, L. and J.W. Larimer: Rev. Geophys. Space Phys. 12(1974), 71-101.

[6] Loper, D.E. and P.H. Roberts, in: Stellar and planetary magnetism (A.M. Soward ed.), Gordon & Breach, New York, 1983, pp. 297-327.

[7] Russell, C.T.: Rev. Geophys Space Phys. 18(1980),77-106.

[8] Schubert, G.; P. Cassen and R.E. Young: Phys. Earth Planet. Inter. 20(1979),194-208.
Schubert, G.; D.J. Stevenson and P. Cassen: J. Geophys. Res. 85(1980),B5,2531-2538.

[9] Stacey, F.D., Physics of the Earth, John Wiley & Sons Inc., New York-London-Sydney-Toronto, 1977, pp. 197-209.

[10] Stevenson, D.J.; T. Spohn and G. Schubert: Icarus 54(1983),466-489.

[11] Young, R.E. and G. Schubert: Geophys. Res. Lett. 1(1974),157-160.

ON THE FOUNDATION OF A THERMAL STATE EQUATION RELEVANT TO THE EARTH'S LOWER MANTLE

Wolfgang Ullmann [1)]

We consider a region of the lower mantle (LM) wherein the material appears to be homogeneous with respect to phase as well as composition. Thus a relation between the field functions generated by the specific volume V, pressure P and either temperature T or specific entropy S exists. Its domain of definition is just that region, and the relation, usually written in the form $P = f(V,T)$, is said to be the thermal equation of state (TES) of the region.

Within the scope of such an approach it is generally accepted to consider the Grüneisen parameter as a function of V only, and constancy of the specific heat C_V throughout the region is taken for granted. Hence, consulting the isobaric cubic expansivity α and the isothermal incompressibility $\varkappa_T$, the TES takes the form

$$P = f(V,T_o) - \alpha(V,T_o)\varkappa_T(V,T_o)(T - T_o) , \qquad (1)$$

where V_o, P_o and T_o are the values of V, P and T, respectively, appropriate to a fixed reference state of the material in question. Note that $f(V,T_o)$ represents the isotherm of the region relating to the reference state. That function is subject proper when handling with the TES.

Nevertheless, for the last two decades, special attention has been paid to the supposition that γ to be directly proportional to V throughout the LM is a reasonable approximation (see, e. g., [1]). If that be so, then $\alpha\varkappa_T$ is constant everywhere in the LM, at least approximately, and therefore (1) is of the so-called Hildebrand type given by

$$P = f(V,T_o) + f(V_o,T) - P_o . \qquad (2)$$

In that case, the term $f(V_o,T) - P_o$ would be proportional to $C_V(T - T_o)$ with $C_V = \text{const}$. However, this conception is at variance with a consistent thermodynamic approach of the region under consideration.

To remedy that inadequacy we introduce the dimensionless quantity

$$\lambda_S := \frac{1}{\gamma T} \left(\frac{\partial \Phi}{\partial S}\right)_V ,$$

where Φ is the seismic parameter. There is every indication that λ_S is negligible with respect to unity, although verification by experimental values from close-packed minerals relevant to the LM is outstanding for the present. However, the same is true of the supposition $\gamma \propto V$

1) Dr. sc. W. Ullmann, Institut für Physik der Erde, Burgweg 11, 6900 Jena, DDR

In justification of it, several test results have been published, but they cannot bear out the idea expressed by $\gamma \propto V$ if only because of the measurement inaccuracies and other defects.

From the reasonable estimation that γ in the LM is close to unity (cf. [3]) and the generally valid relation

$$\lambda_S = 1 + \gamma - \left(\frac{\partial \ln \gamma}{\partial \ln V}\right)_S$$

it follows immediately that

$$|\lambda_S| \ll 1 \tag{3}$$

and $\gamma \propto V$ throughout the LM are contradictory propositions. Henceforth giving preference to the assumption (3) we get the approximate formula

$$\gamma = \frac{\gamma_o}{1 - \gamma_o(\frac{V}{V_o} - 1)} \frac{V}{V_o} ,$$

where γ_o is the value of γ at the reference state.

For obvious reasons let us consider P as a function of V and S by writing $P = P(V,S)$. We then have approximately

$$P = P(V,S_o) + P(V_o,S) - P_o ,$$

where S_o is the reference value of S. This decomposition of $P(V,S)$ is in analogy to (2). The term $Q(S) := P(V_o,S) - P_o$ is given by

$$Q(S) = \varrho_o\gamma_o T_o \int_{S_o}^{S} \frac{dS'}{\vartheta(S')} ,$$

where $\vartheta(S) := T_o/T(V_o,S)$ and $\varrho_o = 1/V_o$. Considering the density $\varrho \equiv 1/V$, acceleration of gravity g, and Φ as functions of depth z and assuming that the hydrostatic equation $dP = \varrho g dz$ comes true, we find that

$$Q(S) = \int_{z_o}^{z} (\varrho g - \Phi \frac{d\varrho}{dz'})dz' ,$$

where the depth z_o corresponds to the reference state. Hence follows that $Q(S)$ is computable for every depth z in the region under consideration, provided that the depth functions ϱ, g and Φ are defined on the interval between z_o and z and available from a regional or global Earth model such as PREM [2].

The same applies to the isentrop of the region approximated by

$$P(V,S_o) = P_o + \int_{z_o}^{z} \Phi \frac{d\varrho}{dz'} dz' .$$

That makes a great difference to the first outlined theory of the TES of the region. In that case the isotherm $f(V,T_o)$ could be computed throughout the region only if T as a function of z would be available. However, that is just the problem which cannot be solved without supplementary supposition, and we are still waiting for a reliable representation of T within that framework. On the other hand, along the lines of our approach we attain automatically to the approximate formula

$$T = T_o \left[1 - \gamma_o \left(\frac{\varrho_o}{\varrho} - 1\right)\right] \left[1 + \frac{Q(S)}{\varrho_o \gamma_o T_o C_V}\right] .$$

Hence follows that T is determined everywhere in the region under study, provided that ϱ and $Q(S)$ as functions of z and, not least, the reference values ϱ_o, γ_o, T_o as well as C_V are available. Unfortunately, as to these values, we must content us with rough estimates up to the present.

Another expression for the isentrop $P(V,S_o)$ emerges in connection with high-pressure experiments at high temperatures. Considering the isentropic incompressibility $\varkappa_S$ as a function of P and S, we have the generally valid approximate formula

$$P(V,S_o) = P_o + \varkappa_S(P_o,S_o)\varphi_n\left(\frac{V}{V_o} ; k_1,\ldots,k_n\right) , \qquad (4)$$

where

$$k_\nu := \varkappa_S(P_o,S_o)^{\nu-1} \frac{\partial^\nu}{\partial P^\nu} \varkappa_S(P_o,S_o) , \qquad \nu = 1,\ldots,n .$$

The $n + 3$ constant quantities P_o, V_o, $\varkappa_S(P_o,T_o)$ and $k_1,\ldots,k_n$ are characteristic of certain constitutive properties of the material in the reference state. The quantity k_{n+1} which is defined by analogy with k_ν can be estimated by an approximate relation written as $k_{n+1} = \chi_n(k_1,\ldots,k_n)$. The polynomial χ_n is unique by the function φ_n. Conversely, if χ_n proves to be reliable for estimating k_{n+1}, then we may be sure that (4) comes true to sufficient accuracy [4].

In practice, however, values of k_n for $n > 2$ are not available from laboratory experiments, and information about k_2 can be gained, if at all, only by way of the polynomial $\chi_1(k_1)$. Therefore interest proper remains confined to (4) with $n = 1$. Let us now turn to the case that the reference state occurs in some place of the region under consideration. Then k_1 can be estimated by the relation

$$\left(\frac{\partial \varkappa_S}{\partial P}\right)_S = \frac{1}{1 + \Delta}\left(\frac{d\varkappa_S}{dP} + \Delta\lambda_S\right) , \qquad (5)$$

where $\Delta := \varphi(d\varrho/dP)-1$. An Earth model such as PREM renders feasible to compute Δ and $d\varkappa_S/dP$ everywhere in the LM. Using PREM we find that, in the LM, Δ ranges between 0.015 and -0.03, and $d\varkappa_S/dP$ is greater than unity throughout.

Hence, making allowance for (3), it follows that the term $\Delta\lambda_S$ in (5) is negligible with respect to $d\varkappa_S/dP$. Therefore the identically valid

(5) can be approximated by

$$\left(\frac{\partial \varkappa_S}{\partial P}\right)_S = 1 + \frac{\varrho}{\Phi}\left(\frac{d\varrho}{dz}\right)^{-1}\frac{d\Phi}{dz} .$$

Based on PREM we thus learn that the values of $(\partial \varkappa_S/\partial P)_S$ are approximately equal to 3.1 as far as the uppermost layer of the LM is concerned. They decrease from 4 at z = 771 km to 3.1 at about z = 2180 km, and from that depth they again increase up to 3.3 at z = 2741 km. In the lowermost subregion of the LM, $(\partial \varkappa_S/\partial P)_S$ continues to increase from 1.5 to 1.7, approximately.

In conclusion it should be pointed out once more that for the present it is not to be expected to gain only just enough information which confirms the fundamental condition (3) under the circumstances of the LM. For all that it cannot be disregarded that both (4) and $(\partial \gamma/\partial S)_V = 0$ lay for the first time the foundation for a consistent thermodynamic approach to the real properties of the LM material.

References

[1] Anderson, O.L.: Evidence supporting the approximation $\gamma\varrho$ = const for the Grüneisen parameter of the Earth's lower mantle. J. Geophys. Res. 84 (1979), 3537-3542.

[2] Dziewonski, A.M., and D.L. Anderson: Preliminary reference Earth model. Phys. Earth Planet. Inter. 25 (1981), 297-356.

[3] Stacey, F.D.: A thermal model of the Earth. Phys. Earth Planet. Inter. 15 (1977), 341-348.

[4] Ullmann, W.: On the interrelation between Earth model and equation of state. Gerlands Beitr. Geophys. 91,(1982) 3, 245-257.

HIGH PRESSURE SYNTHESIS OF TiF_2 WITH FLUORITE STRUCTURE

T. Endo[1], N. Morita[1], T. Sato[1] and M. Shimada[1]

Abstract

A new compound TiF_2 was synthesized bythe conditions of 5-6.5GPa and 700-1400°C for 0.5-5hrs. TiF_3 and Ti were intimately mixed in the stoichiometric proportion of TiF_2, and charged into a platinum capsule prior to assembly in the high pressure cell Recovered sample was slightly gray in color. The powder x-ray diffraction pattern was moderately resembled to that of CaF_2. All diffraction peaks were indexed as the cubic symmetry with the lattice parameter of a=0.51555nm. The structure was the same as the fluorite without distorted symmetry, and its space group was determined to be Fm3m.

Introduction

It is believed that stishovite (SiO_2 with the rutile structure) is one of the most significant materials for understanding the earth's interior. In order to get the more detailed information of the earth's lower mantle, it is necessary to study the high pressure phase transformation of stishovite. However, since it is generally difficult to generate the higher pressure to require the transformation of a denser phase, many studies on the phase transformation of the MF_2 (M=divalent cation) with the same structure of stishovite have been extensively established. According to the previous reports (Vereshchagin et al.), (Jamieson and Wu), (Ming and Mangnan), (Ming et al.) and (Tressaud and Demazeau), the pressure induced phase transformation sequence was summarized as; rutile structure→distorted fluorite structure or fluorite structure→ α-$PbCl_2$ structure. Since the negative dP/dT slope for the phase transformation of rutile structure into distorted fluorite structure and fluorite structure were found in ZnF_2, MgF_2 and MnF_2, it could be estimated that the phase boundary between stishovite and its denser phase might have a similar profile.

Incidentally, in the case of high pressure phase transformation in FeF_2, the reversible phase transformation of the rutile structure into the hexagonal structure was exceptionally demonstrated (Ming and Manghnan). However, at present, it is expected that the FeF_2 would transform into the α-$PbCl_2$ structure under higher pressures, as expected in the other difluorides. It is reasonable that the type of pressure induced phase transformations in the difluorides would be evaluated from the crystallographic points of view.

Among transition metal difluorides of MF_2 except titanium difluoride, the heat capacity anomalies associated with the antiferromagnetic ordering were exhibited (Stout and Boo), but no reports on the synthesis and properties of TiF_2 were found. On the other hand, experimental results on the gaseous molecule of TiF_2 were widely demonstrated. Especially, the measurements of IR-spectra(Hastie et. al), Mass spectra(zmbou and Margrave) were given an useful information on the geometrical conformation around Ti ion. TiOF with rutile structure could be prepared by the

1) Department of Applied Chemistry, Faculty of Engineering, Tohoku University, Aoba, Sendai 980, Japan

reaction of TiO_2 and TiF_3 under high pressure and temperature conditions (Chamberland et. al). However, it has been doubtful whether TiF_2 was crystallized as the opposite end member or not in this system.

The present study was initiated to prepare the $TiO_{2-x}F_x$ ($0 \leq x \leq 2$) solid solutions in order to examine the systematic change in the electrical properties concerning to the valency state of Ti ions. As the results, new compound TiF_2 with the fluorite structure was prepared by the reaction of Ti metal with TiF_3 under high pressure and temperature conditions. We report here the crystallographic studies of TiF_2 with fluorite structure comparing with the results on the crystal structure of the other difluorides.

Experimental

The synthesis of TiF_2 was carried out by using a belt-type high pressure apparatus at 700-1400°C and 5-6.5GPa. TiF_3 and Ti was intimately mixed in the desirable proportion of which corresponded to the chemical composition of TiF_2, and charged into a small platinum capsule in a nitrogen-filled dry box. The closed capsule was inserted into the cell assemblage which was composed of NaCl sleeves and the graphite heater, and was subjected to high pressure-temperature conditions. NaCl was considered to be a favourable pressure medium for substantiating an isostatic pressure condition, because it plastically softened and often melted at the desirable reaction temperature conditions. The magnitude of pressure generated inside the cell was calibrated by the electrical changes on the basis of the phase transformations of Bi(2.55, 2.69 and 7.7GPa) and Ba(5.5GPa). The temperature of the sample was controlled by the supplied electric power through the carbon heater, and calibrated by inserting Pt/Pt-13%Rh thermocouple in the cell. The emf was uncorrected for pressure effects. After maintaining the desired high temperature and pressure conditions for 0.5-5hrs, the sample was quenched to room temperature prior to the release of applied pressure.

The X-ray powder diffraction data of recovered samples were measured using Si as an internal standard, and the lattice parameters of unit cell was preciously calculated by a least-squares refinement of all the observed reflections (two theta [$CuK\alpha$] = 20 - 120 deg.).

Results and Discussion

No reaction between TiF_3 and Ti below 5.0GPa was observed. However, there was somewhat ambiguity to determine whether or not the pressure above 5.0GPa was the thermodynamic condition required for forming TiF_2. Anyhow, all the reaction were conducted to obtain the single phase of TiF_2, but the recovered samples frequently contained TiF_2 and the trace of Ti metal. It had a tendency to increase the yield of a stoichiometric TiF_2 with the decrease of temperature.

X-ray powder diffraction pattern of sample, which was prepared by the reaction of Ti with 2 TiF_3 under the conditions of 5.5 GPa and 900°C for 50 hr, moderately resembled to that of CaF_2. By comparing the present x-ray diffraction patterns with the "in situ" X-ray diffraction patterns of FeF_2 under pressure of 8.2GPa at room temperature as shown in Table 1, all the diffraction peaks of TiF_2 could be indexed as the fluorite structure with the lattice parameter of a = 0.51555 ± 0.00002 nm.

In the X-ray diffraction patterns of TiF_2, since the peaks of (210) and (211)

Table 1 X-ray diffraction data of TiF_2, $FeF_2(II)$ and CaF_2

		TiF_2			$FeF_2(II)$		CaF_2	
		a=0.51555nm			a=0.5064nm		a=0.5463nm	
hkl	I/Io_{obs}	I/Io_{cal}	$d_{obs}(A)$	$d_{cal}(A)$	I/Io	d(A)	I/Io	d(A)
111	100	100	2.978	2.977	100	2.922	94	3.153
200	-	0.3	-	-	40	2.534	-	-
210	-	-	-	-	20	2.264	-	-
211	-	-	-	-	10	2.068	-	-
220	87	57.3	1.8231	1.8227	50	1.789	100	1.931
311	31	30.8	1.5547	1.5544	70	1.526	35	1.647
222	-	0.4	-	-	10	1.463	-	-
400	11	7.0	1.2890	1.2889			12	1.366
331	11	10.5	1.1828	1.1828			10	1.253
420	1	0.8	1.1534	1.1528			-	-
422	22	14.3	1.0524	1.0524			16	1.115
511	11	9.2	0.9923	0.9922			7	1.051
440	6	6.6	0.9115	0.9114			5	0.966

could not be observed, the crystal structure was same as the fluorite without distorted symmetry, and its space group was determined to be Fm3m. The relative intensity of each diffraction was theoretically calculated, and the calculated results are also listed in Table 1.

As the results, the crystal structure of TiF_2 seemed to consist of flattened TiF_8 octahedrons having eight identical Ti-F distances without any Jahn-Teller distortion observed in MnF_2 (Tressaud and Demazeau).

By using the ionic radius of F^- (fourfold coordination) (Shannon and Prewitt), the radius of Ti^{2+} coordinated by 8 fluorines in the fluorite structure could be calculated

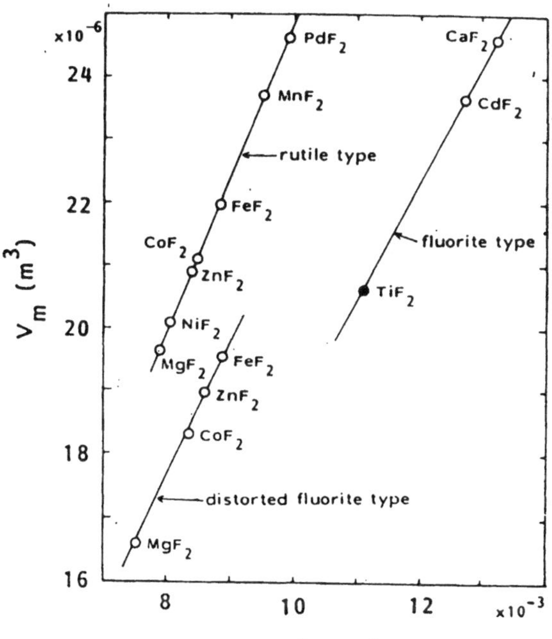

Fig. 1 The ralationship between molecular volume, V_m, and the cubes of the cation-anion distance, r^3, of typical fluorides

to be 0.107 nm. The ratio of $r_{M}2+/r_{F}-$ in TiF_2 was evaluated to be 0.92. On the other hand, the ratio of $r_{M}2+/r_{F}-$ in the first row transition metal fluorides with rutile structure such as NiF_2, CoF_2, FeF_2 and MnF_2 ranged from 0.724 to 0.821. Furthermore, it had been clear that these fluorides transformed to the distorted fluorite structure under high pressure conditions.

Fig. 1 shows the relationship between molecular volumes, V_m, and the cubes of the cation-anion distance, r^3, of the representative fluorides with rutile structure, distorted fluorite structure and fluorite structure. For each type of structure the linear dependence of V_m and r^3 was found, and its coefficient decreased with increasing the coordination number (6 for rutile structure, (2+6) for distorted fluorite structure and 8 for fluorite structure). In this figure, TiF_2 was represented as the solid mark. As seen in this figure, it is considered that TiF_2 with rutile and distorted fluorite structure was hardly formed owing to the restriction of the ratio of $r_{Ti}2+/r_{F}-$. Since the no phase transformation from fluorite structure to rutile structure was occurred by annealing at lower pressure and high temperature conditions and TiF_2 decomposed into TiF_3 and Ti at 900°C in vacuum condition, it is expected that the crystal structure of TiF_2 is only the fluorite structure and anyother structure is not possible.

References

(1) Chamberland, B.L., A.W. Sleight and W.H. Cloud: J. Solid State Chem., 2, (1970). 49.

(2) Hastie, J.W., R.H. Hauge and J. L. Margrave: J. Chem. Phys. 51 (1969) 2648.

(3) Jamieson, J.C. and A.Y. Wu: J. Appl. Phys. 48 (1977) 4573.

(4) Ming, L.C. and M.H. Manghnami: Geophys. Res. Lett. 5 (1978) 491.

(5) Ming, L.C. and M.H. Manghnami: Geophys. Res. Lett. 6 (1979) 13.

(6) Ming, L.C., M.H. Manghnami, T. Matsui and J. C. Jamieson: Phys. Earth. Planet. Inter. 23 (1980) 276.

(7) Shannon, R.D. and C.T. Prewitt: Acta Cryst B25 (1969) 925.

(8) Stout, J.W. and W.O.J. Boo: J. Chem. Phys. 51 (1969) 2648.

(9) Tressaud, A. and G. Demazeau: High Temp-High Press. 16 (1984) 303.

(10) Vereshchagin, L.F., S.S. Kabalkina and A.A. Kotilevetes: Sov. Phys. JETP 22 (1966) 1181.

(11) Zmbou, K.F. and J.L. Margrave: J. Phys. Chem. 71 (1967) 2893.

THE SYNTHESIS OF TRANSITION METAL GERMANIDES UNDER HIGH PRESSURE

M. Shimada[1)], H. Takizawa[1)], T. Endo[1)] and T. Sato[1)]

Abstract

Four new transition metal germanides were synthesized under the conditions of 3-5.5GPa and 600-1300°C for 0.5-10hrs. The high pressure synthesis was carried out by use of a belt-type apparatus after mixing the desired ratio of transition metal and germanium powders. The characteristics of germanides were summarized as follows. MnGe: B20 structure, a=0.4795nm, ρ =4-12 x $10^{-4}\Omega$cm, S=4-12μV/K, T_N=197K. CoGe: B20 structure, a=0.4631nm, ρ =3.8-5.3 x $10^{-4}\Omega$cm, S=-(10-40)μV/K, T_N= 120K. Cr_4Ge_7: Tetragonal structure, a=0.5820nm and c=1.906nm. VGe_2: C40 structure, a=0.4806nm and c=0.6634nm, ρ =1.5-4 x $10^{-4}\Omega$cm, S=-(4-5)μV/K.

Introduction

Many studies have been carried out for the preparation of binary transition metal silicides and germanides. The atomic ratio of silicon/metal and germanium/metal in these compounds except $IrSi_3$ is usually less than 2. First row transition metal monosilicides with the cubic B20 structure show a variety of electrical and magnetic properties. It is expected that the first row transition metal monogermanides with the B20 structure have electrical and magnetic properties similar to those of the silicides, but no systematic investigations have been carried out yet because of little synthesis of compounds. Only two compounds, CrGe and FeGe with the B20 structure were synthesized at ambient pressure condition. A few years ago, CoGe having the monoclinic structure transformed to the B20 structure under high pressure (Larchev and Popova). From the crystal point of view, disilicides and digermanides are classified in several groups. $CoSi_2$ and $NiSi_2$ have the C1 (CaF_2-type) structure, while $FeGe_2$ have the C16 ($CuAl_2$-type)structure. The various crystal structures such as C11b, C40, C54 and the defect disilicide type structure are known as the intermediate structure between C1 and C16 structures. On the other hand, the defect disilicide type structure, so-called Nowotny chimney-ladder structure (Nowtony), is widely found in disilicides and digermanides of 5A-8 group elements. The unit cell in these compounds consisted of the long period superstructure, in which the β-Sn like transition metal subcells were stacked along the c-axis. The number of subcells stacked along the c-axis just corresponded to n, when the chemical formula of the compound was expressed as T_nX_m.

In the present study, VGe_2, Cr_4Ge_7, MnGe and CoGe were synthesized under high pressure-temperature conditions, and their electrical and magnetic properties and effects of pressure on the formation of the defect disilicide type structure were examined.

Experimental

Vanadium, chromium, manganese and cobalt powders (>99.9% in purity) and germanium

1) Department of Applied Chemistry, Faculty of Engineering, Tohoku University, Aoba, Sendai 980, Japan

powder (99.9% in purity) were mixed in the desired molar ratio and uniaxially pressed at 100MPa at room temperature to form pellets. The pellets were put into a cylindrical BN capsule, which was placed in a carbon heater. The assemblage was put into a cell constructed with NaCl and subjected to high pressure-temperature conditions using the belt-type apparatus. The high temperature-pressure reactions were carried out at 3-5.5 GPa and 600-1300°C for 0.5-10hr, and then the samples were quenched to room temperature prior to releasing the applied pressure. The phase of product was identified by means of X-ray powder diffraction analysis. Lattice parameters were determined by a least-squares method using silicon as an internal standard. The electrical resistivity of the sintered sample was measured by a standard four probes method in the temperature range from 80K to 300K. The thermoelectric power mesurement was performed in the temperature range of 80-300K. Magnetic properties were measured using a magnetic torsion balance at 7KOe in the temperature range of 80-300K.

Experimental Results

(A) MnGe was synthesized at 4-5.5 GPa and 600-1000°C for 1-3 hr. All X-ray powder diffraction patterns of MnGe were completely indexed as the cubic B20 (FeSi-type) structure. Lattice constant was calculated to be a=0.4795 nm. MnGe was metastable under ameient pressure condition, and decomposed to $Mn_{11}Ge_8$ and Ge by annealing at 600°C in an evacuated silica tube. The electrical resistivity increased linearly with increasing temperature, which indicated that MnGe was metallic. The thermoelectric power had a maximum value of 12 $\mu V/K$ at 170K, and then decreased with increasing temperature. From the results of magnetic susceptibility measurements, MnGe was antiferromagnetic with the Néel Temperature of 197K. In the paramagnetic region, χ_g obeyed the Curie-Weiss law with the paramagnetic Curie temperature (θ_p) of 187K. The large positive value of θ_p indicated that Mn-Mn ferromagnetic exchange interactions exist in MnGe.

(B) CoGe with the B20 structure was synthesized at 4 GPa and 800-1000°C for 1-3 hr. Lattice constant was calculated to be a=0.4631nm, which was in good agreement with that of reported value (Larchev and Popova). The high pressure phase of CoGe with the B20 structure was metastable under ambient pressure condition and transformed to the monoclinic phase by annealing at 600°C in an evacuated silica tube. The high pressure phase of CoGe with the B20 structure was metallic. The value of the thermoelectric power was relatively larger than that of the monoclinic phase. These results indicated that the concentration of conduction carriers of CoGe with the B20 structure was lower than that of the monoclinic CoGe. From the results of magnetic susceptibility measurements, the high pressure phase of CoGe with the B20 structure was antiferromagnetic with the Néel temperature of 120K. At high temperatures, χ_g obeyed the Curie-Weiss law with the paramagnetic Curie temperature (θ_p) of 132K. The large positive value of θ_p suggested that ferromagnetic exchange interactions are dominant.

(C) Cr_4Ge_7 was synthesized at 4-5.5 GPa and 800-1000°C for 1-3 hr. All X-ray powder diffraction peaks of Cr_4Ge_7 were completely indexed as the defect disilicide type ($Mn_{11}Si_{19}$-type) structure of tetragonal symmetry with a=0.5820nm and c=1.906nm, but was different from those of $Cr_{11}Ge_{19}$ which was formed at ambient pressure. The high pressure product of Cr_4Ge_7 was metastable under ambient pressure condition, and decomposed to $Cr_{11}Ge_{19}$ and Ge by annealing at 800°C in an evacuated silica tube.

(D) VGe_2 was synthesized at 4-5.5 GPa and 800-1100°C for 1-3 hr. All X-ray powder diffraction peaks of VGe_2 were completely indexed as the hexagonal C40 ($CrSi_2$-type) structure. Lattice constants were calculated to be a=0.4806nm and c=0.6634nm. VGe_2 was metastable under ambient pressure condition, and decomposed to $V_{17}Ge_{31}$ and Ge by annealing at 600°C in an evacuated silica tube.

Discussion

A. Consideration of the B20 structure

Transition metal monogermanides had several kinds of crystal structures such as the monoclinic, B20, B31 and B35 structures. Among these structures, the B20 structure was the most closest-packed structure in which two kinds of atoms were 13-fold coordination. The atomic radius ratios (r_M/r_X) in MGe compounds with the B20 structure were limited in a narrow range of $0.90 < r_M/r_X < 1.03$ (Schob and Parthe). The ratio of r_{Mn}/r_{Ge} in MnGe was 0.94. The ratio of r_{Fe}/r_{Ge} in FeGe was 0.90, which was the lowest limit of the stability range of the B20 structure. FeGe had two polymorphs of the B35 and monoclinic structures at high temperature. Since the atomic radius ratio of r_{Co}/r_{Ge} in CoGe was 0.89, at ambient pressure condition CoGe had the monoclinic structure which was isostructural with the high temperature phase of FeGe. Since the ratio of r_{Co}/r_{Ge} in CoGe was very close to the lowest limit of the stability range of the B20 structure, CoGe transformed to the B20 structure under high pressure condition because germanium atom was more compressive than cobalt atom. The phase transformation from monoclinic to B20 structure in CoGe resulted in the increase of the coordination number of the cobalt atoms from 10 to 13 and in the decrease of the cell volume of 8.4%.

B. Crystal structure of the defect disilicide type structure

Among the defect disilicide type structures, the C11b, C40 and C54 structures were composed by the close-packed layers in different stacking sequences. The stacking sequences of ABAB, ABCABC and ACBDACBD led to the C11b (tetragonal), C40 (hexagonal) and C54 (orthorhombic) crystal structures, respectively. The difference in the stacking sequences mentioned above was closely related to the valence electron concentration of the transition metal atom (Nowotny).

The defect disilicide type structure was derived from C54 structure. In this structure, the transition metal atoms formed the β-Sn like subcells stacking along the c-axis by the long period. The formation of the long period superstructure was caused by ordering of metalloid atoms resulting in the metalloid vacancies (Nowotny).

It is expected that the applied pressure plays an important factor to control the stacking number "n" of the defect disilicide type compounds, T_nX_m. The defect disilicide type compounds synthesized at ambient pressure condition generally had large "n" value as found in $Cr_{11}Ge_{19}$, $V_{17}Ge_{31}$ and $Mn_{27}Si_{47}$. While, the compounds synthesized at high pressure conditions had small "n" value as found in Co_2Si_3 (Larchev and Popova), Mn_3Ge_5 (Takizawa et al.) and Re_4Ge_7 (Larchev and Popova). These results indicated that the long period superstructures tended to become more primitive short period under high pressure conditions.

In the present study, chromium defect digermanide of $Cr_{11}Ge_{19}$ ($CrGe_{1.73}$) formed at ambient pressure condition was converted to Cr_4Ge_7 ($CrGe_{1.75}$) under high pressure conditions. In the case of vanadium digermanide, $V_{17}Ge_{31}$ ($VGe_{1.82}$) was synthesized at

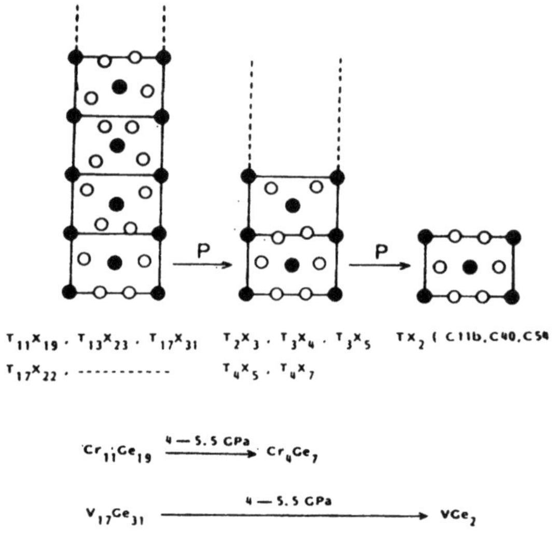

Fig. 1 Effect of pressure on the stacking sequence in the defect disilicide type compounds (figure shows the generating plane ●: T atom, ○: X atom)

ambient pressure condition and VGe_2 with the C40 structure was prepared under high pressure conditions. Since the chemical compositions of disilicides and digermanides synthesized under high pressure conditions ranged from $TX_{1.25}$ to TX_2, the chemical formulae of high pressure products are considered to be TX_2, T_2X_3, T_3X_4, T_3X_5, T_4X_5 and T_4X_7. The possible change in the stacking sequence of tetragonal sublattice is schematically summarized in Fig. 1.

References

(1) Larchev, V.I. and S.V. Popova: J. Less-Comm. Met. 87 (1982) 52.

(2) Larchev, V.I. and S.V. Popova: J. Less-Comm. Met. 84 (1982) 87.

(3) Nowotony, H.: "The Chemistry of Extended Defects in Non-Metallic Solids", North-Holland, Amsterdam (1970) p.223.

(4) Schob, O. and E. Parthe: Acta Cryst. 17 (1964) 452.

(5) Takizawa, H., T. Sato, T. Endo and M. Shimada: J. Solid State Chem. 68 (1987) 234.

The Synthesis, Characterization and Physical Properties of End-member Garnets in the System $(Fe, Mg, Ca, Mn)_3Al_2(SiO_4)_3$

C.A. Geiger, K. Langer, B. Winkler and L. Cemic[1]

Abstract

The synthesis, characterization and physical properties of almandine, pyrope, grossular and spessartine garnet are discussed. Unit-cell dimensions, refractive indices, possible OH^- contents and thermodynamic properties are reviewed and discussed. Optimal values for these properties are proposed, along with characteristic spectra for each of the end-members.

Introduction

The composition of many natural garnets can be described in terms of the aluminosilcate end-members Almandine ($Fe_3Al_2Si_3O_{12}$), Pyrope ($Mg_3Al_2Si_3O_{12}$), Grossular ($Ca_3Al_2Si_3O_{12}$) and Spessartine ($Mn_3Al_2Si_3O_{12}$). Therefore much mineralogical research has been performed on synthetic garnets in an attempt to document and explain their chemical and physical properties. More chemically complex natural garnets can also be better understood if comparisons to well characterized synthetic compositions are made. High pressure mineralogical research (in the range of 1 to 40 Kbar) is presently relatively routine and therefore rigorous and precise characterization of synthetic silicate minerals is possible. We have succeeded in synthesizing aluminosilicate garnets of end-member and solid solution compositions. In addition, our products have been characterized to a degree which has not been previously done. In this paper we present new IR spectroscopic data on almandine (AL), pyrope (PY), grossular (GR) and spessartine (SP) end-members and we review the literature regarding some of their intrinsic properties and present optimal thermochemical values for the four garnet phases.

Results and discussion

Almandine

The synthesis of AL is the most complex of the four end-members, mainly due to oxidation and reduction problems involving Fe^{2+}. Geiger (1986) synthesized polycrystalline AL (~500 mg batches) in a piston cylinder device from a glass of almandine compositon anhydrously in graphite crucibles, which maintain low, but unbuffered fO_2 (Table 1). AL can also be prepared hydrothermally at lower temperatures using an oxide and Fe - metal mix. A reducing buffer such as Fe - FeO must be employed if all iron is to be maintained as Fe^{2+}. The color of AL is a good qualitative indicator of the presence of Fe^{3+}, as end-member AL is green or gray and the presence of Fe^{3+} introduces a red color. Fe^{3+} can be detected

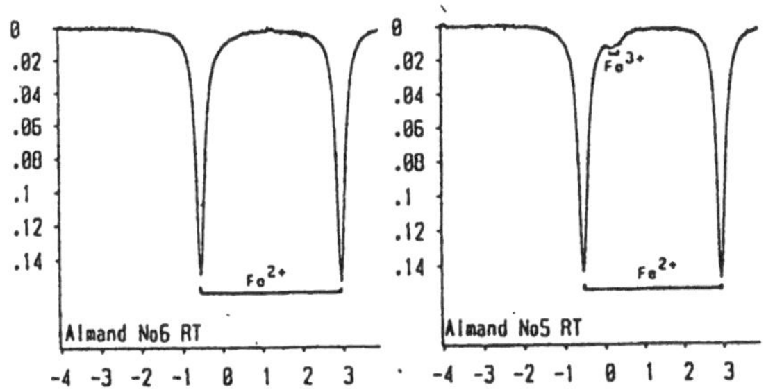

Fig. 1. Mössbauer spectra of two synthetic AL recorded at room temperature (G. Amthauer, pers. comm., 1987). The spectrum on the left shows no Fe^{3+} and has a_0 = 11.528(2) Å. The spectrum on the right contains a little Fe^{3+} and has a_0 = 11.536(2) Å. Both samples were synthesized anhydrously from glasses (Geiger, 1986).

1 Institut für Mineralogie und Kristallographie, Technische Universität Berlin, D 1000 Berlin 12

Table 1 : Physical constants of end-member garnets in the system $(Fe, Mg, Ca, Mn)_3Al_2(SiO_4)_3$

	Synthesis Conditions	a_o (Å)	n	OH^-	V (cm^3/mol)	ρ (g/cm^3)	$\Delta H^o_{f,298}$ (Kcal/mol)	S^o_{298} (cal/K · mol)
Almandine								
anhydrous	a) p = 27 Kbar T = 1173 K in graphite crucible	11.525(1)*	1.832(2)*	none	115.23	4.319(5)	- 12.9 ± 1.2	81.3
hydrous	b) p = 0.5 - 3.0 Kbar T = 773 - 1173 K under different fO_2	11.525 - 11.531(1)	1.829(3)*	small (?)	---			
	c) p = 10 Kbar T = 1073 K in iron capsule	11.528(1)	---	small (?)	---	---		
Pyrope								
anhydrous	a) p = 30 Kbar T = 1673 K	11.454(1)*	1.745(5)*	none	113.12	3.564(5)	- 18.7 ± 0.4	63.6 ± 0.4*
hydrous	b) p = 20 - 25 Kbar T = 1173 - 1273 K	11.456 - 11.459(1)	1.696 - 1.638(3)	small	---	---		63.13 ± 0.12§
Grossular								
anhydrous	a) p = 26 Kbar T = 1473 K	11.850(2)*	1.733(2)*	none	125.26	3.596(5)	-76.6 ± 0.7*	62.2 ± 0.7*
	b) p = 0.2 - 10.0 Kbar T = 1135 - 1489 K	11.845(3)	1.733(1)	none	---	---	- 78.4 ± 0.4§	60.8 ± 0.4§
hydrous	c) p < 7 Kbar T = 573 - 1123 K	≡	≡≡	variable	---	---	- 76.1 ± 1.8§	61.3 ± 0.3§
Spessartine								
anhydrous	a) p = 20 Kbar T = 1373 K	11.615(1)*	1.800(3)*	none	117.95	4.197(5)	no data	no data
hydrous	b) p = 0.001 - 3 Kbar T < 873 K	11.615 - 11.635	1.799 - 1.611(3)	variable	---	---		
	c) p = 20 Kbar T = 1373 K	11.617(3)	1.799(3)	?	---	---		

a_o ≙ unit-cell dimension; n ≙ refractive index; OH^- ≙ qualitative OH^- contents; V ≙ molar volume; ρ ≙ calculated density;
$\Delta H^o_{f,298}$ ≙ enthalpy of formation; S^o_{298} ≙ third law entropy; § values for natural specimens; • preferred optimal values for pure end-member composition.
≡ $a_o = 11.845 + 0.7335 X_{hygro} - 0.16234 X^2_{hygro} + 0.15614 X^3_{hygro} \pm 4 \times 10^{-3}$ ≡≡ $n = 1.733 - 0.08192 X_{hygro} - 0.11488 X^2_{hygro} + 0.05516 X^3_{hygro} \pm 3 \times 10^{-3}$
where X_{hygro} is the mole fraction of hydrogrossular in grossular (Huckenholz and Fehr, 1982)
The sources for the thermodynamic data are listed in Newton et al. (in press). The other data are taken from Mottana (1974), Huckenholz and Fehr (1982) and Hsu (1968).

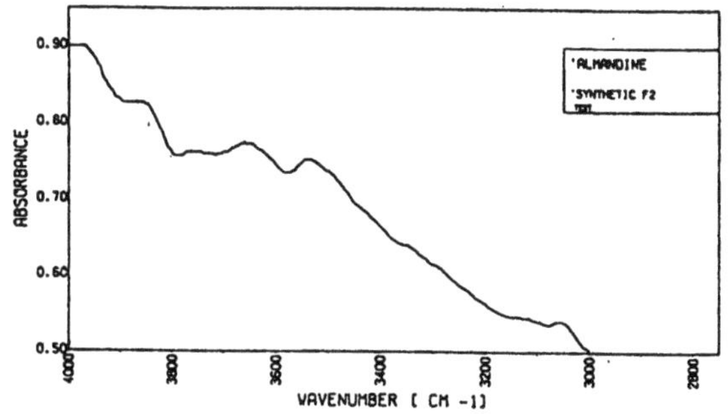

Fig.2. IR spectrum of a hydrothermally synthesized almandine (p = 10 Kbar, T = 1073 K) in the range of 4000 - 3000 cm^{-1}, showing OH^- valence vibrations. The band at approximately 3650 cm^{-1} is probably related to structural OH^- occuring as a hydrogarnet component. The lower energy band at approximately 3500 cm^{-1} is likely due to H_2O held in tiny fluid inclusions. The sloping background is due to intense electronic absorption bands at higher energies (see Fig. 4). The other small bands are artifacts due to the experimental set-up. The crystal is 280 µm thick, the size of the measuring spot was 38 x 170 µm.

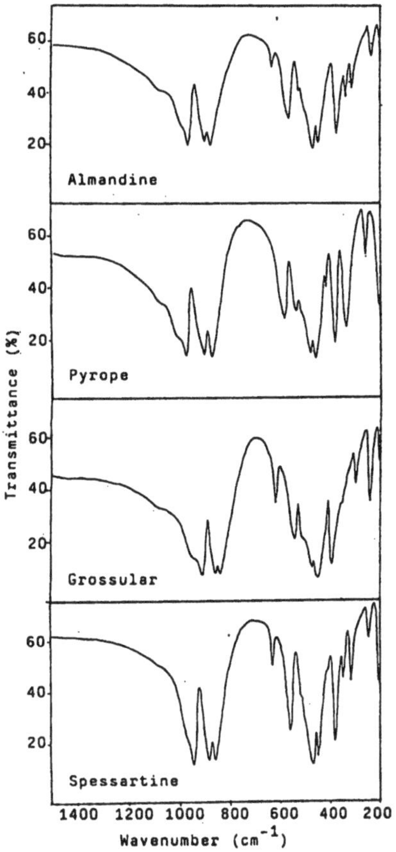

Fig. 3. IR spectra of synthetic end-member AL, PY, GR and SP. The band positions are as follows :
AL : 1074 (sh), 965 (s), 901 (s), 878 (s), 635 (m), 567 (s), 527 (w), 471 (s), 450 (s), 376 (s), 340 (m), 315 (m), 237 (m);
PY : -- , 976 (s), 907 (s), 877 (s), 585 (s), 539 (m), 482 (s), 463 (s), 420 (w), 386 (s), 339 (s), 259 (m)
GR : 1080 (sh), 913 (s), 860 (s), 843 (s), 620 (m), 544(s), 475 (s), 451 (s), 396 (s), 300 (m), 243 (s)
SP : -- , 947 (s), 887 (s), 863 (s), 631 (m), 562 (s), 472 (s), 450 (s), 383 (s), 349 (w), 316 (s), 245 (m)
(sh) = shoulder, (s) = streng, (m) = medium, (w) = weak and (--) = missing

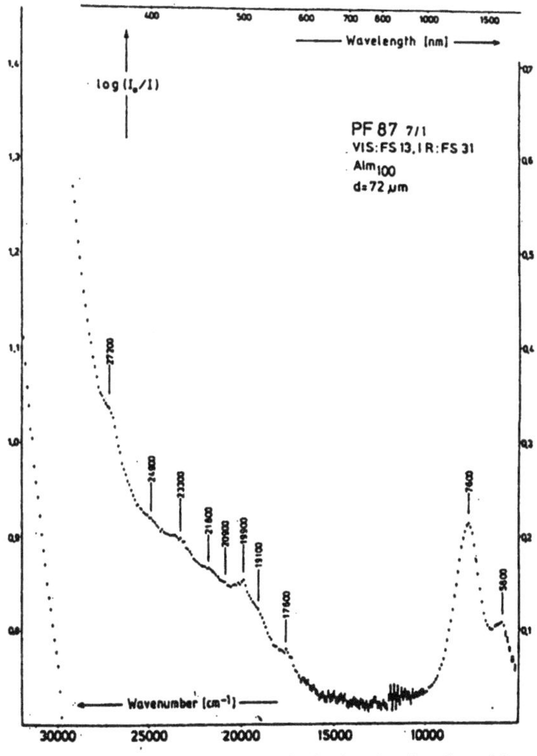

Fig. 4. Absorption spectrum of AL showing band positions of the electronic transitions (Frentrup and Langer, 1982).

quantitatively by Mössbauer analysis. Fig.1 shows the room-temperature Mössbauer spectra of two synthetic almandines, with one containing a little Fe^{3+}. Small amounts of Fe^{3+} also increase the unit-cell dimension over the optimal value of $a_0 = 11.525 \pm 0.002$ Å. Moreover, Geiger (unpublished) noticed that $\Delta H_{AL}^{sol'n}$ was slightly dependent on the presence of Fe^{3+}, although this was not quantified. This could account for the differing values of $\Delta H^o_{f,298}$ that have been proposed for almandine (-12.88 Kcal/mol from calorimetry vs -9.08 Kcal/mol from phase equlibrium work; Table 1, Newton et al., in press). The cell dimension could also be changed, although very slightly, by the incorporation of OH^- in AL. Fig. 2 shows the IR spectra of a hydrothermally synthesized AL with a weak band at ~ 3650 cm^{-1}, which is probably related to the presence of structural OH^- (see other sections below for OH^- in PY and GR). The low energy IR spectra showing the characteristic lattice vibrations and the UV/VIS spectra of synthetic AL are shown in Figs 3 and 4, respectively. A comparison of the spectra in Fig. 3 shows that the IR spectra of AL and SP are similar, indicating that the lattice energies and, hence, the structural make-up of these two species are the most similar of the aluminosilicate garnets. Frentrup and Langer (1982) measured the absorption spectrum above 5,000 cm^{-1}, which forced reassignment of certain spin-forbidden dd transitions related to Fe^{2+} (Fig.4). Such spectroscopic measurements can now be made to test for the presence of Fe^{3+} in natural or synthetic AL.

Pyrope

PY is the only aluminosilicate garnet which is not stable at low pressures. Synthesis must be made at pressures > 16 Kb. PY synthesized anhydrously, which requires even higher p and T, yields polycrystalline material which is relatively fine-grained. Large single crystals (up to 2 mm diameter) can be grown hydrothermally. However, such crystals have intrinsic structural OH^- and often H_2O contained within tiny fluid inclusions (Fig. 5). Intrinsic OH^- contents have been estimated at levels of 0.2 to 0.4 wt % H_2O (Hegeler et al., in prep.). The presence of OH^-, possibly as a hydrogarnet substitution with $(OH^-)_4 = SiO_4^{4-}$, inflates a_0 slightly and decreases the refractive index n (Table 1). OH^- contents change as a function of p and T of synthesis, thus affecting a_0 and n accordingly. The IR spectra of PY at low energies is unique among the aluminosilicate garnets. It lacks a band at approximately 650 cm^{-1}, which is present in the other-end-member compositions (Tarte and Deliens, 1973; see Fig. 3). The intensity of this band can be used analytically as a measure of the PY component in AL-rich garnets.

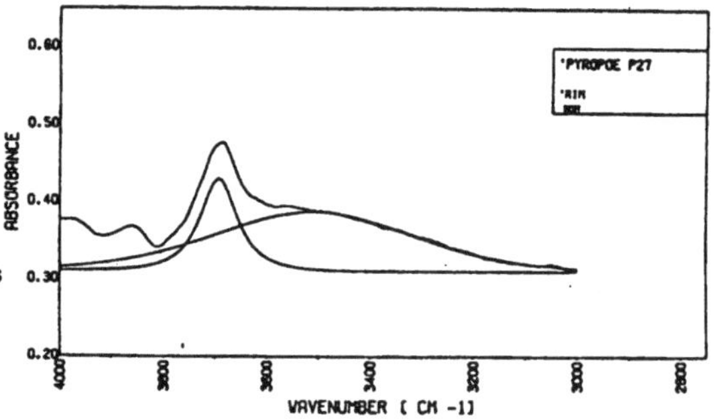

Fig.5. IR spectrum of a hydrothermally synthesized PY (p = 20 Kbar, T = 1173 K) in the range of 4000 - 3000 cm^{-1}, showing characteristic OH^- valence vibrations. The top line is the experimentally determined spectrum. The two smooth curves underlying the spectrum are a two curve LSQ best fit to the spectrum. The high energy band (half band width is ≈ 95 cm^{-1}) has been interpreted as arising from intrinsic structural OH^-, the broad low energy band (hbw ≈ 500 cm^{-1}) as arising from H_2O in tiny fluid inclusions (Ackermann et al., 1983). The crystal is 260 μm thick with a measuring spot of 38 x 170 μm. The bands occuring above 3800 cm^{-1} are artifacts. The given positions are only approximate.

Grossular

GR can be synthesized under anhydrous or hydrous conditions using various gels, glass or oxide mixes. It has long been known that compositions along the join $Ca_3Al_2Si_3O_{12}$ - $Ca_3Al_2(OH_4)_3$ can be found in nature and also be prepared in the laboratory (Huckenholz and Fehr, 1982) . The OH^- content is strongly a function of p and T. Moreover, metastability effects involving the synthetic starting materials and run durations have complicated quantitative calibrations (Huckenholz and Fehr, op cit.). It has not been demonstrated if small OH^- contents are retained in GR at high-temperatures (> 1100 K) at elevated water pressures (p_{H_2O} > 10 Kbar). The addition of a hydrogrossular component in grossular dramatically increases a_0 and decreases n (Table 1). High pressure and high - temperature hydrothermal syntheses appear to give relatively consistent and low values for a_0 and thus suggest that little OH^- is held in GR at such conditions (Table 1). The presence of OH^- in natural GR may account for the discrepancies in the calorimetrically determined $\Delta H^o_{f,298}$ and S^o_{298}, between synthetic anhydrously prepared GR and naturally occurring OH^- bearing material (Newton et al., in press, Table 1). Fig. 3 shows the lower energy region of the IR spectrum for synthetic GR. The small energy separation between the components of the doublet at about 850 cm^{-1} suggests there is, relative to AL, PY and SP, a smaller tetrahedral distortion in GR (Moore et al., 1971). Perhaps partly for this reason and because of the larger Ca^{2+} cation, compared to Fe^{2+}, Mg^{2+} and Mn^{2+}, GR can accept more of a hydrogarnet component than the other end - members.

Spessartine

The synthesis of SP is also relatively simple. Also, it has the largest p,T stability field of the aluminosilicate garnets (Mottana, 1972) and is stable over large ranges in fO_2. Frentrup and Langer (1982) succeeded in growing large single crystals of 600 µm for UV and VIS microspectrometric measurements (Fig 6). Hsu (1968) demonstrated that a_0 and n change sharply for SP synthesized hydrothermally at p < 3 Kbar and T < 900 K (Table 1). Later work demonstrated that hydrospessartines are metastable above 693 K (Hsu,1980). SP grown at higher temperatures shows no change in n and a_0 and, moreover, there appears to be no major differences in a_0 and n in anhydrously vs hydrothermally prepared material at p > 15 Kbar and T > 1300 K. This, as in the case of GR, suggests that little OH^- is incorporated at these conditions. No calorimetric data are available on SP. The characteristic IR spectrum is shown in Fig. 3.

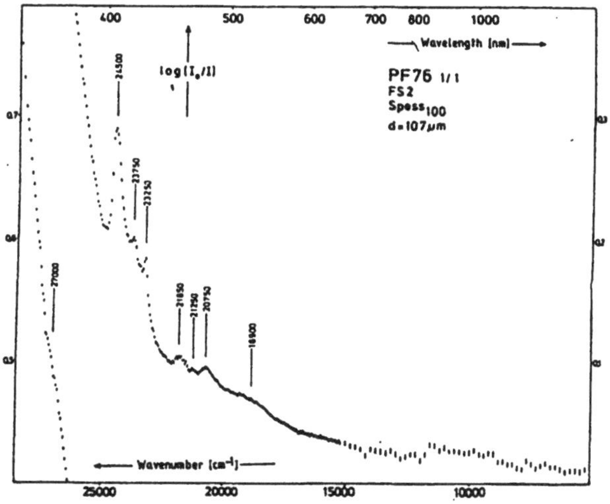

Fig. 6. Absorption spectrum of synthetic SP showing band positions of the electronic transitions (Frentrup and Langer, 1982).

Acknowledgments

We would like to thank Prof. G. Amthauer for collecting the Mössbauer spectra, Prof R. Steudel for use of the IR spectrometer and to Frau Hilarius for technical assistance.

References

Ackermann, L., Cemic, L., and K. Langer (1983) Hydrogarnet substitution in pyrope : a possible location for "water" in the mantle. Earth. Planet. Sci. Lett., 62, 208 - 214

Frentrup,K.R. and K. Langer (1982) Microscope absorption spectrometry of silicate microcrystals in the range 40,000 - 5,000 cm^{-1} and its application to garnet end-members synthesized at high pressures. In : Schreyer, W. (ed.) : High pressure studies in geosciences. Schweizerbart., Stuttgart, 247 - 258.

Geiger, C.A. (1986) Thermodynamic mixing properties of almandine garnet solid solutions. unpub. Ph.D.thesis, Univ. of Chicago, 150 p.
Hegeler, H., Langer, K., Beran, A. and C. Chopin (in prep.) Hydroxyl in pyrope II: A first attempt to determine by IR - microscope spectrometry the "water"- contents in synthetic pyropes and an application to the Doria Maira pyrope.
Hsu, L.C. (1968) Selected phase relationships in the system Al-Mn-Fe-Si-O-H : A model for garnet equilibria. J. Pet., **9**, 40 - 83.
Hsu, L.C. (1980) Hydration and phase relations of grossular-spessartine garnets at p_{H_2O} = 2 Kb. Cont.Min.Petr., **71**, 407 - 415.
Huckenholz, H.G. and K.T. Fehr (1982) Stability relationship of grossular + quartz + wollastonite + anorthite II. The effect of grandite - hydrograndite solid solution. N. Jb. Min. Abh. **145**, 1-33.
Moore, R.K., White, W.B., and T.V. Long (1971) Vibrational spectra of the common silicates : I The garnets. Am. Min., **56**, 54 - 71
Mottana, A. (1974) Melting of spessartine at high pressure. N. Jb. Min. Mh, **6**, 256 - 271.
Newton, R.C., Geiger, C.A., and O.J. Kleppa (in press) Thermodynamics of $(Fe^{2+},Mg,Ca)_3Al_2Si_3O_{12}$ garnet : A review. Korzhinskii Memorial Vol., Springer-Verlag
Tarte, P. and M. Deliens (1973) Correlations between the infrared spectrum and the composition of garnets in the pyrope-almandine - spessartine series. Cont. Min. Pet , **40**, 25 - 37.

ISOTHERMAL COMPRESSION OF THREE SYNTHETIC END MEMBER GARNETS UP TO 25 GPa

J.M.LEGER, C.CHATEAU, A.M.REDON*

The volumes of three synthetic end member garnets (Ca-Al,Mn-Al,Ca-Cr) have been determined at room temperature as a function of pressure up to 25 GPa in a diamond anvil cell by x-ray powder diffraction. Anisotropic stress components were found to have a large effect on volume measurements and must be accounted for to derive reliable volume changes. Experiments in different pressure transmitting media were carried out and bulk moduli were deduced. They are generally lower than those reported from previous x-ray experiments under high pressure but are in good agreement with ultrasonic or Brillouin spectroscopic determinations.

----

* Centre National de la Recherche Scientifique
Laboratoire de Physico-Chimie des Materiaux,
Groupe d'Etudes des Matériaux sous Pression,
1 place A.Briand, 92190 Meudon, France

In the lower crust and upper mantle, garnets are present with variable compositions. It has long been desirable to establish bulk modulus-composition relationships that would allow to derive the bulk modulus of a garnet of arbitrary composition. However, the bulk moduli of pure end member garnets are scarcely known and in addition large discrepancies arise from different experiments.

In this note we report pressure-volume measurements by powder x-ray diffraction analysis as a function of pressure up to 25 GPa for synthetic Mg-Al (pyrope), Ca-Cr (uvarovite) and Mn-Al (spessartine) garnets.

These garnets were obtained as powders by following previously reported syntheses (Naka et al., 1975;Akaogi and Akimoto,1977). The cell parameters were in excellent agreement with the values they obtained as well as the line intensities in the Debye-Scherrer patterns. Under pressure we noticed some differences in the line intensities with respect to normal pressure for pyrope and spessartine; as no further change occured with increasing pressure, this was attributed to preferred orientations.

The pressure-volume measurements have been performed in a diamond anvil cell using molybdenum radiation and a wet film camera (Léger,1984). The pressure was determined through the simultaneous determination of the cell parameter of a pressure marker, silver metal or sodium chloride.

The pressure transmitting medium is critical; below its solidification pressure, the usual 4:1 methannol-ethanol mixture provides a truly isostatic medium, but above this pressure (and any known substance is then solid at room temperature) the applied stress is no longer hydrostatic and uniaxial stress components act onto the sample.

The uniaxial stress components modify the d_{hkl} of cubic structure compounds (Singh and Kennedy,1974) according to the relation :

$$\Delta a/a_o = \varepsilon = \varepsilon_P + S_{12}t + (S_{11}-S_{12}-S_{44}/2)t\,\Gamma(hkl)$$

where P is the effect due to the hydrostatic pressure P, S_{ij} are the elastic compliances of the sample, t the unixial stress and $\Gamma(hkl)$ a function of the Miller indices, zero for h00 lines and maximum,0.33, for hhh lines.

For garnets, at normal pressure, the elastic anisotropy is weak and the last term is very small; so, there should be no dependence of the strain as a function of the observed diffraction line. This turns out to be actually the case. However the term $S_{12}t$ still remains and may be large : the observed volume is then markedly larger than the actual one in absence of anisotropic stress. In this case, a solution is to discard the low pressure data, to extrapolate the high pressure data back to zero pressure and to renormalize the data to the new $(V/Vo)o$ value obtained, which is generally higher by 1%. This procedure is based on the small increase at high pressure, above 5 GPa for NaCl or MgO, of the shear strength of the pressure transmitting medium which is strongly related to the applied anisotropic component (Kinsland,1978).

In this study of the garnets, we systematically performed the experiments in two different pressure transmitting media as use of the liquid medium was not satisfactory. This was especially the case for uvarovite and spessartine; use of a solid pressure transmitting medium, silicone grease or NaCl respectively, allowed to confirm the volume variations by using the above procedure and corrections were then equal to 0.9 % and 1.3 % respectively. For pyrope, results obtained in silicone grease and sodium chloride were in excellent agreement. The volume measurements were very well reproducible and no difference could be noticed when increasing or decreasing pressure.

Once corrected for the effects of the anisotropic stress components as indicated above, the volume variations were fitted to the Birch equation of state. In the case of these garnets, the compression remains small in the pressure range investigated and no attempt to determine the pressure derivative of the bulk modulus was made. We fixed this $B'o$ value to 5, the usual value, which is actually the mean value found from ultrasonic experiments, 5.4 or 4.7 for pyrope and almandite respectively (Bonczar and Graham,1977;Soga,1967).

The compression of pyrope is slightly larger than reported from previous powder x-ray investigations under high pressure (Duba and Olinger, 1972;Takahashi and Liu,1970) but this can be easily accounted for by the different experimental conditions leading to rather high anisotropic components which were not taken into account in these works. It results that the bulk modulus we determined is lower; it is only 165 $\pm$ 7 GPa, compared to 190 GPa. However, this lower value is in excellent agreement with the bulk modulus obtained from ultrasonic experiments (Bonczar and Graham,1977).

The values of the bulk modulus of uvarovite or spessartine, 155 $\pm$ 5 and 170 $\pm$ 8 GPa respectively which were not previously determined by x-ray diffraction under pressure, are in very good agreement with ultrasonic (Babuska et al.1978) or Brillouin scattering (Bass,1986) experiments but differ somewhat from extrapolation of the bulk modulus of garnets with variable composition to pure end members (Leitner et al., Babuska et al.,1978).

The effects of anisotropic stress components revealed extremely important on the values of the bulk modulus of garnets obtained from powder x-ray diffraction under high pressure; values larger by as much as 15 % can be actually found when no account of the uniaxial stress is made. Then it is meaningless to derive any correlation from small variations of the bulk modulus with composition unless special care has been taken to reduce these anisotropic components and account for them. This was not the case in the previously reported works under high pressure and actually, contradictory results about the systematics of the bulk modulus with composition or molar volume were noticed.

Literature

Akaogi,M.,and S.Akimoto : Phys.Earth Planet.Inter.15(1977)90-106.

Babuska,V., J.Fiala, M.Kumazawa, I.Ohno, and Y.Sumino : Phys. Earth Plan. Int. 16 (1978) 157-176.

Bass,J.D.: J. Geophys. Res. 91 (1986) 7505-7516.

Bonczar, L.J, and E.K.Graham: J. Geophys. Res. 82 (1977) 2529-2534.

Duba,A., and B.Olinger: J. Geophys. Res. 77 (1972) 2496-2499.

Kinsland, G.L.: High Temp. High Pressures 10 (1978) 627 and 639.

Léger,J.M.: Revue Phys. Appl. 19 (1984) 815-818.

Leitner, B.J., D.Weidner, and R.C.Lierbermann : Phys. Earth Planet. Int.22 (1980) 111-121.

Naka,S.,Y.Suwa,T.Kameyama : Amer.Mineral.60(1975)418-432.

Singh,A.K., and G.C.Kennedy : J. Appl. Phys. 45 (1974) 4686-4691.

Soga,N.: J.Geophys. Res. 72 (1967) 4227-4234.

Takahashi,T., and L.Liu : J. Geophys. Res. 75 (1970) 5757-5766.

POLYCRYSTALLINE SKELETAL DIAMONDS

*Andrzej JAWORSKI, Michał SUBSTYK**

In this paper there are discused results of statical diamond synthesis in high temperature and pressure which have been achieved in the ASEA's Quintus press. There has been used modified reaction vessel composed of cobalt-graphite sandwich set surrounded by graphite heater. It has allowed to reduce radial temperature gradient. As an effect of the process polycrystals with very different shapes, morphology and internal structure have been formed.

INTRODUTION

For the last ten years there has been observed a growing interest in obtaining mono- and polyphases diamond sinters, used in tool industry for production of grinding and cutting tools. One of such materials are polycrystalline diamonds made in statical synthesis process, from the same components as monocrystalline diamonds. Their origin is not connected with sintering effect, but with nucleation and growth up to creation of coalestent monocrystalline diamonds. Diamond polycrystals which have been obtained so far (NASSAU K., NASSAU J. 1979) are similar in morphology and internal structure to well-known natural diamond types as the ballas or the carbonado. In result of experiments with polycrystalline diamond production, which have been made in IEMT, various kinds of such crystals have been obtained. Because of their unusual morphology and internal structure they are named *the skeletal polycrystals* .

Experiments have been made in the Quintus press in the «Belt» high pressure apparatus (described by HALL (1960) and JAWORSKI (1985)). Reaction vessel is resistance heated by changing current. There can be reached temperatures upto 2300 K and pressures upto 10 GPa. However, real temperature inside the reaction vessel during the process is not directly measurable. The cobalt-graphite sandwich set has been surrounded by graphite heater in purpose to reduce the radial

**Institute of Electronic Materials Technology, 01-919 Warszawa, Wólczyńska 133, POLAND*

temperature gradient (higher temperature in the reaction vessel center) Experiments have been made at temperatures about 1700-2000 K and pressures about 5.7-5.8 GPa. Each process has lasted 23 minutes. In that processes 22-34 ct of diamonds have been obtained and it comprises from 21% to 32% of primary material transformation.

In purpose to isolate the diamonds after synthesis, material has been mechanically crumbled and next chemically purified from extra metal and graphite.

Investigations have been done with various microscope techniques.

CHARACTERISTIC OF SYNTHESIS PRODUCT

Macro-observations of cracked reaction vessel have proved that small grains have been in the center, and great ones on the border with graphite heater. Great grains, upto 6 mm in diameter, forming block shaped polycrystals, very often similar to natural spherolitic polycrystals (crystal brushes) e.g. calcite's or quartz's. Cudgel shaped long diamond crystals have structure like single natural crystals taken out from the crystal brush. Polycrystals are black or dark olive. Small crystals can be yellow and transparent. Great black ones are mat, but small and especially dark olive are glossy. The diamonds have dark colour due to graphite, cobalt, and perhaps cobalt carbide impurites. Their most characteristic feature are their cavities, empty spaces and pores in central part of each grain. Therefore they are named the polycrystalline skeletal diamonds.

On their faces some different structures have been observed.

Rough structures are observed at thin ends of cudgel shaped crystals. They are composed of sharp-edges elements on crystal surface. They come out into band-layer structures .

Band-layer and layer structures are composed of diamond's sine curve bands with similar amplitude. It is observed on side faces of long crystals. That structure comes out of the rough surface at thin end of a crystal and stretches upto aproximately flat surface on thick end. Perhaps this structure represents side surface of one crystal from a crystal brush .

Layer-clouds structures are probably less porous and more dense form of earlier described structures. That structures are also possible to be an effect of polycrystals cracking. Such shape of the structure is very similar to the cirrus clouds composition .

Mosaic-hopper structures are observed only at right crystalographic faces. Face with such structure have hollow center part and elevated outer part. Each step morphology is strong sculptured in mosaic of geometric patterns. Perhaps that forms arise as a result of crystal growth in rather great graphite supersaturation conditions (SUNAGAWA 1984).

Scale structures are mainly observed on large crystal faces. It is like a fish scale. That structure is compound of conformable forms to the face edges, covering the face like tile, one layer over the next one. Sometimes that structure elements have spiral disposing. Each scale surface is not so much sculptured. Its forming is probably connected with overgrowing of the mosaic-hopper structures .

Growth spiral structures. There are large forms - steps about 40 μm wide, and about.5 μm high. Steps are not homogenous, and there can be found remains of the mosaic structures. That structures are observed in no larger than 1 mm crystals. The growth spiral structures arise in conditions of graphite contens decrease around a crystal, as an effect of growth and overgrowth of earlier structures .

Block polycrystals. That is coalescence of less or more fine monocrystals. The monocrystals have fine faces (100) and (111), edges and corners. Parallel free spaces between grains are very characteristic. Grains are conected with small surfaces. Some faces have etching trigons. The block polycrystals have been found in central part of reaction vessel .

On crystals surface there is observed great number of small cavites and pores which have arisen, probably, out of synthesis process during chemical purification, as an effect of imputites etching from surface diamond layer.

In the skeletal polycrystals have been found another very interesting phenomenon - crystals have formed mainly octahedron faces.

Physical and chemical properties of polycrystalline skeletal diamonds are shown in table 1 (according to JAWORSKI 1986).

TABLE 1. Physical properties of the skeletal polycrystals and the MDS* type diamonds.

PROPERTY / TYPE	MASS DENISITY g/cm³	BULK DENISITY g	MAGNETIC SUSCEPTIB. x**	STRENGHT COMPRESSIVE N	STRENGHT IMPULSE imp***	GRAPHITIZATION START °C	GRAPHITIZATION END °C
POLYCRYST.	3.38	1.62	360	48	6000	830	920
MDS	3.50	1.91	170	170	15000	950	1100

* monocrystalline synthetic diamonds

** values in CGS, to recount into SI it should be multiplied by $4\pi 10^6$

*** impuls number which gives 50% of no crumbled fraction

CONCLUSIONS

Using of modified reaction vessel geometry has made change of physical conditions inside reaction vessel, mainly reducing there radial temperature gradient. As the result of it new type of diamonds has been obtained.

Rich morphology of crystals faces and spaced structure observed in the polycrystalline skeletal diamonds have been formed as an effect of crystal growth process. Decreasing of graphite concentration around a growing crystal make change of growth mechanism from continous growth, through two-dimensional nucleation growth to growth spiral (SUNAGAWA 1984). It is well noticed in crystals morphology (mosaic-hopper, scale, growth spiral structures). Details of that crystals genesis is being prepared for publishing by the authors.

The obtained polycrystalline diamonds have not satisfactory mechanical properties. Because of graphite and metal impurites, and great porosity the polycrystals can be used mainly for production of polyphase diamond sinters. It is thought, that proper ratio of the graphite heater mass to the metal-graphite sandwich set mass and choice of heating current power, pressure and process time allow to receive polycrystals with much better mechanical properties.

REFERENCES

HALL H.T.(1960), Ultra High Pressure, High Temperature Apparatus the «Belt». Rev. Sci. Instr. 31.

JAWORSKI A.(1985), Trying of Diamond Sinters Production in the «Belt» High Pressure Apparatus.(in polish). VII Scientific School of Grinding. Koszalin.

JAWORSKI A.(1986), Polycrystalline Skeleton Diamonds as a Product of Static Controllable Synthesis.(in polish). Prace ITME z 20.

NASSAU K., NASSAU J.(1979), History and Present Status of Synthetic Diamond. J. Cryst. Growth 46 .

SUNAGAWA I.(1984), Growth of Crystals in Nature. Materials Science of the Earth Interior. Tokyo.

REACTION MODEL ON THE PHASE TRANSITION OF GRAPHITE AND BORON NITRIDE INTO DIAMOND AND CBN

U. Gerlach 1), H. Vollstaedt 2), E. Wolf 1)

Even after 30 years diamond and cBN synthesis, respectively, very different mechanisms for the high pressure phase transitions are discussed, for example:

- homogeneous nucleation in a metal melt,
- heterogeneous nucleation on metal catalyst or metal-carbon phases with a diamond-similar structure, on dopants or impurities,
- otherwise a solid state transition from the graphitic structure into diamond.

It is well known from nucleation theory that the nucleation work is highest for homogeneous nucleation but lowest for a martensitic solid-solid transition. On the other side, diamond and cBN formation depend strongly on the structural perfection of the source used. Amorphous carbon, diamond-like carbon (i - c), glassy carbon or carbon black can't be transformed into diamond at normal synthesis conditions, only a graphitization process takes place. This graphite can be successfully used in obtaining diamond in a second step of resynthesis /1/.
In the last years increasing by experimental proofs of a solid-solid transition are published, but it is impossible to explain why graphite regrown during synthesis does not transform.
For cubic boron nitride a homogeneous nucleation mechanism is supposed predominantly, but the dependence of wurtzitic or cubic phase on the crystallinity of boron nitride source is not fully understood.
The hitherto existing models of transformation process can't sufficiently explain experimental evidence.

Our considerations start from the general behaviour and formation of inorganic layered structures on pressure and temperature, especially the aspect of the occurence of polytypes. It is known, that there are very close crystallographic relationships betweeen the structures of

- hexagonal diamond and 2H-graphite
- cubic diamond and 3R-graphite
- wBN and 2H-hBN
- cBN and 3R-hBN.

----

1) Central Institute of Solid State Physics and Materials Research
Academy of Sciences of the GDR, Dresden, 8027
2) Central Institute for Physics of the Earth
Academy of Sciences of the GDR, Potsdam, 1561

Hexagonal diamond and wBN are formed from high crystalline graphitic source material by martensitic phase transitions.
Also in other systems different polytypes can form different high pressure structures, for example PbJ_2, in which the 2H- , 4H- and 6H-polytype forms own, diverse high pressure phases, but return these phases transform all into the 2H-polytype /2/.

If we consider the stability of different polytype stacking orders it is shown, that this stacking order depends on certains conditions, as temperature and/or pressure. In the case of $TaSe_2$ DTA-investigations by HAYASHI and KAWAMURA /3/ have shown polytytpe transitions of 2H→3R and 2H⇌4H⇌6R⇌1T by increasing temperature and return by decreasing temperature.
In the carbon system the 3R-structure changed at high temperatures slowly into the 2H-form.

This suggests, that the thermodynamic stable polytype is the 2H-form in those layered structures and higher polytytpes have a more and more metastable character.

A similar energetic behaviour is also expected in systems, in which the transition 2H→3R is not easily possible, as hBN or MoS_2, because additional to the translation of neighbouring layers a rotation movement is necessary for changing the polytype stacking order. Such a transition is hitherto unknown for hBN, but for MoS_2, both types of which occur in nature, similar to graphite.
At 1000 grdC and normal pressure the 3R-structure changed into the 2H-form, but at 2000 grdC and 4 - 7.5 GPa the reverse transformation from 2H into 3R takes place /4/. This suggests, that some metastable polytypes at normal pressure can be thermodynamically stable, not only kinetically at certain HP/HT-conditions.

Unfortunately there are only few investigations on thermodynamical and physical properties of different polytypes. The energy differences between polytypes are small. For the transition enthalpy 3R→2H-graphite COUGHLIN and BOEHM measured a value of 0.586 ± 0.17 kJ/g-atom by using the intercalation of 3R-containing natural graphite /5/. These low energy differences between polytypes can be remarkable at phase transitions, because they must effect different equilibrium lines. In the case of carbon phase diagram the usually used graphite-diamond equilibrium line is the 2H/diamond line. With the enthalpy value of 3R-graphite and a somewhat higher value of entropy a lower equilibrium pressure for the 3R/diamond line of about 3 Kbar can be estimated. Because 3R-graphite will not be stable thermodynamically at HP/HT-conditions, the 3R/diamond line should be established below the 2H-line /Fig. 1/.

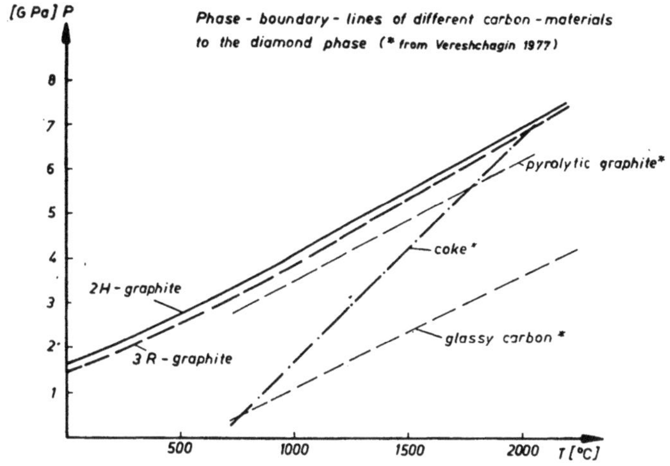

This means that the phase transition 3R-graphite into 3C-diamond is thermodynamically prefered to 2H→3C.
For the boron nitride system an analogeous behaviour can be expected.

This thermodynamical situation causes us to suggest an analogeous way of nucleation in both systems:

1. the formation of 3R polytype and
2. a martensitic solid state phase transition of the 3R-structure into the 3C-structure at HP/HT-conditions.

This way is in accordance with the OSTWALD's step rule and 3R-graphite is also detected in turning back transformation at HP/HT-graphitization of diamond by TITOVA and FUTERGENDLER /6/.

Now let us consider the different ways of 3R-formation for carbon and BN.
3R-carbon can be formed by mechanical treatment of the highly crystalline graphitic structure, for example by pressure generation and increasing in the synthesis chamber because this treatment is not fully hydrostatically. The first detection was on direct conversion by NAKA and coworkers /7/, and we found this phase at usually used conditions of about 6 GPa and 1400 grdC, with a considerable deformation of the source. 3R-graphite is partially formed in the used natural graphite besides

structural defects and stacking faults and is kinetically stabilized also at higher temperatures.
3R-formation is also be possible at graphitization processes in lower crystalline carbon (amorphous or from chemical reactions) by thermal or catalytic graphitization in the solid phase or by volume changing in the graphitic matrix during diamond formation, but not in a liquid solution.

The metal catalyst diffuses between carbon layers and causes an electronic interaction.
- catalytic metal atoms act as electron acceptor and carbon as donor, which causes a shift of the planar sp^2-configuration in the direction of the tetragonal sp^3-configuration with a corrugation of the carbon layers.

This arrangement is considered as a 'preformed nucleus' of martensitic transformation.

The growth takes place above all by diffusion of atomic, positively charged carbon through the liquid phase to the surface of crystallities.

3R-hBN can't be formed directly from 2H-hBN, but by other ways, which are able to create polytypes as CVD-processes from a gas phase, crystallisation from the amorphous state or from a liquid phase. Turbostratic or amorphous BN may form 3R - hBN partially by the graphitization process at HP/HT and this 3R-hBN can be transformed to cubic BN at available low conditions. Crystalline hBN can't change directly and without a catalyst it forms wurtzitic BN. In the presence of a catalyst as alkali or alkaline earth nitrides it is necessary to increase the temperature up to the beginning of eutectic melting and origin of a liquid phase.

This liquid phase dissolves 2H-hBN and can deposit 3R-hBN in accordance with OSTWALD's step rule in a deposition process with homogenous or heterogeneous nucleation. This 3R-hBN afterwards can be martensitically transformed into the dense structure and form a nucleus of cBN. The catalyst acts abouve all as a solvent for nucleation and growth.

Arguments for this model of nucleation can be :
- formation of diamond in solid carbon containing Co-alloys by HOSOMI/8/,
- solid state transformation of GIC /9/, with Fe, Ni or Co
- solid state transformation of ferrocene to diamond /10/,
- mechanism of diamond synthesis from carbon-containing compounds or amorphous carbon via well crystalline graphite,
- 3R-graphite detected in technical graphites , in synthesis process and at HP/HT-graphitization of diamond,

- transformation of plate-like graphite which is shown by TEM and REM-investigations,
- unambiguous shock wave transformations from 2H-hBN to wBN and 3R-hBN to cBN /11/,
- nucleation process takes place at the beginning of synthesis,
- structure dependence of the used carbon or BN-source.

This model of the nucleation process on the uniform way along 2H→3R→3C with a martensitic solid-solid-trasition into the dense phase allows the explanation of experimental results in the processes of:
- direct conversion (graphite and BN)
- catalytic synthesis and mode of different catalytic operations
- formation from chemical reactions along crystalline (C) or amorphous (BN) stage
- shock wave synthesis and multishock treatment of BN-phases.

R E F E R E N C E S

/ 1/ Shul'zhenko, A.A., 5th International Symposium "High pur'ty materials in science and technology", Dresden, GDR, 5.-9. May 1980, part 1, 122

/ 2/ Jayaraman, A., Maines Sr., R.G. and Chattopadhyay, T., Pramana - J. Phys. 27, (1986) 449

/ 3/ Hayashi, K. and Kawamura, A., Mat. Res. Bull. 22, (1987) 11

/ 4/ Silverman, M.S., Inorg. Chem. 6, (1967) 1062

/ 5/ Boehm, H.P. and Coughlin, R.W., Carbon 2, (1964) 1

/ 6/ Titova, V.M. and Futergendler, S.I., Kristallografija 7, (1962) 926

/ 7/ Naka,S., Forii,K., Takeda,Y. and Hanawa,T., Nature 259, (1976) 38

/ 8/ Hosomi, S., Yogyo - Kyokai - Shi 92, (1984) 146 and 93, (1985) 96

/ 9/ Kalashnikov,J.A., Timofeyevna,N.V., Farafontov,V.I.,Goldanski,V.I., Stukan,R.A., Nefedyev,V.A., Volpin,M.E., Novikov,Ju.N. a. Lapkina,N. Synthetic Metals 4, (1982) 363

/10/ Soloshenko, V.L. et al., Procc. Symp. Appl. High Pressure, 25.-27.07.1986, Moscow 1986

/11/ Sato,T., Ishii,T. and Setaka,N., J. Am. Ceram. Soc. 65, (1982) 162

ON THE METHODS OF CALCULATING THE EFFECTIVE TRANSPORT COEFFICIENTS IN COMPOSITE MATERIALS - AN APPLICATION TO PLANETOLOGY

Janusz Eluszkiewicz [1/]

ABSTRACT: Based upon a comparison with the results of Monte Carlo calculations it is argued that the self-consistent approach provides a good approximation to the effective thermal and/or electrical conductivity of heteregeneous materials. Examples are given where the usage of this approximation /instead of the volumetric average/ should lead to an improvement in the existing models of planetary bodies: medium-sized satellites of the outer planets /for which porosity might be an important structural factor/, asteroids composed of phases with considerably different electrical conductivities, and the binary mixture metallic hydrogen + helium. The self-consistent formula might also be of some help in a quantitative treatment of the results obtained in high-pressure experiments /e.g. in rock physics/.

From the molecular standpoint all planetary bodies are multicomponent systems. However, some of them are multicomponent at the macroscopic level, i.e. are composed of a few different phases. For example, the "icy" saturnian satellites are usually modelled as ice-rock mixtures /Ellsworth and Schubert 1983, Federico and Lanciano 1983/ - in this case ice and rock constitute two phases which might, in principle at least, be distinguished macroscopically /they are thought to form grains/. This work is devoted to those planetary materials which are multicomponent in the macroscopical sense. /such materials are also called heteregeneous or composite/. Their molecular structure will be neglected, i.e. they will be treated as continuous media. As a consequence, the relevant physical processes will be described by macroscopical equations with some effective coefficients which depend on the properties and geometry of mixed phases. For instance, in the case of heat flow one has

$$\vec{q} = - k^* \operatorname{grad} T , \qquad /1/$$

where $\vec{q}$ is the heat flux, T denotes the temperature, and k^* is the effective thermal conductivity. In general k^* depends not only on the conductivities and the volume fractions of the constituent phases but also on the details of their geometry /see e.g. Beran 1967/. Other

[1/] University of Warsaw
Institute of Geophysics
ul. Pasteura 7, 02-093 Warszawa
Poland

transport processes are described by equations similar to Eq./1/, e.g.

$$\vec{j} = -\sigma^{*} \operatorname{grad} V \ , \qquad /2/$$

where $\vec{j}$ is the current density, V denotes the electrical potential, and σ^{*} is the effective electrical conductivity.

While for a truly random arrangement of phases it is not possible to determine effective constants in terms of volume fractions alone, it is common in material science to use approximate expressions of this kind in view of the fact that more detailed information on the phases' geometry /i.e. correlation functions of higher-order/ is difficult to obtain /Beran 1967/. In modelling planetary interiors the volumetric average of single-phase properties is usually taken /see e.g. the works on saturnian satellites cited above/. However, while such an approximation may be accurate for the effective specific heat C_p^{*} and the effective thermal expansivity α^{*}, it always overestimates the effective transport coefficients k^{*} and σ^{*}, corresponding to the artificial geometry of conductors in parallel. In this paper I will argue that a better approximation to the effective conductivities is provided by the self-consistent approach /Landauer 1952/ and will describe some planetary situations in which this improvement might be important.

The most direct way of determining the effective conductivities of a heteregeneous material would consist in performing the relevant measurement in the laboratory. However, in application to planetology this method is often impractical due to the necessity of working at extreme p-T conditions. Where direct measurement is not possible recourse to theory and/or numerical experiments must be had. One can, for example, solve the relevant transfer equation for a random arrangement of components. In practice this must be done numerically and to this end one has to discretize the continuous problem - the random continuous medium is replaced by a network of randomly distributed conductivities /Kirkpatrick 1973/. For a mixture of N phases the distribution of conductivities in the network has the form

$$P/k/ = \sum_{i=1}^{N} v_i \delta/k-k_i/ \ , \quad v_1 + v_2 + \ldots + v_N = 1 \ , \qquad /3/$$

where v_i and k_i denote the volume fraction and conductivity, respectively, of the i-th phase. I have employed this method in order to determine the effective thermal conductivity of the system ice-rock-pores. A porous ice-rock texture has been proposed for Mimas as a possible explanation for the low value of the moment of inertia of this satellite /Eluszkiewicz and Leliwa-Kopystyński 1987/. In this case N = 3 and $k_1/\mathrm{Wm^{-1}K^{-1}}/ = 488.12/T + 0.4685$, $k_2 = 4.2\ \mathrm{Wm^{-1}K^{-1}}$, $k_3 = 0$ are the conductivities of ice, rock, and pores, respectively. In the calculations the volume fraction of pores $v_3 \equiv \Phi$ /porosity/ was treated as the free variable and the volume fractions of ice v_1, and rock v_2 were then determined so as to give the assumed silicate mass fraction C

$$\frac{v_2 \varrho_2}{v_2 \varrho_2 + v_1 \varrho_1} = C . \qquad /4/$$

In the above equation ϱ_1 and ϱ_2 are the densities of ice I and rock respectively. The values ϱ_1 = 940 kg m^{-3} and ϱ_2 = 3000 kg m^{-3} were used. Since the value of C for icy satellites is poorly constrained the computations were repeated for C = 0.3, 0.5, and 0.7. Two values of the overall temperature were used /corresponding to two different values of the thermal conductivity of ice/: T = 60 and 120 K /this is roughly the temperature range expected in the interior of Mimas/. Plots k^* vs. Φ for C = 0.5 at the two temperatures are shown as dashed lines in Fig.1a. For $\Phi \gtrsim 75\%$ conductivity goes abruptly to zero - the percolation threshold is observed. Also plotted in this figure are results obtained by taking the volumetric average and from the self-consistent approach.

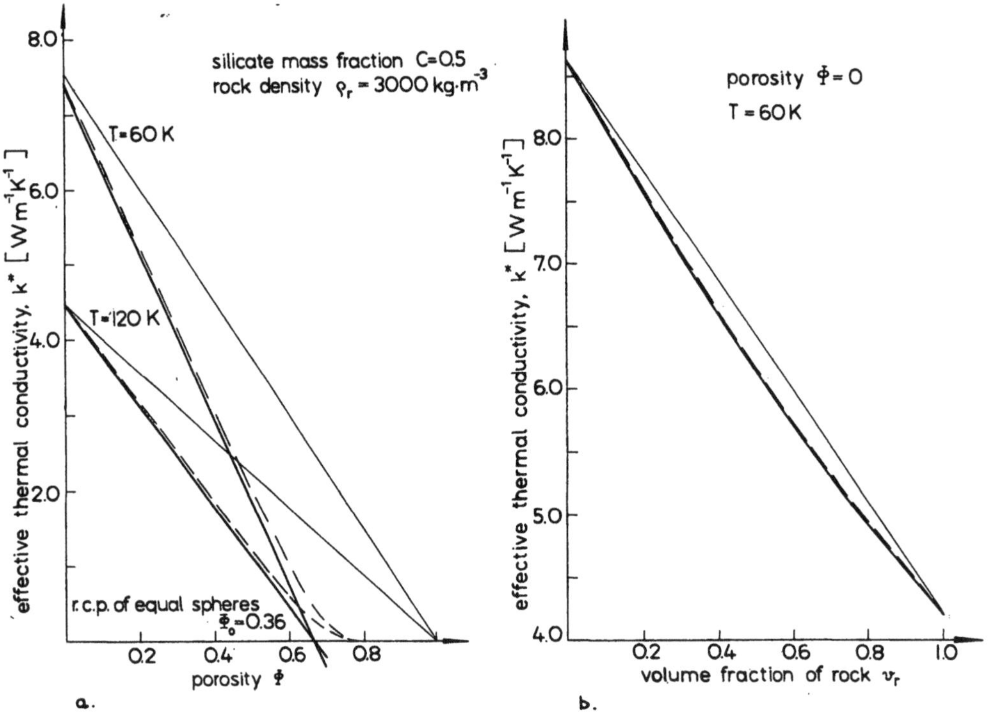

Fig. 1

a. The effective thermal conductivity k^* vs. porosity Φ in the mixture ice-rock-pores: volumetric average /thin solid line/, mean-field theory /bold solid lines/, Monte Carlo calculations /dashed lines/. The value Φ = 0.36 is the porosity of random close packing of equal spheres and also corresponds roughly to the initial porosity in sedimentary rocks on the Earth[6].

b. The effective thermal conductivity k^* vs. silicate volume fraction v_2 in the mixture ice-rock.

The determination of the effective conductivity by the above method is a time-consuming task. For each run k^* vs. Φ /twenty values of were used/ averaged over ten samples at fixed values of C and T about 25 minutes CPU at the Soviet R-60 computer were required. In the models of planetary bodies' interiors the variables upon which k^* and/or σ^* depend /such as Φ, C, T/ might take on many values or even change continuously and it is therefore desirable to have a simple approximate formula which could be used instead of the Monte Carlo technique. Such a formula is provided by the self-consictent approach /mean-field theory/ /Landauer 1952, Beran 1967/. In this method one calculates the effective constant /such as k^*/ for a material composed of N phases by considering a spherical inclusion of one phase in a continuous matrix of the other N-1 phases; the matrix having a constant effective constant k^*. One then arrives at the following equation

$$k^* = k_N + \sum_{i=1}^{N-1} v_i/1 - \frac{k_N}{k_i}/ \frac{3k_i k^*}{k_i + 2k^*} \quad , \qquad /5/$$

from which k^* may be found. Plots k^* vs. Φ obtained by means of Eq./5/ for the ice-rock-pores mixture are presented as bold solid lines in Fig.1a. Also shown in this figure /as thin solid lines/ are results obtained by taking the volumetric average

$$k^* = \sum_{i=1}^{N} v_i k_i \quad . \qquad /6/$$

It is seen from Fig.1a that the mean-field approximation agrees well with the Monte Carlo results except in the vicinity of the percolation threshold /for binary mixtures of conductors and insulators this fact is known ever since Kirkpatrick's work/. At the same time the volumetric average always overestimates Monte Carlo results, especially for large porosities. I have verified that these conclusions are true for all values of C and T studied. It should be noted, however, that the agreement between the volumetric average and Monte Carlo method becomes better when the conductivities of constituent phases do not differ markedly. This may be appreciated from Fig.1b where results of the three methods for the mixture ice-rock at T = 60 K are shown. This fact justifies the use of Eq./6/ in those models of icy satellites in which the presence of pores is not taken into account.

It might seem that the slight improvement over the volume-weighted average provided by the mean-field approximation should play a minor role at the present stage of modelling of the planetary interiors because the conductivities of individual phases are usually poorly known /in the models of icy satellites referred to above it is assumed that k/rock/ = 4.2 $Wm^{-1}K^{-1}$ and the dependence of k on temperature and composition of "rock" is ignored/. However, there are at least two reasons why the self-consistent formula should be preferred. The first is a methodological one. Why should one not use Eq./5/ instead of Eq./6/ when both relations are based upon the same amount of information about single phases /k_i, v_i/, both are computationally simple, and at the same time the former is certainly more accurate than the latter? The second reason is that there are planetary situations where a more detailed knowledge of the effective thermal /or electrical/ conductivity might be important. Two such situations are described below.

A. The knowledge of a more accurate value of the thermal conductivity might be crucial in thermal history models of the icy outer-planets satellites /Hubbard 1983/. This may be illustrated by the following example. According to the calculations by Ellsworth and Schubert /1983/ the internal temperature in Mimas never exceeded 115 K if the satellite accreted cold. This and other results obtained in their model would probably not be changed by the use of the self-consistent formula instead of the volumetric average for the effective thermal conductivity because the two approaches give very close values for k^* if the presence of pores is ignored. However, at such low temperatures /and the pressures $< 10^2$ bar expected in Mimas' interior/ a porous structure should be expected /Smoluchowski and William 1983/. A significant porosity would in turn lead to a decrease in k and therefore to greater internal temperatures - eventually it could have other physical consequences such as shrinking and/or migration of pores /Smoluchowski and William 1983/, onset of convection or even melting /in the latter case the porous picture would break down!/. The issue of porosity in satellites' interiors can only be consistently addressed to by performing thermal history calculations in which the presence of porosity is assumed as an initial condition. Clearly, for such calculations Eq./5/ is more suitable than Eq./6/ /for $\Phi \lesssim 70\%$/.

B. The effective electrical conductivity of some asteroids can probably be determined in a spacecraft mission /Ip and Herbert 1983/. If an asteroid is an undifferentiated body composed of a few phases whose electrical conductivities differ significantly, e.g. silicates + chondritic material + metallic iron, and the value of σ^* for it were known, then Eq./5/ could be used to put a constraint on possible interior models.

Besides the above examples there are other planetary materials which could probably be described as composite media. For example, at the conditions expected in the interiors of Jupiter and Saturn helium is expected to be immiscible with hydrogen. Hydrogen is predicted to undergo a metallization transition at pressure ~ 3 Mbar whereas helium remains an insulator for pressures < 100 Mbar /Hubbard 1983/. The effective electrical conductivity of the binary mixture /metallic H + He is certainly more precisely calculated by means of Eq./5/ than by Eq./6/. As the differentiation of hydrogen and helium goes on even the use of percolation-theory results might become necessary.

The self-consistent approach can also be taken in order to determine effective elastic constants of a composite medium /Beran 1967/. In this respect the self-consistent formula might be of some value in a quantitative treatment of the results obtained in high-pressure experiments on rocks.

REFERENCES

Beran, M.: Statistical Continuum Theories. Interscience, N. Y. 1967.

Ellsworth, K., and G. Schubert: Saturn's icy satellites: thermal and structural models. Icarus 54 /1983/ 490-510.

Eluszkiewicz, J., and J. Leliwa-Kopystyński: A model of the porous structure of icy satellites. Lunar and Planetary Science XVIII /1987/ 260-261 /abstract/

Federico, C., and P. Lanciano: Thermal and structural evolution of four satellites of Saturn. Ann. Geophys. 1 /1983/ 469-476.

Hubbard, W.: Planetary Interiors. Van Nostrand, N. Y. 1984.

Ip, W.-H., and F. Herbert: On the asteroidal conductivities as inferred from meteorities. The Moon and the Planets 28 /1983/ 43-47

Kirkpatrick, S.: Percolation and conduction. Rev. Mod. Phys. 54 /1973/ 574-588.

Landauer, R.: The electrical resistance of binary metallic mixtures. J. Appl. Phys. 23 /1952/ 779-784.

Smoluchowski, R., and A. McWilliam: Structure of ices on satellites. Icarus, 58 /1984/ 282-287.

HIGH PRESSURE ORGANIC SYNTHESIS. POLYCYCLIC ADDUCTS VIA [2 + 2 + 2] CYCLOADDITION REACTIONS

G. Jenner[1] and M. Papadopoulos[2].

Introduction

The success of high pressure organic synthesis is intimately dependent on the value of the volume of activation $\Delta V^{\neq}$. The later or more "product-like" the transition state, the more pressure-accelerated the reaction rate. Number of reactions were investigated in this way, especially cycloadditions [1] which are usually characterized by $\Delta V^{\neq}$ values of at least -30 $cm^3.mol^{-1}$. Less studied reactions include [2+2+2] cycloadditions involving free independent reacting systems. Such additions cannot experience concertedness, unless at least two reacting double bonds are present in one reactant and thus the cycloadditions are bimolecular.

Two kinds of [2+2+2] cycloadditions are considered in the present paper : the $[\pi^2+\pi^2+\pi^2]$ cycloadditions termed as the homo-Diels-Alder reaction and the $[\pi^2+\sigma^2+\sigma^2]$ cycloaddition termed as the bis-homodiene reaction.

1. [2+2+2] Monoaddition reactions.

a) The homo-Diels-Alder $[\pi^2+\pi^2+\pi^2]$ cycloaddition of a 1,4-diene with a suitable dienophile is of important synthetic value since it produces two new bonds and a cyclopropane ring in one step. However, the feasibility of the reaction is rather poor at ambient pressure and limited practically to bicyclo [2.2.1] heptadienes [2]. As previously demonstrated, the reactivity can be induced or enhanced by operating under higher pressures according to the general equation (E = CO_2CH_3, E' = H or CO_2CH_3) [3] :

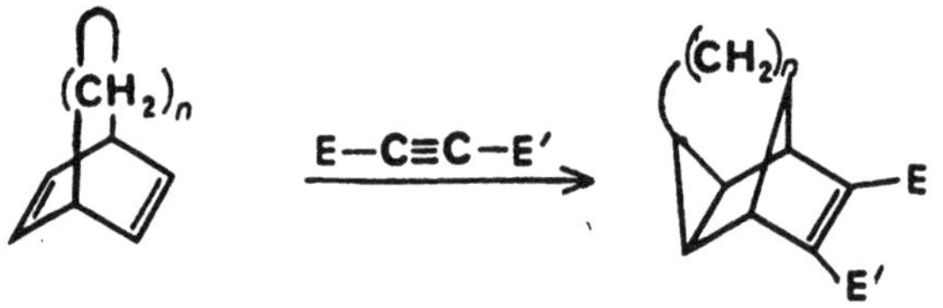

1) Laboratoire de Piézochimie Organique, Chimie Organique Appliquée, EHICS, 1, rue Blaise Pascal, 67008 Strasbourg.

2) University of Crete, Department of Chemistry, P.O. Box 1470, 711 10 Iraklion (Greece)

Table 1 portrays some representative reactions performed at 100°C under 10 kbar (reaction time : 24 hours). No reaction occured at atmospheric pressure.

Table 1
High pressure $[\pi^2+\pi^2+\pi^2]$ reactions

Diene	E'	Yield (%)
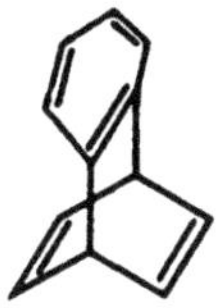	CO_2CH_3	96
	H	51
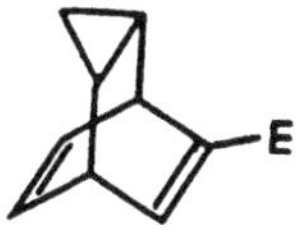	CO_2CH_3	52
	H	28 (t : 65 h)

As evidenced in Table 1, the use of high pressure extends the scope of the homo-Diels-Alder reaction. The effect of pressure is essentially to accelerate reaction rates (kinetic effect) rather than to modify the geometric arrangement of the molecules.

b) The $[\pi^2+\pi^2+\pi^2]$ cycloaddition requires the polycyclic hydrocarbon, quadricyclane, which adds to suitable dienophiles according to the reaction [4] :

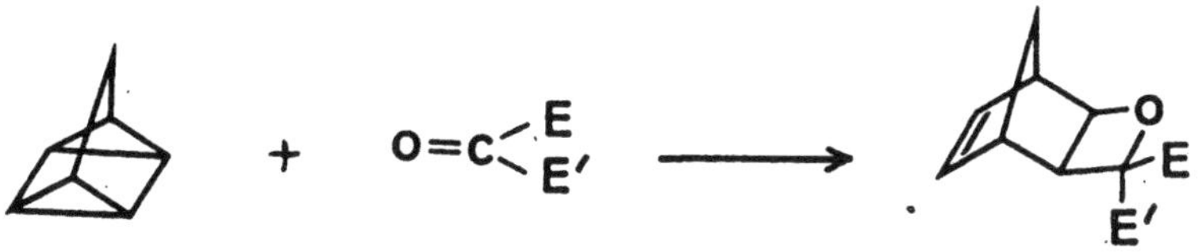

Quadricyclane is extremely reactive, however, only with strongly activated unsaturated compound. With activated carbonyl bonds, only mesoxalates ($E=E'=CO_2R$) cycloadd to quadricyclane. The use of high pressures allows the extension to other carbonyl bonds with variable yields (Table 2).

Table 2
High pressure $[\pi^2+\sigma^2+\sigma^2]$ reactions (8 kb, 100°C)

E	E'	Yield (%)
CO_2Me	Ph	60
CO_2Et	Me	90
COPh	Ph	60
$CHCl_2$	Cl	20

2. Double [2+2+2) reactions.

The considerable kinetic effect of pressure on [2+2+2] cycloadditions prompted to consider possible double [2+2+2] processes. This was attempted in the reaction of either quadricyclane 2 or norbornadiene 1 with methyl propynoate 3.

a) Double $[\pi^2+\sigma^2+\sigma^2]$ process

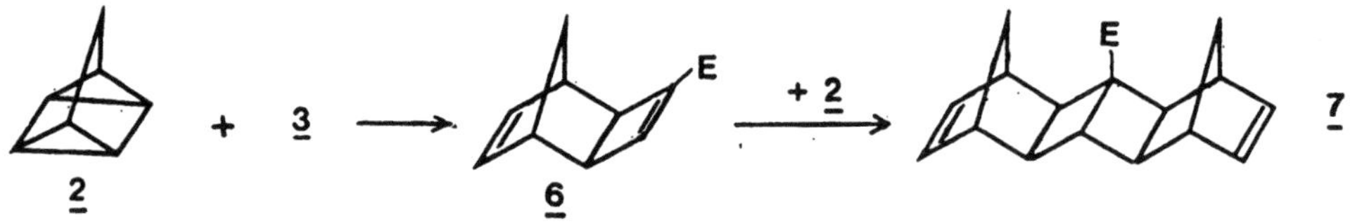

Table 3
Reaction of quadricyclane 2 with methyl propynoate
(initial composition 2 : 3 = 33 : 67)

Conditions	Yield	Monoadduct 6	Bisadduct 7
1bar, 100°C, 24 h	80%	100%	0%
10 kbar, 100°C, 24 h	100%	23%	77%
10 kbar, 100°C, 65 h	100%	4%	96%

The high pressure reaction yields the bisadduct in nearly quantitative yield.

b) Bisadducts via $[\pi^2+\pi^2+\pi^2]$ and sequential $[\pi^2+\pi^2+\pi^2]$ and $[\pi^2+\sigma^2\sigma+^2]$ processes.

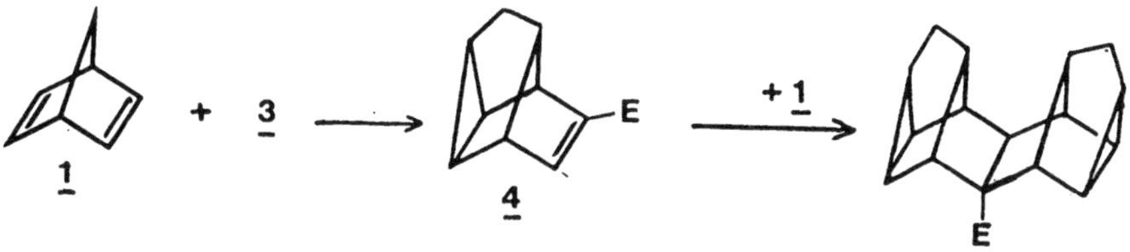

5

The monoadduct 4 is produced in 100% yield even at 1 bar, whereas the bisadduct 5 is formed only at 10 kbar (T : 110°C) in 68% yield after 48 h.

Monoadducts 4 and 6 can be subjected to cross addition with quadricyclane and norbornadiene respectively yielding the same bisadduct 8, however only under high pressure (yield 80%).

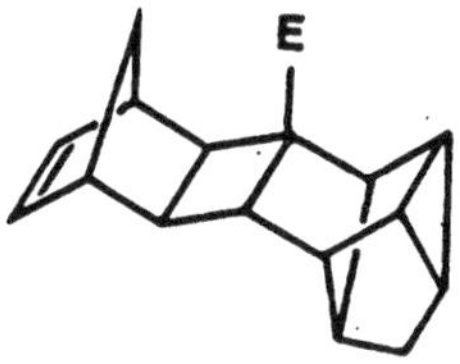

8

Conclusion

The present results demonstrate the usefulness of high pressures in accelerating [2+2+2] cycloaddition reaction and producing new polycyclic compounds. They represent an extension of the considerable synthetic possibilities of the pressure parameter. Reactions with transition states inaccessible for steric or electronic reasons under normal conditions can occur at high pressures.

Literature.

[1] T. Asano and W.J. le Noble, Chem.Rev. 78 (1978)4, 407-489.
[2] G. Jenner and M. Papadopoulos, Tetrahedron Lett. 23 (1982)42, 4333-4336.
[3] G.N. Fickes and T.E. Metz, J. Org. Chem., 43 (1978)21, 4057-4061.
[4] M. Papadopoulos and G. Jenner, Nouv. J. Chim., 7 (1983)8, 463-464.

THE INFLUENCE OF PRESSURE ON THE PROCESS OF FORMATION OF STRUCTURE IN THE PROCESS OF SINTERING OF NITRIDE CERAMICS FROM AlN

P.S.Kisly, M.A.Kuzenkowa and I.P.Fesenko 1)
H.Recht, H.Vollstädt 2)
G.S.Olejnik 3)

In the present work were examinated the influence of the pressure of Nitrogen and high pressure on the kinetics of densification and the formation of structure of ceramics from ultradisperse AlN powders (UD)and relatively coarse technical AlN powders (T). The chemical composition and specific surface of these powders are given in the table.

	chemical		composition, mass%			specific surface
AlN	$Al_{tot.}$	N	Al_{free}	O	Fe	m /g
UD	65,4	33.7	0,16	1,5	not est.	19,5
T	65,0	33,2	0,80	0,6	0,6	1,8

The microstructure of UD-AlN is shown in Fig.1.
Sintering was carried out in the laboratory gasostat under Nitrogen pressure of 0,1 MPa; 10 MPa and 100 MPa and in the high pressure apparatus of the "Toroid"-type at pressures of 4 to 9 GPa in a temperature of 1300 °C to 2000 °C. The shrinking of the sintered samples is shown in Fig. 2, their density in Fig. 3 and their mass lost in Fig.4.
As the diagrams show, the samples from coarse powder sinter badly and the increase of the Nitrogen pressure up to 100 MPa shows practically no influence on the velocity of the densification of the sintered material.
The sintering of samples from UD-powder under Nitrogen pressure of 0,1 MPa occures actively in the temperature range from 1300°C to 1850 °C, where shrinking reaches 28 %, and after sintering the density approaches the theoretical density.
With increasing Nitrogen pressure up to 100 MPa the shrinking is somewhat smaller (24-26 %), the density also does not approach the theoretical value but reaches 2,95 g/cm^3. The mass lost of the samples is the smaller the higher the Nitrogen pressure is.

1) Institute for Superhard Materials, Kijev, USSR
2) Institute for Physics of the Earth, Potsdam, GDR
3) Institute for Problems of Materials Science, Kijev, USSR

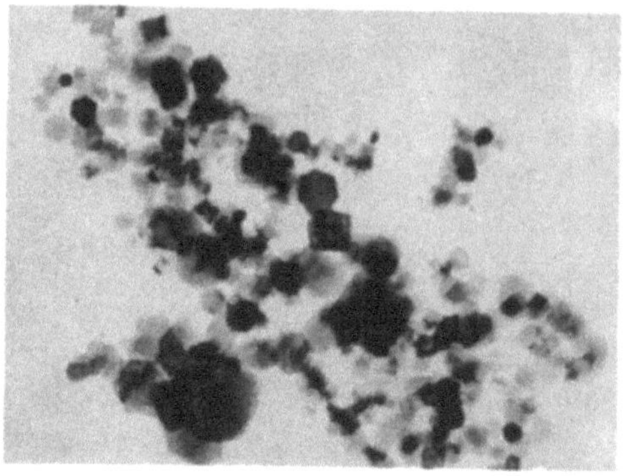

Fig. 1: Morphology of the particles of UD-powder of AlN x 60000

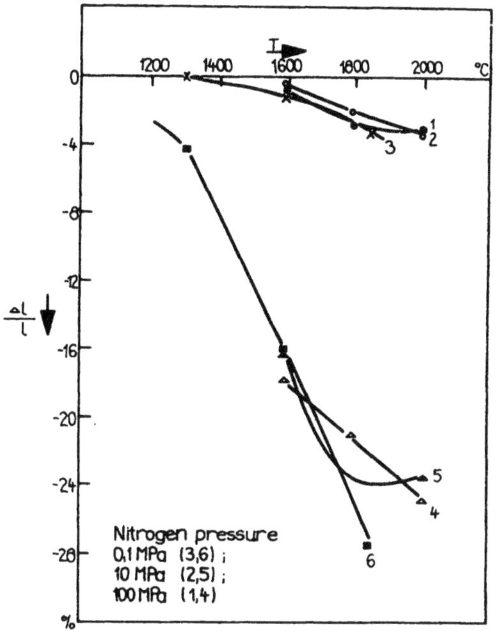

Fig. 2: Temperature dependence of the shinkage(Δl/l) at the sintering of AlN-T(1-3)and AlN-UD(4-6) under Nitrogen pressure time of reaction- 60 min

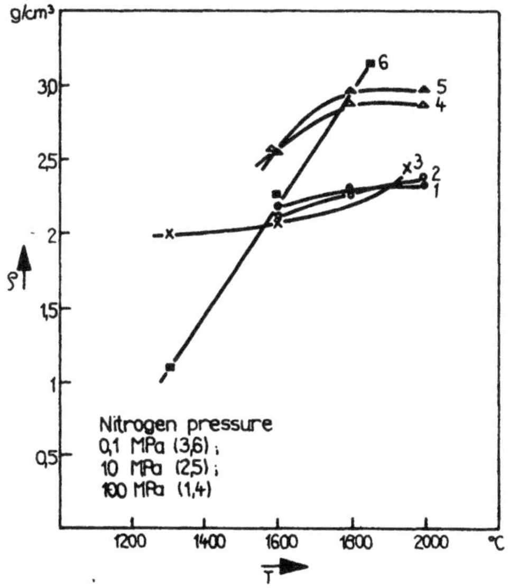

Fig. 3: Dependence of density (ρ) of the samples on the sintering temperature of AlN-T (1-3) and AlN-UD(4-6)

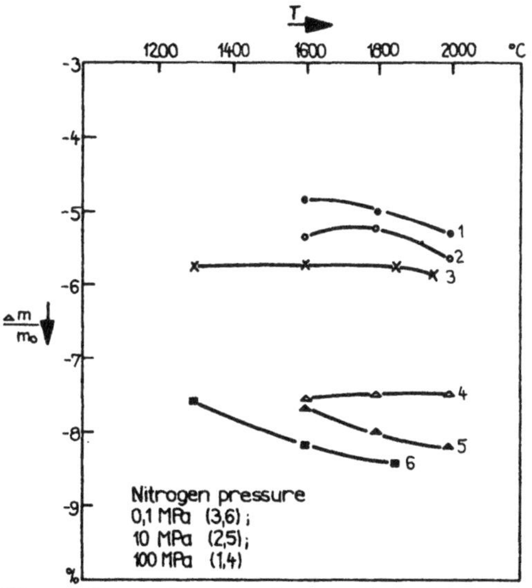

Fig. 4: Dependence of mass lost(Δm/m) of the samples on the sintering temperature of AlN-T(1-3) and AlN-UD(4-6)

Fig. 5: Formation of structure of ceramics from AlN-UD-powders in different stages of sintering

5a) 0,12MPa N_2
1600°C, 90 min.

1μm

5b) 100 MPa N_2
1600°C, 60 min.

1 μm

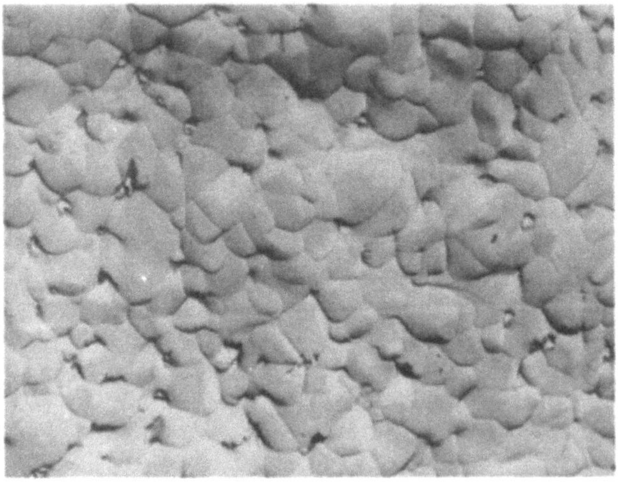

5c) 9GPa
1750°C, 90 sec

1 μm

Electron microscopic investigations of the microstructure permit to explain the following morphological and structural transformations. In the samples which were pressed from UD-powder, the Oxygen adsorbed on the surface of the particles formes an Oxynitride phase at temperatures above 1200°C, which is localized at the triangular boundaries of the particles. [1] At temperatures above 1600° C a quick densification of the starting particles takes place forming aggregates, in which monocrystalline particles originate afterwards. The Oxynitride layer dissolves in the particles of the main phase by forming a solid solution of Oxygen in Aluminium Nitride. At a temperature above 1700°C there takes place a growth of the particles, while packing defects in the (0001)-plane appear and manylayering polytypes arise at 1800°C to 1900°C from which particles of the plate type (Fig. 5a) are formed. The increase of the Nitrogen pressure causes the retardation of the processes described above and the slowing down of the recristallization. At the particles boundaries there appear pores between the grains, which are probably filled with gas and are preserved up to the highest temperatures (Fig. 5b). The recrystallization growth of the particles retards even more in those samples, which were sintered in the in the high pressure chamber (Fig. 5c). There are no polymorph transformations in the structure, the particles of an octahedral from grow larger in their dimension.

Conclusions:

The application of high pressure permit the regulation of structural changes and recrystallization processes, which take place in the process of sintering ultradispers Aluminium Nitride powders and thus affect the physical and mechanical qualities of Nitride ceramics.

Reference

[1] P.S.Kisly; M.A.Kuzenkova; G.S.Olejnik; A.N.Piljankevich: Recrystallising sintering of high dispersion Aluminium Nitride. International Pulvermet. Tagung, DDR, Dresden, 1985, pp. 51-57

ON THE DYNAMICAL SCALING BEHAVIOUR IN A ONE-DIMENSIONAL MODEL FOR NUCLEATION AND GROWTH

B. Lorenz 1)

1. INTRODUCTION

The theory of first order phase transitions has been a subject of increasing attention /1,2/ because of its significance for various phenomena, e.g. primary recrystallization, polymorphic transitions in solids or industrial crystallization. The late stage of the transition characterized by the process of domain growth and coarsening /3,4/ has attracted special theoretical interest due to the validity of dynamical scaling laws relating the time and space dependences of the scattering structure function (see /5/ and references therein).
In the early stage of the transition and for thermodynamic conditions not too far from the equilibrium line the stable phase evolutes from the metastable initial phase by the nucleation and growth process /6 to 8/. For the case of constant nucleation rate (homogeneous nucleation) it has been shown very recently that the structure functions of the one- and two-dimensional nucleation and growth models obey the scaling relations /9/. However, the validity of the scaling law for arbitrary nucleation kinetics is not clear.
Besides the special case of homogeneous nucleation systems with preferred nucleation sites due to impurities or defects are frequently considered /7,10 to 12/. The kinetics is specified as follows. At time t=0 all nuclei are created randomly with density n_0 in the initially homogeneous state. No further nucleation takes place at t>0. The nuclei grow with the constant velocity v until two neighbouring nuclei impinge to each other. As a result of the nucleation and growth process there arises a distinct grain pattern which changes with time. The time-dependent size distributions of grains and clusters of impinged grains are characteristic quantities of the nucleation kinetics. In the one-dimensional case the results are obtained analytically by Schulze /12/. The validity of scaling laws in that model is discussed in the following.

---

1) Central Institute for Physics of the Earth,
Academy of Sciences of the GDR, Potsdam, 1561

2. TEST OF SCALING HYPOTHESIS

The normalized spatial correlation function G(x,t) which is related to the structure function by a Fourier transformation is defined as /9/

$$G(x,t) = 1/L \int (x'-x)\ W(x',t)\ dx' \tag{1}$$

W(x',t) denotes the size-distribution function of the substructure contributing to a coherent scattering and $L=\int x'\ W(x',t)\ dx'$ is the mean length in the substructure. In /9/ it is assumed that the grains are not correlated to each other. Then W(x',t) is given by the grain-size distribution $W_g(x',t)$. If the grains belonging to one and the same cluster are strongly correlated W(x',t) is substituted by the cluster-size distribution $W_c(x',t)$. Both cases are considered in the following.

CASE A: Uncorrelated grains

The grain-size distribution in terms of the dimensionless time and space variables $\tau = 2n_0vt$ and $x = n_0r$ is given by /12/

$$W_g(x,\tau) = \begin{cases} 4\ x\ \exp(-2x) & , \quad 0 \leq x \leq \tau/2 \\ [\ \delta(x-\tau) + 4(1-x+\tau)]\ \exp(-2x) & , \quad \tau/2 < x \leq \tau \\ 0 & , \quad x > \tau \end{cases} \tag{2}$$

$\delta(x)$ denotes the Dirac δ-function.

The grain autocorrelation function (1) is obtained by integration with the mean grain size $L = 1 - \exp(-\tau)$ as

$$G_g(x,\tau) = \begin{cases} [(1+x)\ \exp(-2x) - \exp(-\tau)]\ /\ [1-\exp(-\tau)]\ , & 0 \leq x \leq \tau/2 \\ (\tau - x)\ \exp(-2x)\ /\ [1 - \exp(-\tau)]\ , & \tau/2 < x \leq \tau \\ 0\ . & x > \tau\ . \end{cases} \tag{3}$$

The short-time behaviour ($\tau \ll 1$) of (3) is given by the linear dependence

$$G_g(x,\tau) = 1 - x/R(\tau) \quad , \quad 0 \leq x \leq \tau \tag{4}$$

with $R(\tau) = \tau$ ($\tau \ll 1$). Fig. 1 shows the fit of (3) to the dependence (4). Obviously, for different times there is a distinct spread in the curves and the scaling behaviour detected for homogeneous nucleation is not observed in the present case.

CASE B: Correlated grains

The normalized cluster-size distribution according to /12/ is obtained as

$$W_c(x,\tau) = \exp(-\tau) \begin{cases} \delta(x-\tau) & , \quad x \leq \tau \\ 1 & , \quad \tau < x \leq 2\tau \\ g(x,\tau,k) & , \quad k\tau < x \leq (k+1)\tau , \quad k \geq 2 \end{cases} \tag{5}$$

with $$g(x,\tau,k) = 1 + \sum_{j=1}^{k-1} (-1)^j \exp(-j\tau)\{ \frac{1}{j!} [x-(j+1)\tau]^j + \frac{1}{(j-1)!} [x-(j+1)\tau]^{(j-1)} \}$$

The correlation function (1) is obtained from (5) by integration and the results for $G_c(x,\tau)$ are evaluated numerically.
The mean cluster length $R_0(\tau) = \exp(\tau) - 1$ diverges for $\tau \gg 1$. This implies that the appropriate scaling length is of the order of $R_0(\tau)$ at large τ. Expanding $G_c(x,\tau)$ for $\tau \gg 1$ we get $G_c(x,\tau) = \exp\{-x/R_0(\tau)\}$ which indicates the scaling form of $G_c(x,\tau)$ to be

$$G_c(x,\tau) = A \exp\{-x/R(\tau)\} \tag{6}$$

The scaling length $R(\tau)$ is determined by fitting $G_c(x,\tau)$ to the exponential dependence (6). The results are shown in Fig. 2 for various times τ. The scaling behaviour is evident for $\tau > 1.5$ in accord with the general scaling arguments /5/. The characteristic length scale $R(\tau)$ is always less than $R_0(\tau)$ and approaches $R_0(\tau)$ for $\tau \to \infty$.

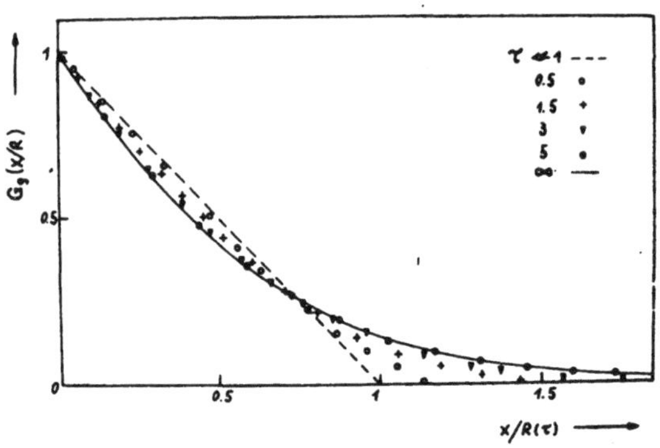

Fig. 1
Grain autocorrelation function in dependence on the scaled length $x/R(\tau)$ for various times.

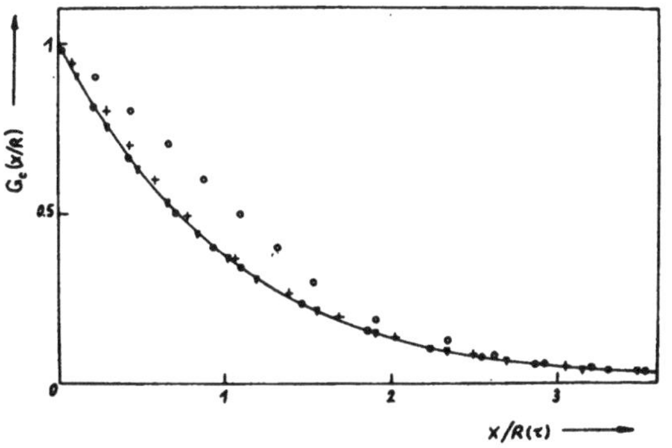

Fig. 2
Scaling behaviour of the cluster autocorrelation function. The symbols denote the same times as in Fig. 1. The solid line indicates the scaling function for $\tau \rightarrow \infty$

3. DISCUSSION

The dynamical scaling theory for phase transitions and phase separation /5/ predicts the scaling laws to be valid in the late stage of domain growth and coarsening. For the process of homogeneous nucleation and growth the grain autocorrelation function (1) indicates the scaling hypothesis to be valid in a large time interval /9/. This result, however, depends strongly on the nucleation kinetics and cannot be generalized to arbitrary nucleation and growth reactions. In the present case scaling is observed only in the long time behaviour of the cluster correlation function $G_c(x,\tau)$. The validity of dynamical scaling in dependence on the nucleation kinetics should be a subject of further investigations.

REFERENCES

/1/ Christian, J. W.: The Theory of Transformations in Metals and Alloys, Pergamon Press, Oxford, 1965.

/2/ Binder, K. and D. Stauffer: Adv. Phys. 25 (1976), 343.

/3/ Cahn, J. W.: Trans. metall. Soc. AIME 242 (1968), 166.

/4/ Lifshitz, I. M. and V. V. Slezov: J. Phys. Chem. Solids 19 (1961), 35.

/5/ Furukawa, H.: Adv. Phys. 34 (1985), 703.

/6/ Kolmogorov, A. N.: Bull. Acad. Sci. USSR, Phys. Ser. 3 (1938), 355.

/7/ Avrami, M.: J. Chem. Phys. 7 (1939), 1103; 8 (1940), 212; 9 (1941), 177.

/8/ Johnson, W. A. and R. F. Mehl: Trans. Am. Inst. Mining Metall. Engrs. 135 (1939), 416.

/9/ Axe, J. D. and Y. Yamada: Phys. Rev. B 34 (1986), 1599.

/10/ Meijering, J. L.: Philips Res. Rep. 8 (1953), 270.

/11/ Gilbert, E. N.: Ann. Math. Stat. 33 (1962), 958.

/12/ Schulze, G. E. W.: J. Cryst. Growth 62 (1983), 7.

CLUSTER-SIZE DISTRIBUTION IN NUCLEATION AND GROWTH PROCESSES

B. Lorenz and I. Orgzall 1)

1. INTRODUCTION

The theory of first order phase transitions has been a subject of increasing attention because of its significance for various phenomena, e.g. primary recrystallization, polymorphic transitions in solids or industrial crystallization. The kinetics of the transformation near equilibrium conditions is determined by the nucleation and growth process. After a short transient phase the stage of steady state nucleation characterizes the evolution of the stable phase and the KOLMOGOROV model [1] yields an adequate description.

According to the physical mechanisms of nucleation and growth the model parameters are the mean nucleation rate I(t) and the linear growth velocity v (for interface controlled growth the rate v is constant). If two growing nuclei impinge they continue to grow with the same rate except at their common interface. The nucleation rate I(t) usually depends on the time t according to the physical nucleation process. Two limiting cases are frequently considered:

(a) Site saturation case

In the presence of preferred nucleation sites due to impurities or grain boundaries all nuclei are formed in a short time intervall at the beginning of the transition. The nucleation rate is given by

$$I(t) = n_0 \delta(t); \tag{1}$$

n_0 is the density of nuclei and $\delta(t)$ is the DIRAC δ- function.

(b) Homogeneous nucleation

If nucleation takes place as a consequence of thermal fluctuations the rate is constant,

$$I(t) = I_0 = \text{const.} \tag{2}$$

The agglomeration of the growing grains results in the formation of larger clusters differing in shape and size. The time dependent size distribution of the clusters W(R,t) is a characteristic function of the

1) Central Institute for Physics of the Earth,
Academy of Sciences of the GDR, Potsdam, 1561

nucleation process as far as the net transformation rate x(t) does not exceed the percolation threshold $x_c \approx 0.3$. For $x > x_c$ the microstructure is determined by large polycrystalline clusters.
In the following the cluster-size distribution is calculated from an analytical theory and a numerical simulation procedure. The results are discussed with respect to the investigation of nucleation kinetics in crystallization phenomena.

2. THEORY OF CLUSTER-SIZE DISTRIBUTION

The theoretical investigation of the microstructure resulting in a nucleation and growth reaction has been performed extensively in the one-dimensional model [2,3,4]. In the three-dimensional case, however, there are no exact results for the grain- or cluster-size distributions. The average grain diameter and its variance have been shown to depend significantly on the dimension [5]. For lack of analytical results numerical simulation techniques are used to determine the microstructure of crystallization products [6,7].
An analytical theory for the cluster-size distribution function is developed as follows. The probability W (R,t) that a spherically growing nucleus with radius R does not impinge with any other nucleus at time t is given by

$$W_0(R,t) = \exp\{-4/3\,\pi \int_0^R [(R+r)^3 - (R-r)^3]\, n(r,t)\, dr - 4/3\,\pi \int_R^\infty (R+r)^3\, n(r,t)\, dr\}, \quad (3)$$

with $$n(r,t) = 1/v\ I(t-r/v)\ \Theta(t-r/v).$$

$\Theta(x)$ is the step function and I(t) has to be replaced by (1) or (2) for the nucleation kinetics (a) and (b), respectively. With (3) the probability distribution of two impinged nuclei $W_2(R,t)$ is calculated and reduced to the functional form (3) by an approximate renormalization (for details see [8]). Repeating this procedure a recursion relation for the probability distribution $W_n(R,t)$ is derived and the total cluster-size distribution W(R,t) is obtained as the sum of all W_n. The result is shown in Figs. 1 and 2 (dashed line) for a transformation rate of x(t)=0.1. Obviously, there are clear differences in the analytic structure of W(R,t) for the two nucleation mechanisms discussed above.
Besides the cluster-size distribution the numbers of clusters consisting of n nuclei z_n and the corresponding volume fractions x_n are of special interest. The time dependence of these quantities shows a characteristic maximum. It is worth noting that the maximum portion of single and ideally shaped grains x_1 is below 5 per cent of the total volume.

3. COMPUTER SIMULATION

To avoid the influence of approximations we have performed a computer simulation of the nucleation and growth reaction in a cubic volume with periodic boundary conditions [9]. Thereby, 100 to 200 nuclei are created at random positions according to the nucleation kinetics (a) and (b). The nuclei are growing with constant velocity and the cluster-size distribution and other physical quantities are evaluated numerically. In order to decrease statistical fluctuations this procedure is repeated 500 to 1000 times and the results are averaged. In comparison with the exact results for x(t), $z_1(t)$, and $x_1(t)$ the simulation yields sufficiently accurate results.

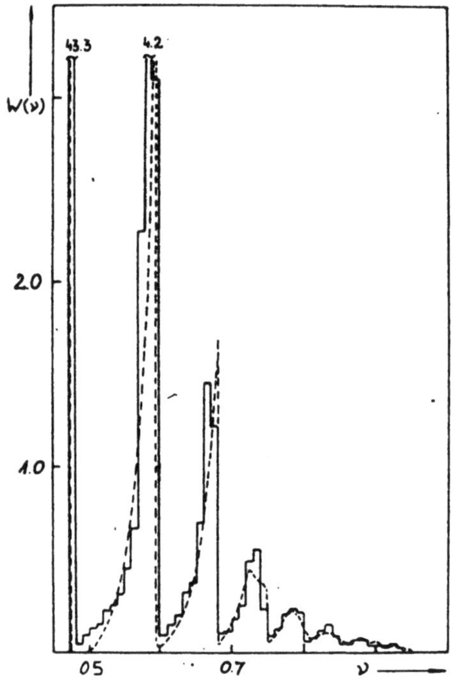

Fig. 1
Cluster-size distribution function (Case a, x=0.1) in comparison with analytical results (dashed line).

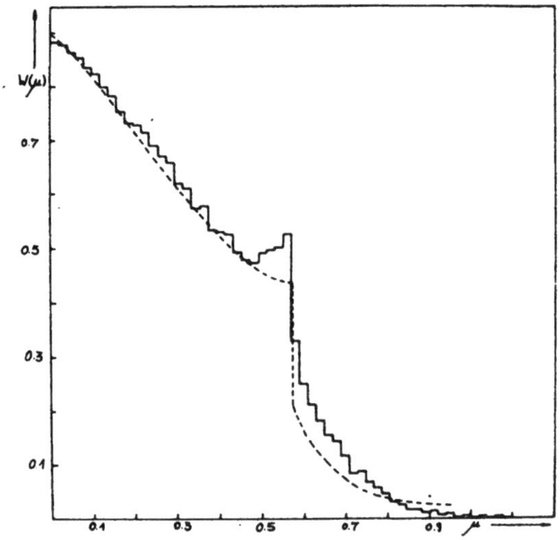

Fig. 2
Cluster-size distribution function (Case b, x=0.1) in comparison with analytical results (dashed line).

4. RESULTS AND DISCUSSION

The results of the analytical and the numerical calculations are given in terms of scaled time and length variables. The overall agreement is good for the site saturation case (a). For homogeneous nucleation (b) there are larger deviations near the maxima of x_n, z_n due to the approximations introduced in the analytical treatment. The fractional numbers z and x show a typical maximum which is shifted to larger times with increasing n. The cluster-size distribution for case (a) is characterized by a number of pronounced peaks (Fig.1). The analytical results are confirmed by the numerical simulation (solid line). With increasing time the hight of the peaks decreases and the weight of the distribution is shifted to larger radii.

On the other hand, for homogeneous nucleation W(R,t) is a continous function with the exception of a remarkable step (Fig.2). The simulation results deviate from the analytical theory near this step indicating a distinct shoulder as shown in Fig.2.

The cluster-size distributions for the nucleation models (a) and (b) have a completely different functional form (Figs.1 and 2). Especially, there are qualitative differences which are much more pronounced than those in the grain-size distributions [6]. The experimental investigation of the cluster-size distribution in the early stage of the nucleation and growth process should yield substantial information on the nucleation kinetics.

REFERENCES

[1] Kolmogorov, A. N.: Bull. Acad. Sci. USSR, Phys. Ser. 3 (1938),355.

[2] Schulze, G. E. W.: Z. Metallkde. 69 (1978),605;
J. Cryst. Growth 62 (1983),7.

[3] Sekimoto, K.: Physica 125A (1984),261.

[4] Axe, J. D. and Y. Yamada: Phys. Rev. B34 (1986),1599.

[5] Gilbert, E. N.: Ann. Math. Stat. 33 (1962),958.

[6] Mahin, K. W., K. Hanson and J.W. Morris Jr.: Acta metall. 28 (1980),443.

[7] Saetre, T. O.,O. Hunderi and E. Nes: Acta metall. 34 (1986),981.

[8] Lorenz, B.: Cryst. Res. Technol. 22 (1987),869.

[9] Orgzall, I. and B. Lorenz: to be published in Acta metall.

PHILOSOPHY AND REALITY OF STRENGTH DESIGN OF THE BELT CHAMBER BY FINITE ELEMENT METHOD

J.Vrbka [1), Z.Knésl [2)

1. Introduction

One of the most successful devices for production of superhard materials under high pressures (up to 10 GPa) and high temperatures is the Belt chamber. It consists of the high pressure container (compound vessel) with matrix made of sintered carbide WC-Co and supporting steel rings and two stamps moving against each other to produce required pressure in the working space of the chamber - see Fig. 1.

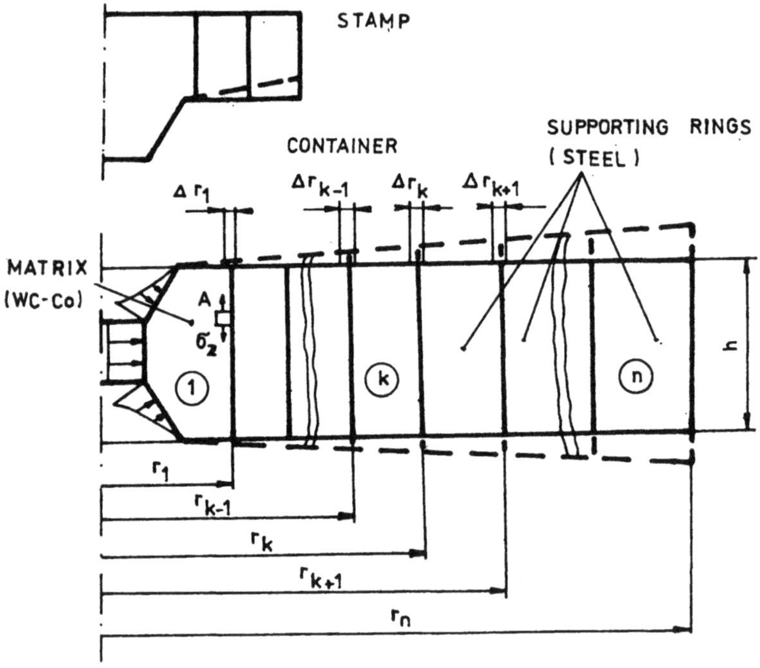

Fig. 1. Scheme of the Belt chamber

Many realized strength calculations show, that the most loaded part of the Belt chamber is the container (under the assumption that the pressure loading is axially symmetric). In our contribution we shall limit our

1) Technical University of Brno, Brno, Czechoslovakia
2) Institute of Physical Metallurgy, Czech.Acad.Sciences, Brno, Czechoslovakia

analysis to the container only. In many practical cases the strength design problem can be considered as an optimization one [4] , defined by the design variables (number n of components, contact surface radii r_k and overlaps Δr_k - see Fig. 1), geometrical restrictions of explicit type (outer shape of container is given by process technology and dimensions of used press), the stress restrictions of implicit type (we want to keep the loading of supporting steel rings in elastic state, which permits their economical repeated application). The objective function is the safety of the matrix, which is the most loaded part of the container.

2. Mathematical formulation of the problem and its solution

The stress and strength calculations are made on the proper computational model, which is closely connected with the computational possibilities. For the modeling of pressure loading we assume that the pressure distribution in working space is uniform. Its value p_o in the cold state can be measured making use of the electric resistance wire (Bi, Tl, Be). The pressure distribution (1) in the sealing conical part is given on the basis of the direct analogy with the pressure relations in pyrophylite loaded in the Bridgman anvils

$$p(s) = p_o \cdot e^{-\frac{2\nu}{t} \cdot s} \qquad (1) \qquad\qquad F_p = F_o' + F_o'' \qquad (2)$$

where s is the position coordinate (Fig. 2), ν is the inner friction coefficient of pyrophylite and t is the sealing thickness. Relation (1) is graphically shown on Fig. 2. The curve A corresponds to the state at the synthesis beginning, curve B to the state during the synthesis. In the latter case stamps move further against each other to keep the pressure necessary for synthesis and the pyrophylite sealing is further more compressed. Applied pressure distribution is checked by the equilibrium equation (2) on the stamp in axial direction z (Fig. 2). F_p is the measured axial force on the press.

The deformation loading at the cylindrical contact surfaces (Fig. 3) is taken into account by the deformation relations (3) and (4) in doubled points i, j. Corresponding stresses have to fulfil relations (5) and (6). Shear stress is restricted by condition (7) based on the Coulomb law of friction [5] .

$$u_{r_k}(j,k+1) = u_{r_k}(i,k) + \Delta r_k \qquad (3) \qquad\qquad \left[\frac{\partial u(j,k+1)}{\partial z}\right]_{r_k} = \left[\frac{\partial u(i,k)}{\partial z}\right]_{r_k} \qquad (4)$$

$$/\sigma_r(j,k+1)/ = /\sigma_r(i,k)/ \quad (5) \qquad /\tau_{rz}(j,k+1)/ = /\tau_{rz}(i,k)/ \quad (6)$$

$$/\tau_{rz}(i,k)/ = f./\sigma_r(i,k)/ \quad (7) \qquad w(j,k+1) = w(i,k) \quad (8)$$

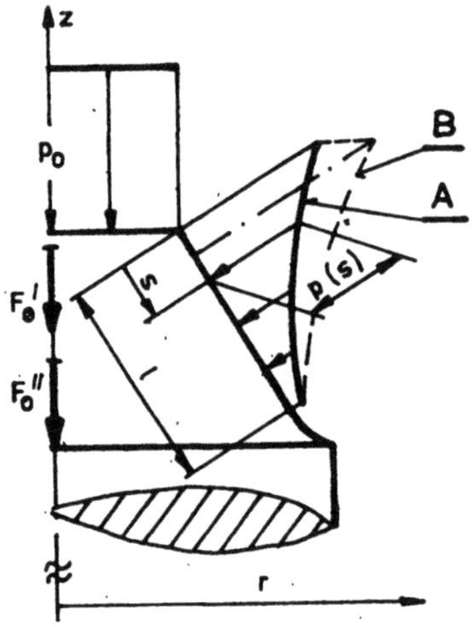

Fig. 2. Pressure loading of the stamp

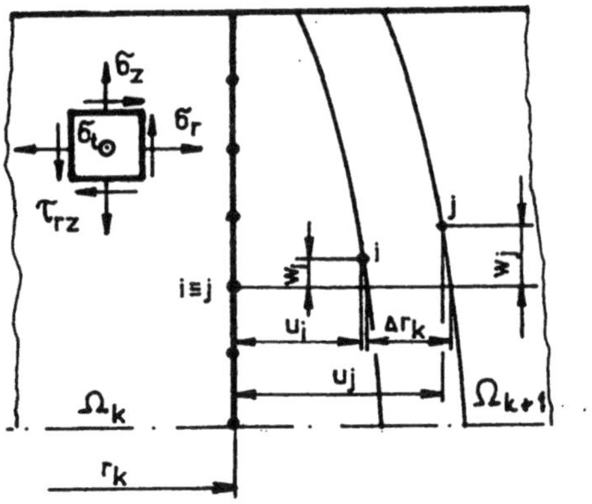

Fig. 3. Deformation loading at the contact surface

The optimum strength design of the container is made on two accuracy levels. If we transform the matrix to the cylindrical ring (Fig. 4) with the same average rigidity. [6] , for the calculation we may use the well known solution of the cylindrical compound vessel, based on Lamé's analytical relations for an elastic isotropical material. Thus the optimum number n of vessel components and optimum radii r_k and overlaps Δr_k can be calculated. As an example we present here an expression (9) for contact radii

$$r_1 = r_0^* \sqrt{\frac{n\left[\sigma_{D1} + p_0^*(1-\varkappa)\right]}{\sum_{j=1}^{n} \sigma_{Dj} - p_0^*(1+\varkappa)}} \qquad r_k = r_{k-1} \sqrt{\frac{n.\sigma_{Dk}}{\sum_{j=1}^{n} \sigma_{Dj} - p_0^*.(1+\varkappa)}} \quad (9)$$

$$\varkappa = \sigma_{Rt}/\sigma_{Rd} \quad (10) \qquad k \in \langle 2, n \rangle$$

where p_0^* is the statically equivalent inner pressure, r_0^* is the equivalent inner radius (from the point of view of matrix stiffness - Fig. 4), σ_{Dj} is the allowed effective stress of j-th component. σ_{Rt} and σ_{Rd} are tensile and compressive rupture strengths of sintered carbide. Optimization condition of the minimum outer radius r_n together with the strength conditions (Mohr's brittle fracture condition for matrix of sintered carbide and maximum shear stress condition for the supporting steel rings) on the inner radius of each component were applied to the derivation of all relations. For calculation of basic parameters p_k (contact pressure on surface r_k), r_k and Δr_k the program VMPRES in FORTRAN IV for computer ICL 2950/10 has been performed. The calculated values are then taken as input parameters for the stress and strength calculations making use of the finite element method (FEM). Verification of such cylindrical compound vessel model was done by FEM calculation of contact pressures p_k - see Fig. 4.

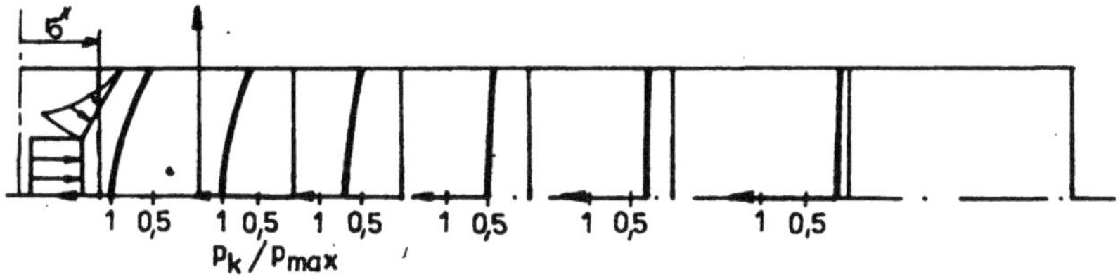

Fig. 4. Distribution of the contact pressure on the contact surfaces

The second computational level represents the stress-strength calculation of the chamber making use of FEM. For the stress analysis the code PROKOP [1] is used. The applied cubic Hermitien approximation function for displacements describes very well the great loading gradients. The double noding technique permits to realize the deformation loading at the contact surfaces, modeled by relations (3) - (6). The problem is solved for the whole body, the friction can be taken into account [5], which has substantial influence on the stress state (especially at the area near the contact surface). For example, the value of axial tensile stress in the dangerous place A (Fig. 1) for smooth surface (without the friction) is σ_z = 1080 MPa and for adhesion simulated by equation (7) σ_z = 415 MPa. On the basis of FEM the optimum calculation may be realized [4].

3. Accuracy of the computational model

The accuracy of the whole computational model is substantially influenced by its least accurate part. The natural requirement is that all the parts would have approximately the same accuracy. The contemporary reality is as follows

<u>Physical model</u> (model of the material, loading etc.). The errors are up to 20 %.

<u>Mathematical and numerical model</u>. By the application of the FEM the errors are of order 1 %.

Model of actual limiting states for postprocessing calculations of safety and reliability. In the case of matrix made of sintered carbide the errors are of order 100 % (calculated safety of the matrix which is successfully used in practice is for Mohr's hypothesis k_R = 0,26, for Balandin's hypothesis k_R = 0,60).

Our present effort is concentrated on reduction of the greatest discrepancies. To make the agreement by the material model (constitutive relations) better, the material inverse elastic-plastic problem has been solved [3] for an identification experiment on the thick cylindrical ring of similar dimensions as the matrix. The input parameters are measured strains. Calculated material characteristics E (Young's modulus of elasticity), ν (Poisson's ratio) and relation $\sigma_i(\varepsilon_i)$ (σ_i is the stress intensity, ε_i is the strain intensity). Evaluated characteristics differ from those obtained on the basis of the conventional 1-D stress-strain test because of great size factor. Expecially for E the differences were significant ($E = 4{,}35.10^5$ MPa compared to $E = 4{,}83.10^5$ MPa by 1-D test).

The discrepancies by the safety evaluation mentioned above show that the applied limiting state theories are not correct for sintered carbide WC-Co. It seems that rupture conditions here are strongly stress state dependent. We would like to try to formulate the new limiting state condition for WC-Co, which should agree better with the reality. Contemporary safety calculations are based on the stress analysis of the matrix used successfully in the production. The determined maximum effective stress is then taken as a strength material characteristics for the safety calculation of the new proposed device. We also assume, that the applied Mohr's brittle fracture hypothesis is valid only relatively in the narrow loading range. Isolines of the safety coefficient obtained directly from the computer in graphical form are demonstrated for one calculated case on Fig.5.

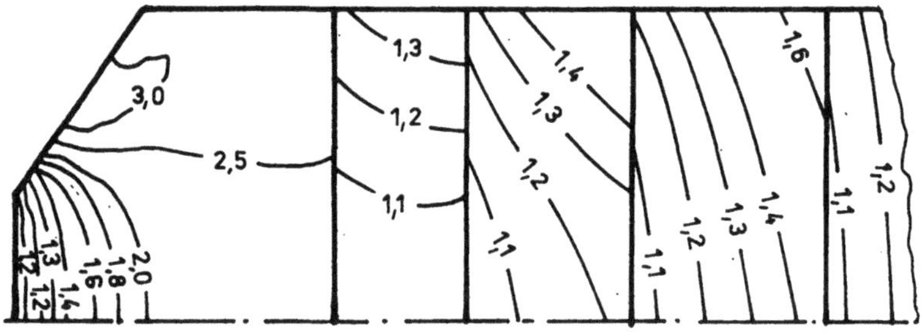

Fig. 5. Isolines of safety coefficient

All described calculations were made for the matrix material without defects. The containers with the cracked matrix often work in practice. The linear fracture mechanics has been applied to the lifetime estimation of the cracked matrix [2] . The stress intensity factors K_I and K_{II} were evaluated making use of the double noding technique for displacements at he vicinity of crack tip. Some results of calculations will be presented at the conference.

As the conclusion it can be said that in the present time the accuracy of the strength calculations of the Belt chamber is mostly determined by the accuracy of the proposed physical model (including the limiting state conditions). The accuracy of the numerical calculations by FEM is sufficient.

References

[1] Holuša, L. et al: PROKOP - version 1 (research report). Regional Computer Center VUT, Inst. of Physical Metallurgy Czech.Acad.Sciences Brno, 1983 (in Czech)

[2] Knésl, Z., Vrbka, J.: Solution of problems of linear fracture mechanics by package PROKOP. Proc.Conf.Mathematical Methods in Engineering, Czech.Science and Technical Society, Karlovy Vary, 1986, pp. 647-652 (in Czech)

[3] Vrbka, J.: One elastic-plastic material inverse identification problem. Sent for publication.

[4] Vrbka, J., Knésl, Z.: Optimized design of a high pressure compound vessel by FEM. Comp.Struct., 1986, Vol.24, No 5, pp. 809-812

[5] Vrbka, J., Knésl, Z.: Modeling of constraints in a contact surface - FEM solution. Proc.Conf.Mathematical Methods in Engineering, Czech. Science and Technical Society, Karlovy Vary, 1986, pp. 153-158 (in Czech)

[6] Vrbka, J., Moráček, M.: Dimensional optimization of compound cylindrical vessel, strength analysis by FEM. Proc.Conf. Physic and Technique of High Pressures, Czech.Science and Technical Society, Tábor 1980, pp. 29-32 (in Czech)

INVESTIGATION OF PRESSURE DISTRIBUTION IN WORKING SPACE BETWEEN DIAMOND ANVILS BY CHANGE OF PROPERTIES OF A DEFORMABLE SPACER

B. I. Beresnev, B.M. Efros[I)]

The present paper deals with investigation of phenomena which form the basis of plastic deformation of spacer (gasket) materials under superhigh pressures realized by the diamond-anvil method.

At the diamond-anvil plant under superhigh pressure plastic deformation of gaskets-containers of stainless steel T-30I and heat-resistant alloy Inconel X-750 was carried out. Loading of the tested device with pressures ca. 6,5 GPa is accompanied by appearance of a deformation zone around the compression area. In the range of pressures 8,5-40GPa around the compression area one observes the pronounced zone with a ring-like collar. Further loading - up to pressures ca. 65 GPa - alters no more shape and dimensions of this zone. In Fig. I the features of the plastic deformed zone for a gasket of T-30I steel are shown. They were studied by electron microscopy using a REM JSM-35 microscope at the accelerating voltage 50 kv.

In Fig. 2 the experimental pressure P_o acting upon the gasket axis vs. its deformational level $\mathcal{E} = h_o - h/h_o \cdot 100\%$, $h_o = 250\ \mu m$ is given. The obtained data permit to distinguish three main steps of gasket deformation: elastic-plastic, plastic and quasielastic.

Features of plastic deformation determined by the diamond-anvil method show the formation of polycrystalline material structure of a gasket. It is different in comparison with gasket material structure produced by the traditional deformation. The obtained difference is important for the deformed material properties.

Effect of superhigh pressures on residual strength characteristics of a gasket material is investigated. Microhardness is chosen as a measure of the latter. It shows the level of material strength.

Let us now consider the material deformation in the compressed zone as uniform and neglect pressure at the periphery of this zone in comparison with pressure along the axis. In this case the value of residual barohardening may be estimated from the difference $\Delta H\mu$ of microhardness along the axis and the periphery of the compressed zone.

In Fig. 3 the plots are given for the $P_o - \Delta H \mu$ relation. They may be used to recover the pressure distribution in the compressed zone using the values of distribution of the material microhardness in this zone [I].

In conclusion we compare direct and inherited effects of pressure on the level of deformational resistance of the gasket materials. Deformational resistance σ_S under pressure we estimate by model identifica-

I) Donetsk Physico-Technical Institute, Academy of Sciences of the Ukrainian SSR, 340II4 Donetsk, ul. R. Luxemburg, 72 U S S R

tion [I] in the experiment.

The relation σ_S/H_μ vs. P_o is given in Fig. 4. Here H_μ is the mean microhardness over the cross-section. At low pressures the value of σ_S/H_μ is ca. 0,3 ÷ 0,4, that corresponds to the conclusions of the paper [2]. The fact that this relation increases with pressure indicates the prevailing direct influence of pressure on the value of deformational resistance.

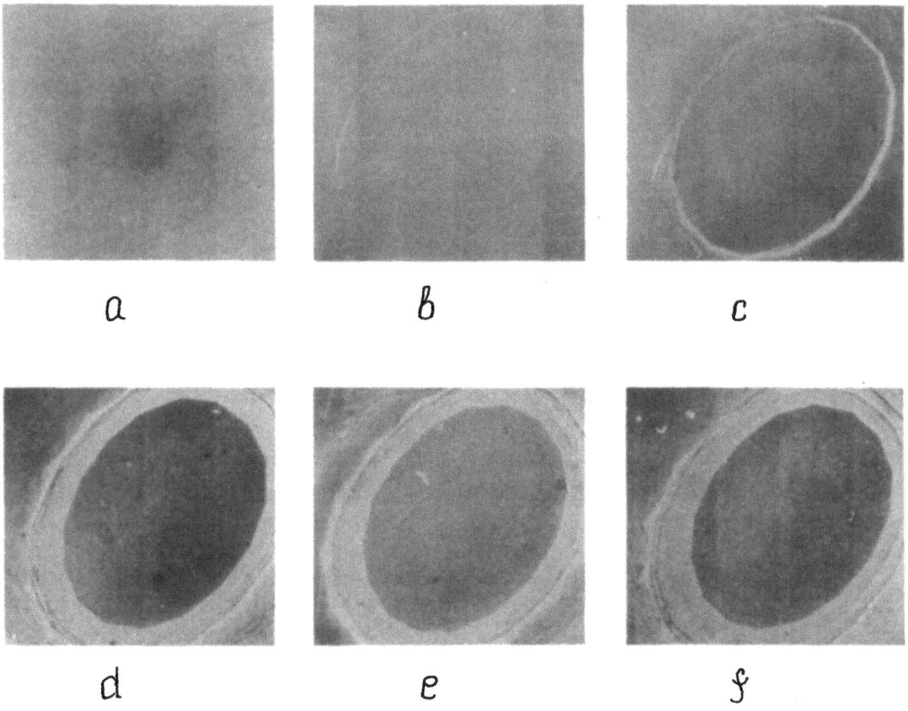

Fig. I. View of a deformed gasket made of T-30I stainless steel after loading in a pressure plant with different loads:
a - 6,5 GPa, b - 8,5 GPa, c - I4,5 GPa,
d - 40 GPa, e - 45,5 GPa, f - 65 GPa.

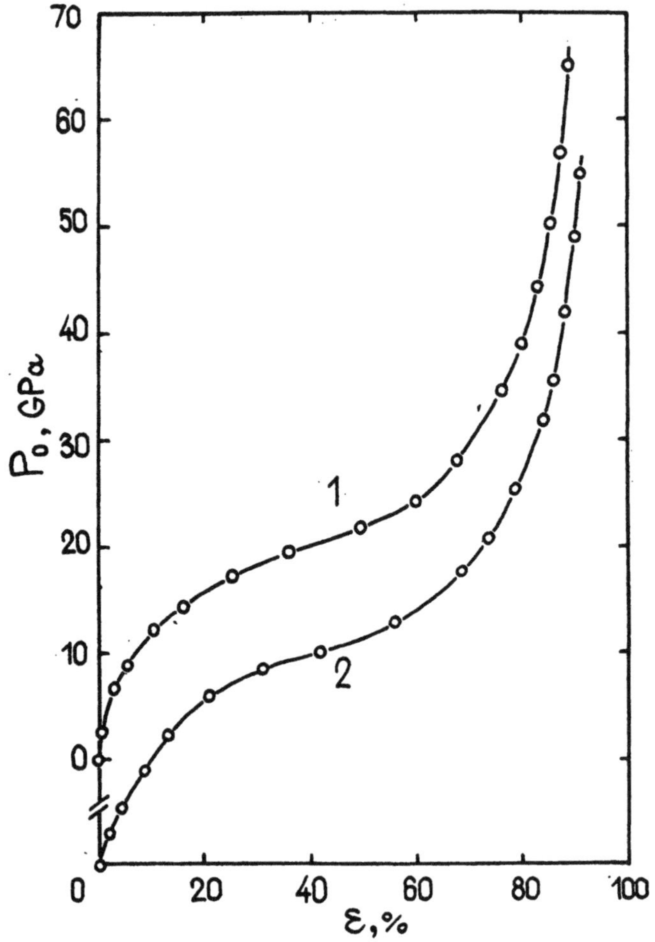

Fig. 2. Pressure P_0 along the gasket axis vs. deformation level ε for gaskets made of different materials:
I - T-30I, 2 - Inconel X-750.

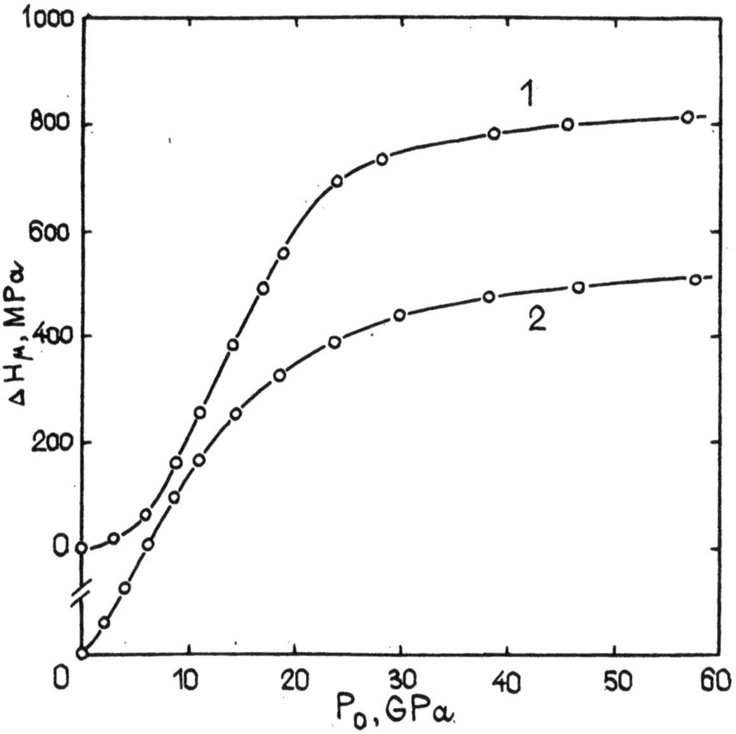

Fig. 3. Level of barohardening ΔH_μ of gasket material vs. pressure P_0 along the gasket axis for the different gasket materials: I - T-30I, 2 - Inconel X-750.

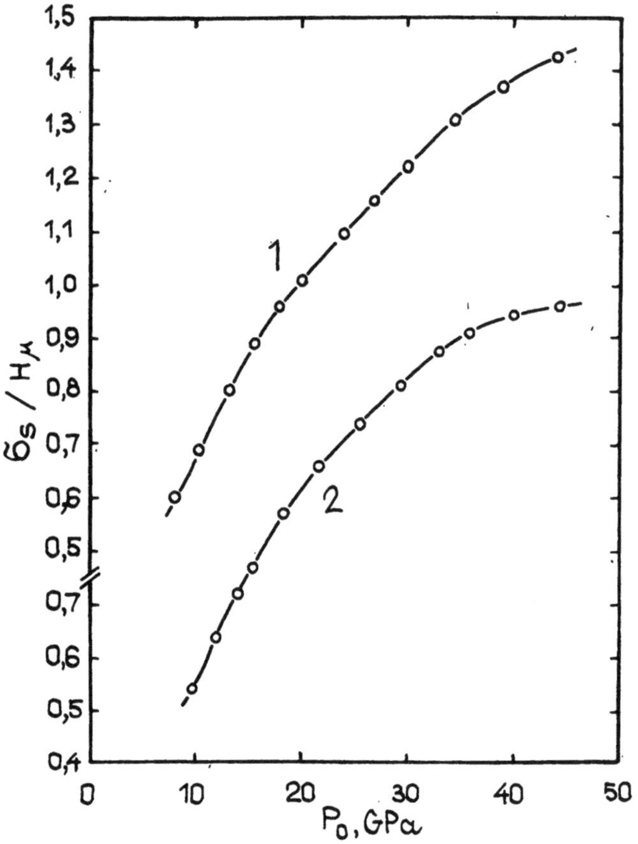

Fig. 4. Stress relation σ_S / H_μ vs. pressure P_0 along the gasket axis for different materials:
I - T-30I, 2 - Inconel X-750.

Literature

[I] Beresnev, B., B. Efros: Pressure generation in a diamond-anvil apparatus. Physica. 139 (1986), 910-915.

[2] Del; G. D.: Determination of stresses in a plastic zone by hardness distribution. Isd. Mashinostroyeniye, Moscow 1971, (in Russian).

THEORETICAL RESEARCH OF THE PROCESS CONTRACTING AND WELDING TUBE BY EXPLOSION

Han duc Kim and Trinh dinh Thang 1)

Presently, the requirements resolving problems of contracting and welding tube by explosion appears in order to reproduce a lot of wearing steel parts in the cylindrical shape and manufacture many products of the industry.
In this arcticle we introduce theoretical research of the process contracting and welding tube by explosion.
In the part one of the article, the mathematical formation of the problem in contracting metal tube by contact explosion is put forward and based on it, the plastic deformation of tube is calculated.
Part two says about looking into the tube welding diagram, in which using of explosive energy. The conditions of metal welding force are assumed and optimal distance between tubes, indispensable quantity of explosive for carrying this process out are estimated.

I. The tube with external radius R_1 and internal radius R_2 contacts with explosive tube having thickness H. Under the action on explosive pressure the tube deformed plastically (Fig. 1).
The motional equation of deformed medium in case of symmetrical deformation follows (1)

$$\frac{\partial \sigma_r}{\partial r} + \frac{\sigma_r - \sigma_\varphi}{r} = \rho_1 \cdot \frac{\partial^2 r}{\partial t^2} , \qquad (1)$$

where: $r = r(r_0, t)$ - radial coordinate of the mass point in the motional process; t - time; r_0 - radial coordinate of the mass point at the beginning moment; σ_r , σ_φ - radial and circle stress; ρ_1 - density of material of tube.
Considering the Treska plastic condition $\sigma_r - \sigma_\varphi = 2K$ [1], from equation (1) we have

$$\frac{\partial \sigma_r}{\partial r} + \frac{2K}{r} = \rho_1 \cdot \frac{\partial^2 r}{\partial t^2} \qquad (2)$$

1) Hanoi Machinery Research Institute, Vietnam

Supposed that material is not able to press, the continuous equation is as follows

$$\frac{\partial r}{\partial r_0} = \frac{r_0}{r} \qquad (3)$$

Take integration (3) we receive

$$r = \sqrt{r_0^2 - c^2(t)} \qquad (4)$$

where C(t) some equations of time.

Replaced (4) into (2), taken integration on the received equation according to from R to R , after some transformations, we receive

$$\frac{d^2(R_1^2)}{dt^2} \cdot \frac{1}{4} \ln \frac{R_1^2}{R_2^2} - \left[\frac{d(R_1^2)}{dt}\right]^2 \left(\frac{R_{10}^2 \quad R_{20}^2}{8R_1^2 \cdot R_2^2}\right) - \frac{1}{\mathit{S}_1}\left[P_1 - K \ln \frac{R_1^2}{R_2^2} - P_2\right] = 0 \qquad (5)$$

Non-pressurized condition: $R_1^2 = R_2^2 + (R_{10}^2 - R_{20}^2)$; (6)

where R_{10} and R_{20} beginning external and internal radius of tube.

$P_1(t)$ and $P_2(t)$ pressure, acted on external and internal side of tube.

In this case $P_2(t) = 0$

By passing the "toughness" of material, supposing K = 0

$$(\sigma_r = \sigma_\varphi = -P)$$

i.e. the tube deformes as ideal non-pressurized liquid, we receive Derebas' equation [2,3] from equation (5) .

Supposed that the explosive waves act directly on the surface of the tube and thickness of the tube is less than average radius of the tube.

Derebas received the equation of explosive pressure [2,3]

$$P_1(t) = \frac{16}{27}\frac{\mathit{S}_0}{D}\left[\frac{H + R_{10} - R_1(t)}{t + H/D} + \frac{dR_1(t)}{dt}\right]^3 \qquad (7)$$

where S_0 , H - density and thickness of the explosive tube

D - speed of explosion.

The constant K in (5) is contact yield strength of the material.

Because of deforming speed of the tube under the action of extremely great contact explosive pressure is high deforming speed of the tube K dependent function of deforming speed material. In this case we take $K = K_d$ where K_d - dynamical contact yield strength.

Equations (5), (6), (7) are integrated with beginning conditions

$$R_2(0) = R_{20} \qquad \frac{dR_2(0)}{dt} = 0$$

Example I.

The tube with its dimension $R_{10} = 56$ mm ; $R_{20} = 50$ mm

made of steel 45 $\rho_1 = 7,8$ g/cm^3; $K_d = 62$ Kg/mm^2 ;

Explosive is 6 * B ; $\rho_0 = 0,9$ g/cm^3; H = 9 mm

For small thickness H, explosive speed is calculated by the formula (4)

$$D = 4,5\left(1 - \frac{2,44}{H^{0,79}}\right)$$

D expressed in mm/μs and H in mm. The calculated results are introduced in table 1 [according to the diagram of one side explosion].

Table 1:

t	Diagram of one side explosion		Diagram of two side explosion	
	R_2 (mm)	dR_2/dt (mm/μs)	R_2 (mm)	dR_2/dt (mm/μs)
0	50,00	0,00	50,00	0,00
1	49,97	-0,046	49,99	-0,019
2	49,91	-0,067	49,96	-0,037
3	49,83	-0,077	49,91	-0,056
4	49,75	-0,083	49,84	-0,074
5	49,67	-0,085	49,76	-0,093
6	49,58	-0,086	49,85	-0,111
7	49,50	-0,086	49,53	-0,130
8	49,41	-0,085	49,40	-0,126
9	49,33	-0,084	49,28	-0,123
10	49,24	-0,082	49,16	-0,120
20	48,55	-0,056	48,12	-0,089
30	48,15	0,00	47,39	-0,056
38	48,04			
47			46,92	

The one side explosion diagram is pointed out above. It is fact that a two side explosive diagram exists, i.e. contracting tube is carried out by the explosive wave sliding over the surface. Because the explosive speed is very fast and after the happened explosive process, the deformed degree to axial side of the tube is smaller than to radial side very much, we can consider that it is a one-side explosive diagram. In this case it has small deviations. However, the two-side explosive diagram is different with the one side.

The pressure in the case of the diagram, in which the explosive waves should be slided is smaller, but exists longer and changes monotonously according to time. To replace (7) we would like to put forward a manner to form approximately $P_1(t)$ in this case of sliding explosive wave with the two-side explosive diagram of plates, we receive the formula of maximum explosive speed [3,5]:

$$V_{max} = \frac{1{,}2D\sqrt{1+\frac{32}{27}q}-1}{\sqrt{1+\frac{32}{27}q}-1}$$

where q is the ratio of the explosive mass with mass of the plate should be made explosion.

The impulse pressure of explosion acts on an area unit of the plate in the form of displacement, therefore:

$$\int_0^{\infty} P_1(t)\,dt = m\,V_{max} \tag{8}$$

Thus, we are able to define approximately the time happened when the explosive pressure should act on the predetermined point of the plate: T = 2H/D. By then P (t) can be established approximately as follows:

$$P_1(t) = \begin{cases} max \cdot \frac{m}{T} \; ; & 0 \leq t \leq T \\ 0 \; , & t > T \end{cases} \tag{9}$$

Because of the thickness of the tube is smaller than average radius, the formular (9) can be used in our problem for element of the tube.

The results of the solving equations (5),(6),(9) is introduced in table 1 (two-side explosive diagram).

The calculating results show that final parameters (final shrinkage R_2, maximum speed $\frac{dR_2}{dt}$ and distance ΔR in which it reaches the maximum speed) of two explosive diagrams are only different a little bit. The second diagram gives the values higher than that of the first as we expect. Because of this, we can accept all of two diagrams but results of the first is more accurate than of the second.

II. Welding process of the tubes

External metal tube with density ρ_1 and initial radius R_{10} and R_{20} is enclosed by an explosive tube with a thickness of H.

Internal metal tube has density ρ_2, external radius R_3. We express $\Delta R = R_{20} - R_3$ (Fig.2). Under the action of explosion pressure the external tube has a speed, which reaches the maximum value through a short time. After that, due to the action of plastic stress of this material and the weakening of explosive pressure, the speed of displacement reduces. It is proved in the table 1. Therefore, the optimal distance ΔR corresponded with predetermined quantity of explosive is the indispensable distance for the external tube displacing and reaching maximum impact speed.

Example II:

External tube made of steel 08 KP, having $\rho_1 = 7,82\ g/cm^3$; $K_d = 47\ kg/mm^2$

$R_{10} = 108$ mm ; $R_{20} = 100$ mm

The explosive is 6*B ; having $\rho_0 = 0,9\ g/cm^3$; $H = 12 \div 20$ mm.

The calculating results are shown in table 2.

The external tube impacts to the internal tube in speed V. The impact pressure can be calculated as follows [5,6]

$$P_v = \frac{V \rho_1 . \rho_2 . C_1 . C_2}{\rho_1 C_1 + \rho_2 C_2} \qquad (10)$$

Here C_1, C_2 are corresponding speed of longitudinal wave in material of external and internal tube.

Table 2

H (mm)	$V_{max} = -[dR_2/dt]_{max}$ (mm/µs)	ΔR (mm)
12	0,119	1,15
14	0,148	1,65
16	0,175	2,2
18	0,203	2,8
20	0,230	3,5

We suppose that these tubes should be welded each other if the impact pressure is enough to form plastic yield in the contact area of material of tube having higher "toughness".

According to the theory of sliding line of flat tough-plastic deformation [1] we receive the indispensable pressure made the plastic yield:

$$P_v = (2 + \pi) K_d \qquad (11)$$

where K_d is the dynamic contact yield strength of material of tube having higher "toughness", as introduced above:

when $K_d = \frac{\sigma_d}{\sqrt{3}}$ $\qquad P_v = (9 \div 12)\, \sigma_T$

where σ_d -dynamic yield strength and for ordinary solf steel

$\sigma_d \approx (3 \div 4)\, \sigma_T$; $\qquad \sigma_T$ - static yield strength.

This criterion corresponds to which observed on [6].

From (10) and (11) the indispensable impact speed for welding is

$$V = \frac{\rho_1 C_1 + \rho_2 C_2}{\rho_1 . C_2 . C_1 . C_2} (2 + \pi) K_d \qquad (12)$$

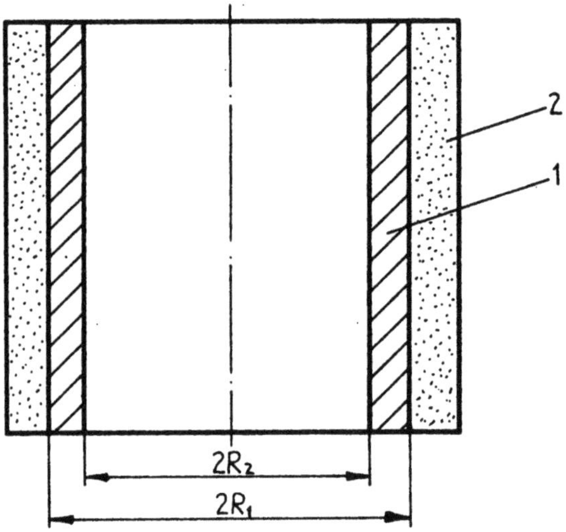

Fig. 1
1- Tube
2- Explosive

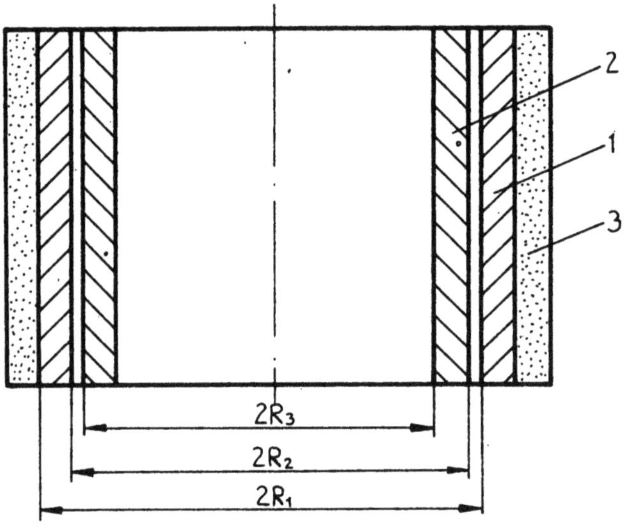

Fig. 2
1- External metal tube'
2- Internal metal tube
3- Explosive

Example 3:

All as example 2

Supplementary data are as follows:

Material of the second tube is aluminium alloy with $\vartheta_2 = 3$ g/cm^3

$C_2 = 5,2$ mm/µs $C_1 = 4,5$ mm/µs

Due to the formula (12) we receive V = 0,22 mm/µs

Table 2 gives the optimal distance Δ R = 3,5 mm and H = 20mm.

In fact, if we are not able to choose the optimal distance R, we have to pick out the value of H to satisfy the condition (12).

References:

[1] Robotnov, J.N. - Mechanics of deformed solid boddies, 744 C "Science" 1979, (in Russian)

[2] Derebas, A.A. - Physics of consolidation and welding by explosion, 188 C, "Science" 1972 (in Russian)

[3] Krupin, A.B. - Deformation of metals by explosion, 415 C, "Metallurgy" 1975, (in Russian)

[4] - High speed deformation of metals, 221 p, "Minsk" 1978 (in Russian)

[5] Hellmann, A.C. - Plating of steel by explosion, 189 p, "Machineengerneering" 1978, (in Russian)

[6] - Principle and practic of explosive metalworking, 210 p, 210 p, "London", 1973

PARAMAGNETIC CENTRES IN HARD MATERIALS UNDER UNIAXIAL PRESSURE

R. Stößer[1)], A. Rericha[2)], R. Lück[3)]

Hard materials like Al_2O_3 and diamond contain usually paramagnetic centres (PC), which even at low concentrations are qualified to characterize the state of the system. Especially they can serve as sensitive indicators for structural changes. Pressure induced structural changes are in general also reflected by the variation of the EPR parameters of the PC's contained.

The goal of the present investigations is to characterize the magnetic and spectral behaviour of hard materials under the influence of uniaxial pressure with respect to an application of the material for abrasive purposes. To monitor the magnetic properties at the action of pressure on hard materials is one actual trend of the pressure application in spectroscopy. For this purpose conventional EPR technique equiped with a suited pressure apparatus (Fig. 1a) can be used. Generally one observes changes of the magnetic and spectral parameters under the influence of pressure. This behaviour can be reflected by a suitable modified spin-hamiltonian |1, 2|. On the one hand pressure effects can help to solve assignment problems of PC's in crystals and powders (the identification and localization of the centres). On the other hand, especially from a technical point of view one is searching for definite connections between the pressure induced magnetic (and structural) effects and the mechanical properties of the material like the state of strain |3, 4| the hardness and the brittle strengh.

The in situ EPR measurements using uniaxial pressure were carried out with a X-band EPR spectrometer (E4, Fa. Varian). Use was made of a pressure equipment which is schematically displayed in Fig. 1b. The apparatus allows the variation of uniaxial pressure up to the GPa range and the rotation of the upper piston inclusive.

In principle all terms in the spin-hamiltonian change at the application of pressure; distinct effects can be expected for the fine structure tensor (for systems with $S' \geq 1$, parameters D, E). Whereas under isostatic conditions |5, 6| one obtains differential quotients like $(\frac{\partial D}{\partial P})_T$ for the zfs and other parameters respectively, under uniaxial pressure conditions at higher pressures a distribution $(\frac{\partial D_i}{\partial P})_T$ results. In practic the distribution depends on the space-coordinates of the PC's in the sample. It means, e. g. D can vary from site to site in the lattice of the crystal and can yield an image of the state of strain in the crystal.

Objects of the in situ investigation were selected crystals of

---

1) Humboldt-Universität Berlin, Sektion Chemie, 1080 Berlin, Bunsenstr. 1

2) VEB Rationalisierung Potsdam

3) Zentralinstitut für anorganische Chemie der AdW der DDR, 1199 Berlin, Rudower Chaussee 5

$MgO/Mn^{2+}/Fe^{3+}/Cr^{3+}$, topaz ($Al_2F_2SiO_4/Fe^{3+}$), ruby ($Al_2O_3/Cr^{3+}$), sapphire ($Al_2O_3/Fe^{3+}$) and synthetic diamond (AC 65 T/ ^{14}N).
The cubic MgO represents a good model system for pressure investigations |5, 6, 7|. Results of investigations under uniaxial pressure are not available so far.
A distinct pressure dependence results for the Fe^{3+}-ions at cubic sites in MgO yielding at ≈120 K line shifts of $6.8 \cdot 10^{-9}$ T/Pa and of $1.85 \cdot 10^{-9}$ T/Pa at pressures p > 0.5 MPa (Fig. 1c). The largest p dependence was observed for Fe^{3+}- and Cr^{3+}-ions at non cubic sites in (and near the surface of) the MgO crystal. The line shifts and the intensity decrease are remarkable. The last effect appears as a result of a drastic increase of fine structure contributions to the line width at the enlargement of P. Every line of the six ^{55}Mn hfs lines from Mn^{2+}-Ions at cubic sites gives a response to the p influence in a different way (Fig. 1d). The intensity distribution of a given spectrum (at p = const.) is a function of the actual zfs-values (mainly of D, Fig. 1e) and their local distribution.The overall pressure dependence (Fig. 1d) which is responsible for the surprising pattern represents the solution of the equation $(\frac{\partial Di}{\partial P})_T$ = f (P). Reversibility of pressure dependence was only reached for the cubic sites at p < 1 MPa. Irreversible changes of fine structure pattern corresponds to the results obtained at external pressure treatments of MgO |7|.
In topaz crystals the substitutional Fe^{3+}-ions at the four different Al^{3+}-sites cause together with ^{19}F-shfs contributions fine structure transitions in the g'≈ 4.2 region with a distinet anisotropy. The pressure influence is above all indicated by an intensity variation of the transitions and only at higher p by a line shift (Fig. 1f). These effects can be used for an assignment of the fine structure transitions.
Corundum crystals doped with Fe^{3+} and Cr^{3+} were used in an arrangement like Fig. 1g. As the pressure apparatus allows to rotate the upper (u) piston together with the Al_2O_3 (u) anvil one can simultanously observe the pressure influence on the two anvils at different orientations with respect to the magnetic field. So it is possible to monitor different PC's in both anvils (Fig. 1h) under the influence of pressure.
At selected samples of synthetic diamond (AC 65 T, SU) only small pressure effects in situ at 300 K were observed. In external low temperature treatments (77 K, GPa-region) and subsequent low temperature EPR-measurements (at p ≈ 0) significant changes in the ^{14}N h/s were registrated. At higher temperatures (77 ⟶ 300 K) the crystals undergo some mechanical relaxation processes |s. a. 9| but seldom reachs their original state. Generally, from the investigations on synthetic diamonds of different origin one can conclude, apparently by reasons of failing crystals perfection, fracture processes appear at relatively low pressures without detectable spectral indication. From this behaviour one can deduce conditions for the synthetic and also for the subsequent processes of quality selection.

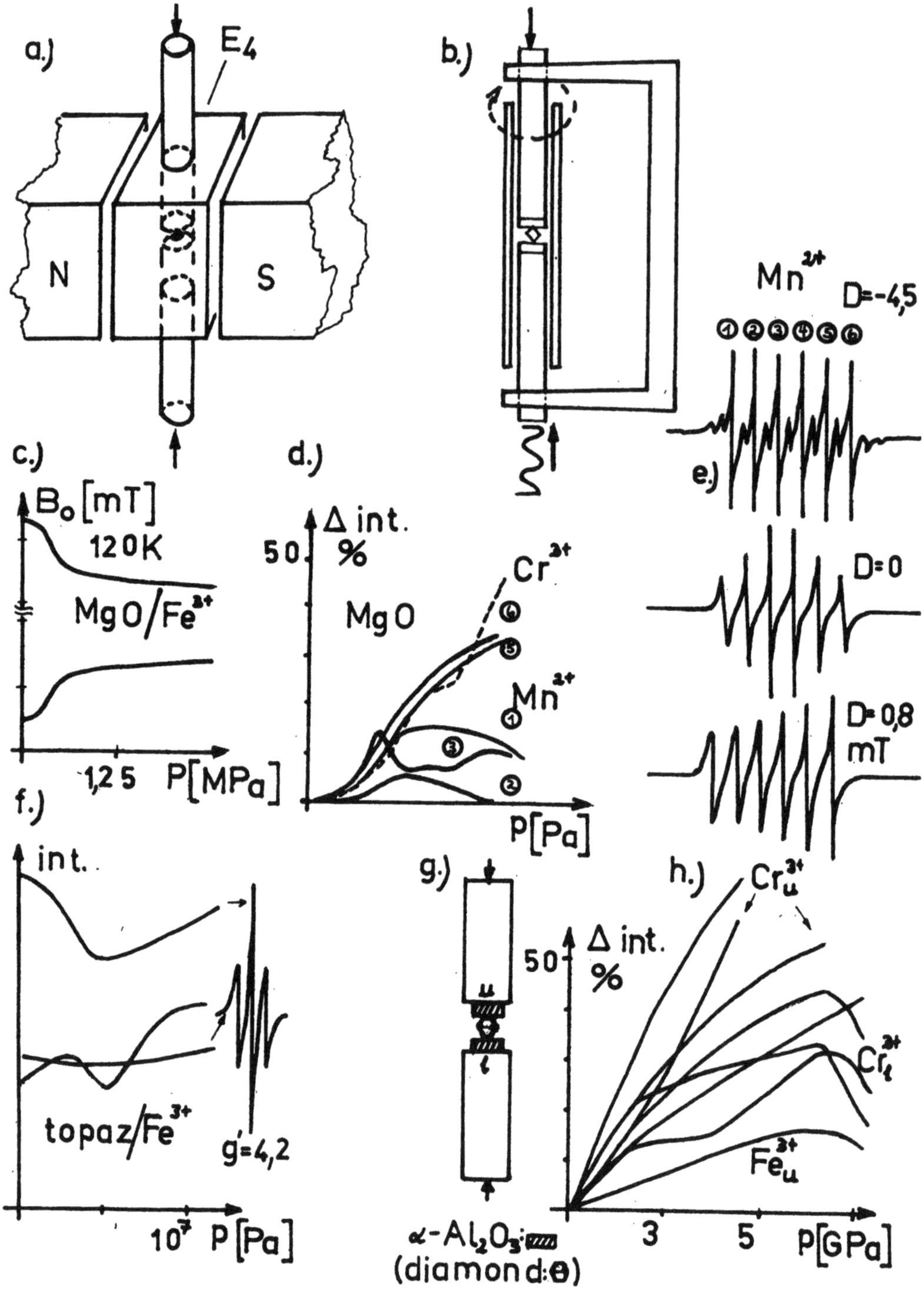

Figure 1. EPR pressure apparatus a.),b.); pressure influence on the EPR line shift of Fe^{3+} in MgO c.); on the intensity of Mn^{2+} hfs in MgO d.); hfs pattern of Mn^{2+} as a function of D e.); p influence of Fe^{3+} intensity at $g' \approx 4.2$ in topaz f.) two anvils arrangement for pressure induced intensity change in Al_2O_3 g.) and h.)

In connection with all our present experiments we can point out, that there exist certain relations between magnetic behaviour of a crystal and his state of stress deformation also under conclitions of abrasive working.

|1| Abhvani, A. S., Bates, C. A., J. Physics C, Solid State Physics 14 (1981) 2617
|2| Vasjukov, V. N., Samsonova, G. Ja., Jinzadze, G. A., Solid State Physics 28 (1986) 808
|3| Gravaix, A., J. Phys. C.: Solid State Physics 16 (1983) 6031
|4| Ammerlaan, C. A. J., van Kemp, R., J. Phys. C.: Solid State Physics 18 (1985) 2632
|5| Walsh Jr., W. M., Phys. Rev. 114 (1959) 1473; 122 (1961) 762
|6| Low, W., Phys. Rev. 105 (1957) 801
|7| Vlasova, M. V., Gorbachuk, S. I., Kakazei, N. G., Ristic, M. M., J. Mater. Sci. 18 (1983) 245
|8| Jones, P. M., Bruker EPR application note 105 (1986)
|9| Ammerlaan, C. A. J., Burgemeister, E. A., Industrial Diamond Rev. 4 (1980)

SOME CRITERIONS OF EVALUATION OF HIGH PRESSURE APPARATUSES FOR INDUSTRIAL USE

J. Flašar 1/

INTRODUCTION

High Pressure Apparatuses for industrial super hard material synthesis are optimized with respect to requirement of minimization of cemented carbide specific consumption. The specific cemented carbide consumption is given by limited service life of extremely loaded parts - the die and the punches.

The service lives of the dies or punches are limited by resistances of their materials against repeated loading. We were interested in influence of service life upon loading magnitude, especially in the influence of service life of the die upon the prestress magnitude.

Proper loading of the extremely loaded parts is superposition of the constant prestress induced by prestress rings and repeated quasi static work loading (press, temperature).

Value of the specific consumption of cemented carbide as a material of the die and the punches (at production scale) is further dependent on making use of exact design method, on using of high fatique resistant material with low dispertion variance, on synthesis optimalization with pressure and temperature minimization and on strict following the optimized technologic and operating instructions.

THEORY

The die life dependence on the level and scattering of the work loading has been followed up on an optimized industrial high pressure apparatuse "belt" having work volume of 50 cubic centimetres. The work loading was simulated by working use of superposition of mounting prestress and loading by synthesis conditions inside a chamber and a gasket (press, temperature).

The following physical model of the "belt" high pressure apparatuse was used:

Model of loading

- The work loading in the chamber is statical, stress distribution in conical gasket area is simulated by using of analogy of press distribution inside a compressed disc.

1/ Pramet, Powder Metallurgy Works, National Enterprise, Sumperk, CSSR

- The deformation loading in contact surface of prestress rings is dated at magnitudes of single interferences of rings and is simulated by using of duplicated parameters. For force conditions there is used transformation of stress relations into deformation relations (the deformation variant of finite element method).

Model of geometry

- The examined body is symmetrical rotary body with cylindrical contact surfaces.

Model of stress state

- The state of stress is axially symmetrical, for small deformations we can use the first order theory:

$$T_{\sigma} = \begin{vmatrix} \sigma_r(r) & \tau_{rz}(r) & 0 \\ \tau_{zr}(r) & \sigma_z(r) & 0 \\ 0 & 0 & \sigma_t(r) \end{vmatrix}$$

- The proper solution utilises Lagrange's variation principle and numerical solution by finite elements method for rotary symmetrical problem.

We contemplated a set of dies where the mounting prestress of the dies was set up in the range of 100 ÷ 115% of the optimized work maximum reduced stress. The changes of prestress were realized by the magnitudes of interferences of prestress rings. Scattering intensity of the work loading was kept in the range of ± 2%. This value reduced to the maximum reduced stress of the die was in the range up to ± 1%.

RESULTS

There was contemplated the set of 200 dies. The die lives were estimated with respect to the kind of damage. On the base of the hitherto experience there was supposed steep dependence of the die life on the prestress magnitude.

Results of the die lives in the explored region ($\sigma_{red_{opt}} \div 1{,}15\ \sigma_{red_{opt}}$) did not verified this assumption.

In the region of $\sigma_{red_{opt}} \div 1,15\ \sigma_{red_{opt}}$ there was no signifioant decrease of die lives in the regions of the minimum, average and maximum die lives.

DISCUSSION

In our opinion we can explain this result in several ways:

- $\sigma_{red\ max}$ of the die, in our case $\sigma_{red} = \max(\sigma_1 - \varkappa\sigma_3)$ - Mohr's brittle strenght condition MOS is not sufficiently accurate condition of limiting state of WC - Co cemented carbide

- $\sigma_{red_{max}}$ of the die is not the only one limited condition of the die life

- on base of the fractographic observation we can presuppose significant influence of the tensile components of the stress state

- the very important factor for the die life scattering is scattering of structure characteristics of WC - Co as a material of the dies, especially geometrical and topological arrangement of defects.

BIBLIOGRAPHY

Novotný: Strenght of material I, SNTL Praha, 1984
Vrbka, Flašar, Lakomý: Mechanical characteristics of the WC - Co Hardmetals from the Material Inverse Problem Solution, VII. Internationale Tagung Pulvermetallurgische in der DDR vol.1, 109, Dresden, 1985.

SIMPLE TWO STAGE HIGH PRESSURE APPARATUS FOR HYDROSTATIC PRESSURES TO 5.0 GPa MADE FROM USUAL STEELS

S.M. Filipek [1)]

INTRODUCTION

The Bridgman's [2] idea of two - stage high pressure apparatus allowed to reach pressures higher than 5.0 GPa [3]. The experiments were however restricted to solids also because of small space, not sufficient for packing arrangement necessary for liquids. Hydrostatic pressures up to 5.0 GPa could be obtained when liquid was closed in a soft metal container [4] and in that case the accurate measurement of pressure was not possible. Thereafter two - stage high pressure apparatuses were described by several authors [5,7] but in all cases the solids were used as high - pressure transmitting media.

In this paper the two stage piston - cylinder apparatus, characterized with a very simple construction and relatively large working volume, suitable for liquids /and for gases after appropriate modification/ is presented. The actual version made from usual steel /only main piston was from tungsten carbide/ has been proved up to about 5.0 GPa at 25°C. For the first stage the piston - cylinder apparatus of Baranowski and Bujnowski [1] has been applied. In this case the pressure, supporting the second stage, can not be higher than 3.0 GPa. Of course the application of higher pressure in the first stage makes possible to increase pressure limit also in the second stage. In this aspect the development of normal one stage piston - cylinder apparatus [6,8,9] creates promising perspectives for our actual design.

APPARATUS

1. Details of design.

The design of our two - stage high pressure apparatus is presented on Fig. 1. As the first stage the vessel 1, composed from three coaxial cylinders, is used. Total dimensions of the first stage vessel are: internal diameter 25 mm, external diameter 210 mm, height 210 mm. From bottom side the vessel 1 is closed by the piston - stopper 9/i.e. piston for the second stage, stopper for the first stage/ through which six /or more/ insulated electrical leads enter directly to the second stage working volume. The second stage high pressure vessel 5 /int. diam. 12 mm, height 60 mm/ has a small hole 7, which enables uniform increase of pressure outside and inside vessel 5 during the initial period of work. Movable piston 2 of the first stage is used for the pressure icrease in both first and second stage pressure vessels. When the piston 2 moves

1) Institute of Physical Chemistry, Polish Academy of Sciences, 01-224, Warsaw, Poland.

down in contact with the vessel 5 the geometry of all assembly guarantees that for unit displacement of the piston 2 the relative volume decrease outside the vessel 5 is at least two times smaller than inside of it.

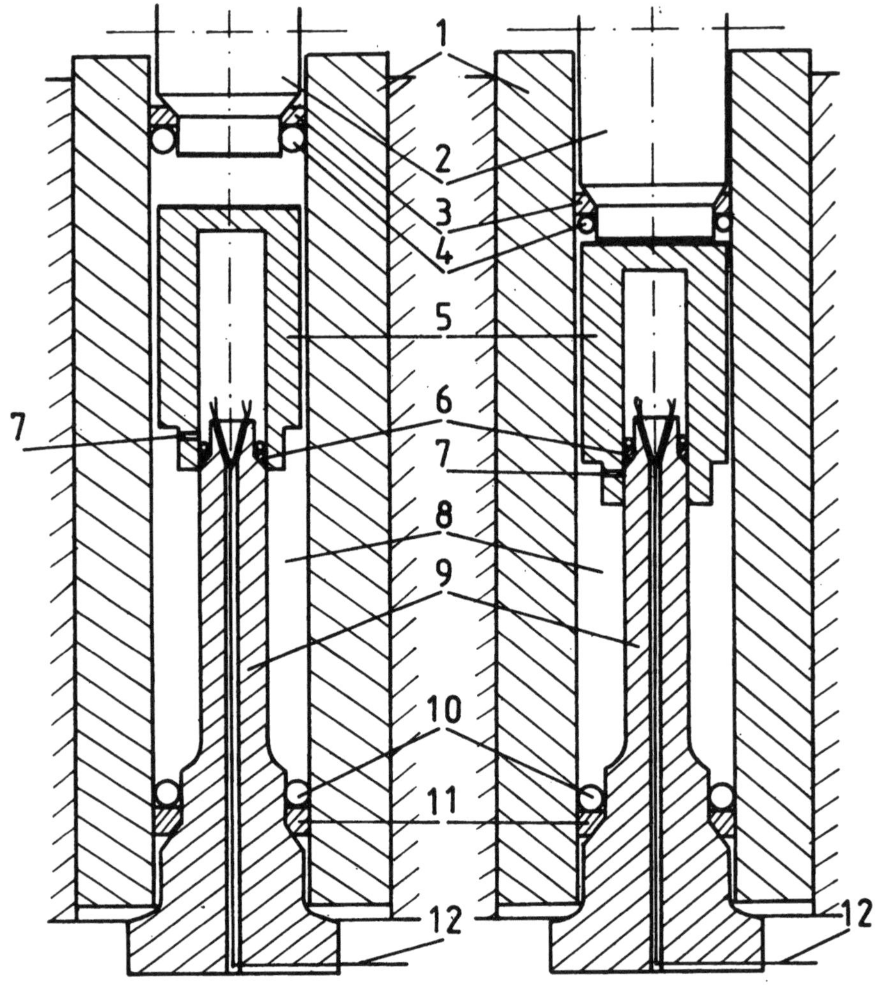

Fig. 1. Two - stage piston - cylinder apparatus during the 2-nd /left/ and 3-rd /right/ period of work.

1. External /first stage/ high pressure vessel
2. Movable piston
3. Metal sealing
4. Rubber O-ring
5. Second stage high pressure vessel
6. Second stage sealing
7. 0.3 mm hole
8. High pressure medium
9. Piston - stopper
10. Rubber O-ring
11. Metal sealing
12. Electrical leads

The first stage high pressure vessel 1 and vessel 5 were made from steel 45HNMF /according to Polish Standards/, for a piston - stopper 9 the bearing steel was used.

The piston 2 was a unique element made from tunsten carbide.

2. Operational principle.

During whole working cycle three following periods can be distinguished:

Period 1.

The vessel 5 is filled with a suitable pressure transmitting liquid /for example ethanol - methanol mixture/, then the vessel 5 is closed by the piston - stopper 9 /without closing the hole 7/ and placed inside the first stage high pressure apparatus 1.The empty volume of the apparatus 1 is then filled with the same /or different/ high pressure transmitting liquid. After that the first - stage high pressure cylinder 1 is closed with the piston 2. Now the apparatus is ready to work.

Period 2.

During this period the piston 2 is moved, by the force of a hydraulic press, into the cylinder 1 and due to the hole 7 the pressure medium is compressed uniformly inside and outside of the vessel 5 /position of the vessel 5 doesn't change/. A Bourdon manometer of the hydraulic press can be used to give approximate pressure values. Accurate pressure in indicated by a manganin gauge mounted on the stopper - piston 9. During compression the piston 2 moves towards the vessel 5 and finally both these elements contact each other. From that moment the vessel 5 starts to move down together with piston 2. The second period of work is finished when sealings 6 close the hole 7. Pressure reached at this moment can be modified by changing the starting volume of the liquid used to fill the working space of the apparatus 1.

Period 3.

During the period 2 the total assembly works as a normal one - stage piston - cylinder apparatus. From the moment in which internal volume of the vessel 5 became closed /the moment when position of the hole 7 is below sealings 6/ the pressure inside the vessel 5 /in the second stage/ increases faster than outside of it. The pressure inside the vessel 5 /the pressure of the second stage/ can be always measured directly by manganin gauge from which the electrical leads 12 go out passing only by one conical feed - through. The pressure, supporting the vessel 5 from outside, can be evaluated from indications of the manganin gauge and values of the total force applied to the apparatus by hydraulic press. Pressure release occurs in the opposite sequence.

CONCLUSIONS

The peculiarities of the presented design and the experience gained so far may be summarized as follows:

1/. Actual apparatus, made - excluding the piston 2 - from usual /non maraging/ steels works satisfactorily up to 4,5 GPa.
For a short time the pressures near 5.0 GPa can be also reached. Much higher pressures seems to be available if all elements would be made from stronger materials /maraging steel and tungsten carbide/. Especially the stronger vessel for the first stage is of great importance. A new apparatus, besed on the high quality materials is now under construction.

2/. Preliminary compression during the 2-nd Period of work allows to have final pressure in relatively large volume /in our case about 6 cm^3/.

3/. The pressure supporting the second stage vessel 5 increases proportionally to the pressure inside this vessel /but with smaller rate/.

4/. The pressure in the second stage vessel 5 can be measured directly by manganin gauge or any other pressure sensor.

5/. The piston - stopper 9 and its sealings 6 are effectively supported by the first - stage pressure. For this reason the piston - stopper charged by the axial stress exceeding 4.5 GPa was not damaged /normally it brokes about 2.5 GPa/.

6/. Introduction of six or more leads directly /by one conical feed - through/ into the second stage space allows to perform "in situ" electric measurements and / or internal heating.

ACKNOWLEDGMENTS

The author wishes to thank C. Zaręba M.Sc. from Huta Warszawa, M.Tomczyk M.Sc. and G. Christoph M.Sc. from Huta Baildon for providing steels and tungsten carbide elements.

REFERENCES

[1]. Baranowski B., Bujnowski W.: Roczniki Chem., 44, 2271 /1970/.

[2]. Bridgman P.W.: Phys. Rev. 57, 342 /1940/.

[3]. Bridgman P.W.: Am. Scientist 31, 1 /1943/.

[4]. Bridgman P.W.: J.Chem. Phys. 5, 964 /1937/.

[5]. Kennedy G.C. and Newton R.C.: Solids under pressure, Mc Graw - Hill, New York /1962/; also Newton R.C., Jayaraman A. and Kennedy G.C., J.Geophys. Res. 67, 2559 /1962/ and Kennedy G.C., Jayaraman A. and Newton R.C.: Phys. Rev. 126, 1363 /1962/.

[6]. Lavergne A. and Whalley E.: Rev. Sci. Instrum. 49 /7/, 923 /1978/.

[7]. Lorent R.E.: Rev. Sci. Instrum. 44, 1691 /1973/.

[8]. Pugh H.D.: Mechanical Behavior of Materials under Pressures /Elsevier, New York 1970/, p.87.

[9]. Whalley E. and Lavergne A.: Rev. Sci. Instrum. 47, 136 /1976/.

INDENTATION HARDNESS TESTING OF ULTRAHARD MATERIALS

H. Richter 1); M. Petzold 2); G. Zahn 1); P. Grau 2)

INTRODUCTION

The indentation hardness measurements are carried out by using conventional and recording methods. In the conventional hardness testing the diameter of the indentation pit is measured after loading. This method is applied especially to ductil materials. In the recording hardness test the penetration depth of the indenter is registered during loading. This method is applied to brittle hard solids but rarely to ultrahard materials, e. g. [1].

We have made investigations into the performance of the recording hardness test for polycrystalline diamond composite material. For comparison conventional hardness tests are carried out and the effects resulting from the indenter depression on the microstructure of the specimen are discussed.

EXPERIMENTAL DETAILS

The recording hardness test were performed by means of the equipment of FROEHLICH et. al. [2] with Vickers indenter. The load and the penetration depth were continously registered by the x-y-t-recorder. The penetration depth was electronically amplified by a factor up to 100,000. The load ranges were 2 N, 10 N and 50 N. The Vickers hardness under load was calculated according to

$$LVH = \frac{1.854 x L}{49 x p^2}$$

(L = load; p = penetration depth).

For the conventional hardness measurements a commercial microhardness

1) Central Institute for Physics of the Earth,
Academy of Sciences of the GDR, Potsdam, 1561

2) Section of Physics, Martin-Luther-University,
Halle/Saale, 4000, GDR

tester with Vickers indenter was used. The load ranges were 20 N and 50 N. The Vickers hardness VH was calculated by using equation

$$VH = \frac{1.854xL}{d^2}$$

(L = load; d = length of the pit diagonal).

All the experiments were made on samples of Soviet diamond-based polycrystalline tool material SKM. The indentation surface of these compact specimens was metallographically prepared using the grinding-lapping-method corresponding tool preparation [3]. The surface roughness was determined by means of the ME 10 profilometer. The values were R_t = 0.5 µm; R_q = 0.05 µm.

RESULTS AND DISCUSSION

In fig. 1 is shown a typical load-penetration depth curve. (a) represents "loading", (b) represents "unloading" and (c) the elastic recovery. The latter was found between 75% and 85%. That means that the depression of the indentation surface is predominantly elastic.

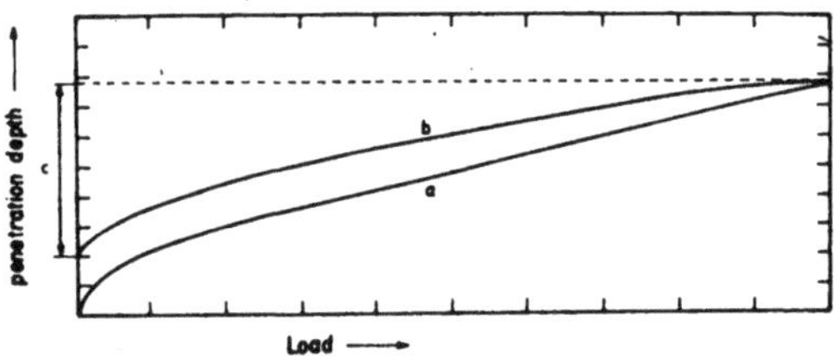

Fig. 1
Recorded loading-penetration curve, schematically

The recording hardness LVH as a function of load is plotted in fig. 2. The dotted line illustrates this fact without considering the elastic indenter deformation. The solid line represents corrected hardness values with regarding to the elastic deformation of the indenter. This correction was done according to the conception suggested in [4]. The LVH values were determined with a precision of about +/- 10%.

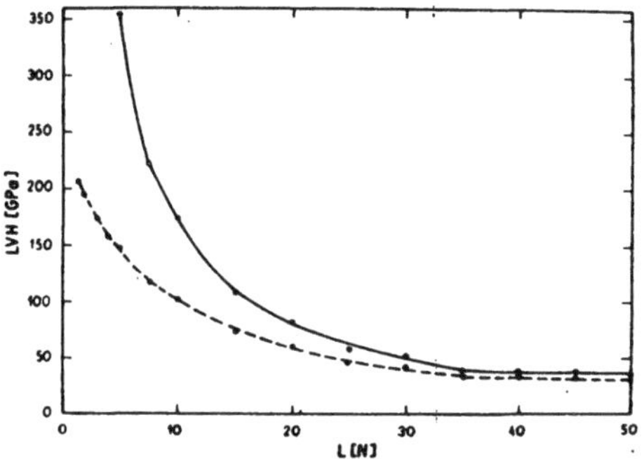

Fig. 2
Recording hardness LVH as a function of load for SKM (dotted line = non corrected, solid line = corrected LVH values)

A significant dependence of the hardness on the load was found. For loads larger than 35 N the recording hardness is nearly independent on the load and was evaluated between 31 GPa and 35 GPa. The increase of the hardness with decreasing load is known and particularly pronounced for materials with covalent bonding [5]. In comparison with the recording method the conventional hardness turns out higher and ranged from 40 GPa to 70 GPa related to the load of 50 N.
In according to FROEHLICH et. al. [2] it is possible to derive a load independent hardness number L_2VH from the slope of the L/p-p - function. Fig. 3 shows that the formulation of the hardness number L_2VH for SKM is only permissible for definite load ranges with constant slope.

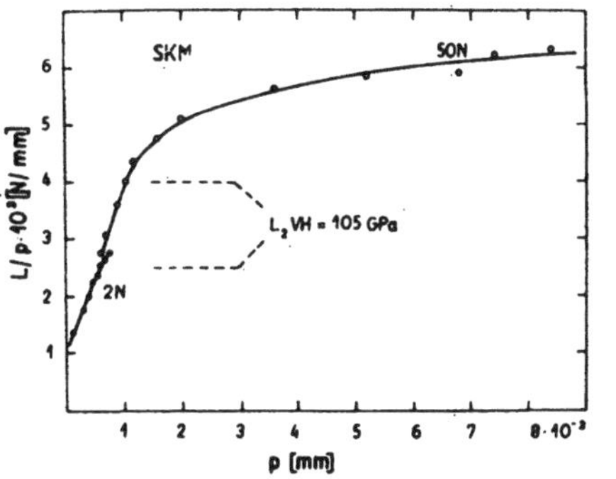

Fig. 3
Load to penetration depth ratio versus penetration depth for the load ranges of 2 N and 50 N

The L_2VH amounts to 105 GPa. This seems to be too high. This fact and the typical course of the function L/p-p will be interpreted as follows: The hardness value L_2VH of 105 GPa is due to the penetration depth of about 1 μm, that means an indentation square which corresponds to the size of a diamond single crystal within the bulk of the compact. Therefore the hardness of a diamond crystal has to be considered. According to KRISCH et. al. [6] the conventional Vickers indentation hardness of single crystal diamond is nearly 150 GPa. The significant increase of the indentation depth with increasing load is due to the compression of diamond single crystals within the deformable matrix of the polycrystalline material.

The scanning electron micrograph of an indentation pit into SKM confirms the above mentioned connections. This micrograph is shown in fig. 4.

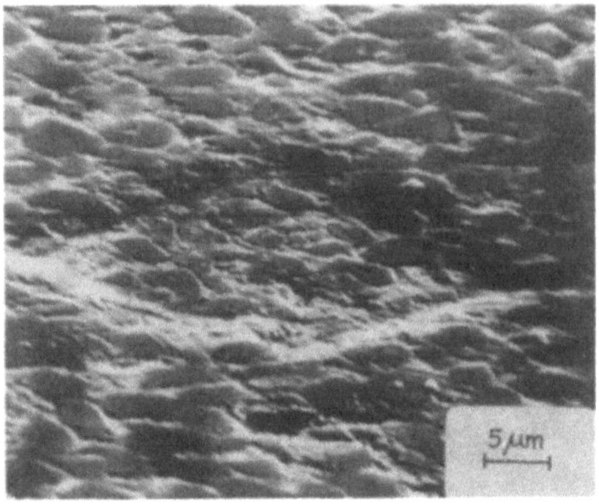

Fig. 4
SEM micrograph of the indentation pit into SKM using the conventional method under load of 50 N (tilt angle = 60° about the horizontal axis of the micrograph)

There are no visible cracks around the pit. The diamond grains seem to be pressed into the bulk material. The pit is not ideally pyramid-shaped but rather like a very flat stump of a pyramid. That illustrates the very high portion of the elastic recovery of the depression after reloading.

For the Meyer exponent of the polycrystalline material SKM our investigations yielded a value of 1.55, which well corresponds to literature estimations [5].

LITERATURE

1) V.V. Berdikov, O.J. Puschkarev, Ju.J. Artemeva, G.A. Adler; Sverchtverdye Materialy 26 (1983), 5, 21
2) F. Froehlich, P. Grau, W. Grellmann; phys. stat. sol. (a) 42 (1977), 79
3) W. Naumann; Teilabschlussbericht V 5, Forschungszentrum der Werkzeugindustrie Schmalkalden, Karl-Marx-Stadt, 1987
4) H.-J. Weiss, A. Krell, B. Mueller; phys. stat. sol. (a) (93) (1986), 509
5) C.A. Brookes, W.A. Lambert; in: Ultrahard Materials Applications Technology, De Beers Industrial Diamond Division, Vol. 1 (1982), 128
6) A. Krisch, V.V. Gasilin, V.V. Kunchenko; Materialpruefung 24 (1982), 4, 130

UNIVERSAL MICROPROCESSOR CONTROLLED MEASURING SYSTEM FOR MATERIAL SYNTHESIS AT HIGH PRESSURES AND TEMPERATURES

Schnell, E. 1) ; Meyer, H. 1) ; Rothe, V. 1) ; Lorenz, H. 1)

Introduction

A well developed high pressure high temperature technique and measuring system are main demands for good results in material synthesis. Reproducible pressure and temperature conditions are necessary for solid experiments. In order to manage the great quantity of experiment data the use of microcomputer technique is indispensable. Therefore we developed an universal microprocessor controlled measuring system for material synthesis at high pressures and temperatures with a B E L T - apparatus.

Measuring System

The schematic diagram of the measuring system for material synthesis at high pressures and temperatures is shown in Fig. 1.

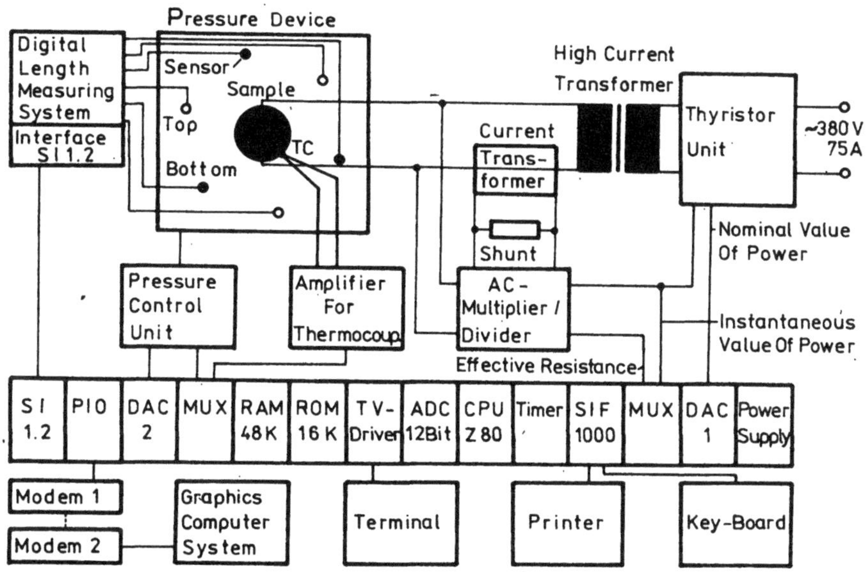

Fig. 1 Schematic diagram of the measuring system

---

1) AdW der DDR, Zentralinstitut fuer Physik der Erde, Telegrafenberg, Potsdam, DDR - 1561

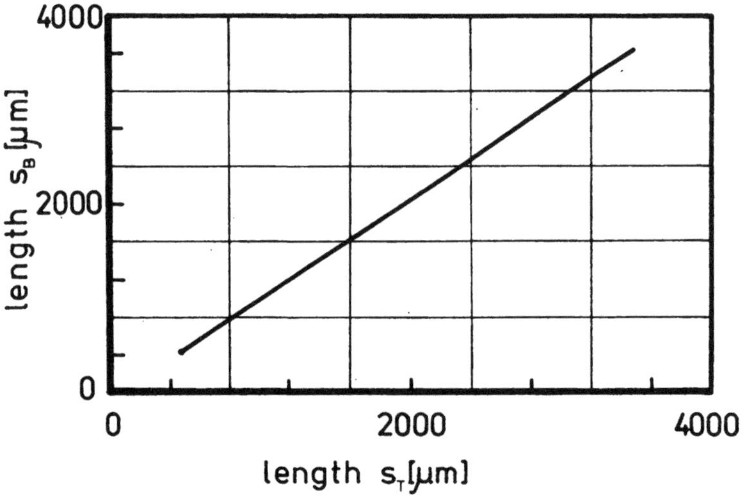

Fig. 2 Symmetry of the two anvils ways

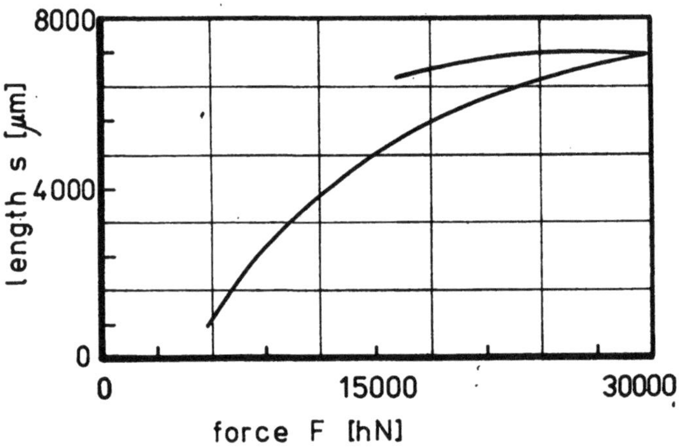

Fig. 3 Hysteresis of pyrophyllite in a deformation experiment

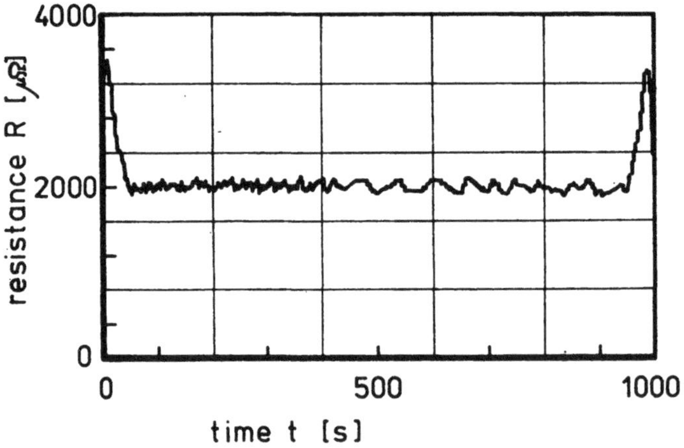

Fig. 4 Resistance of the indirect heater in a synthesis

Fig. 5 shows an example of a power -force -time -arrangement for a synthesis.

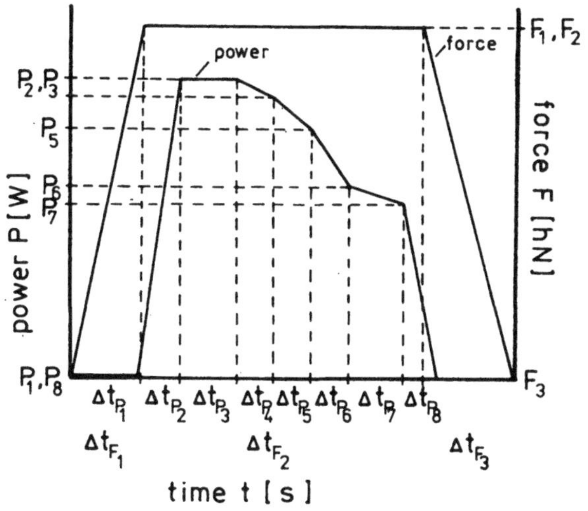

Fig. 5 Power - force - time arrangement

Some of the output - options are :

* printer tables of all storaged data,
* online data transfer to the graphics computer system,
* analog or digital PI - control element,
* calculation of the temperature which is measured with
  a) Ni / Cr - Ni
  b) Pt / Rh - thermocouples by means of a polynomial without pressure correction
  c) only thermo - voltage registration.

Conclusions

The presented microprocessor controlled measuring system for material synthesis yields good reproducible results and a high efficiency in interpretation of the great data quantities. A new and more powerfull system which will be designed is characterized by :

* heating with AC or DC,
* T - controlled or power - controlled heating,
* possibility of arbitrary heating and pressure functions,
* 16 Bit graphics computer for experiment interpretation,
* 1 micrometer lenght resolution.

The measuring system consist of :

* a 20 mm - Belt - apparatus,
* a 400 t - press with a PI - control element,
* a microcomputer system (Zilog 80 CPU) with 10 Bit analog output for the thyristor unit and the press control unit and 32 channels of 12 Bit analog input with a conversion time of 20 microseconds,
* standard computer periphery,
* online - coupled graphics computer system,
* digital six sensors linear length measuring system with a resolution of 25 or 10 micrometers,

  thereby we measure : - the total way of anvils
  - the single anvil ways and
  - the slope of the die for a danger indication,
* a thyristor - unit with a PI - control element,
* a high - current - transformer (5 V , 2000 A , AC),

  (The heating power and the effective resistance are calculated with an ICL 8013 - circuit).

Software Capabilities

The complete system allowed a full automatically control of heating power and force and the storage of all values which were measured during the process time (time, power, resistance, length, temperature). The maximum memory capacity for the process data is 46 kByte. Using the coupled graphics computer it is possible to make diagrams of all requested physical dependencies. Three examples (hard - copies) are shown in Fig. 2 to Fig. 4. Fig. 2 shows the symmetry of the anvil ways . Fig. 3 shows a hysteresis of pyrophyllite in a deformation experiment and Fig. 4 sets an example of the progression of the resistance of an indirect heater.

The realtime control and storage system for the synthesis as well as the coupling software is written in Assembler - language. Graphics software is written in BASIC and Assembler - language. Input values for the synthesis are :

* nominal values of power and force,

  (there are 8 data pairs for power and force possible),
* time ranges, constants, factors,
* distance of data storage (between 0.2 s and 255 s),
* wait, continue, danger indication.

LIST OF PARTICIPANTS

AKIMOTO, S.-Okayama University, Institute for Study of Earth's Interior, Misasa, Tottori-ken 682-02, Japan

ALLAKHVERDIEV,K.-Academy of the Azerbaijan SSR,Institute of Physics,Narimanova 33,Baku,USSR

BAKUN-CZUBAROW,N.-Polish Academy of Sciences,Institute of Geological Sciences, Zwirki i Wigury 93 02--89 Warsaw, P.R. Poland

BAUER,H.-J.-Universität München, Sektion Physik,Schellingstr.4, 8000 München 40, FRG

BAUTSCH,H.-J.-Humboldt Universität Berlin, Museum für Naturkunde, Invalidenstr.43,1040 Berlin,GDR

BERESNEV,B.-Academy of Sciences of the Ukrain. SSR, Physico-Technical, ul. R. Luxemburg 72, 340114 Donetsk, USSR

BILAL,A.- Hahn-Meitner Institut Berlin, Glienicker Str. 100, Berlin (West)

BIRKS,A.-The Queen's University of Belfast, Dept. of Mechanical Engineering, Belfast, Great Britain(Northern Ireland)

BOEHLER,R.-Max-Planck-Institut für Chemie, Abt. Geochemie, Saarstr. 23, 6500 Mainz, FRG

BOETCHER,R.-Akademie der Wissenschaften der DDR, Zentralinstitut für Physik der Erde, Telegrafneberg, 1561, Potsdam, GDR

BUDZINSKI,H.-Akademie der Wissenschaften der DDR, Zentralinstitut für Physik der Erde, Telegrafenberg, 1561, Potsdam, GDR

CARROLL,E.-BOART EUROPE Shannon Industrial Estate, Tungsten Carbid Division, Shannon, Countrry Clare, Ireland

CHUDINOVSKICH,L.-Academy of Science USSR,Institute of Experimental Mineralogy,142432 Chernogolovka, Moscow District, USSR

DAHM,K.P.-Akademie der Wissenschaften der DDR, Zentralinstitut für Physik der Erde, Telegrafenberg, 1561, Potsdam, GDR

DIBLIK,J.-PRAMET National Enterprise Powder Metallurgy Works, 78753 Sumperk, Czechoslowakia

DZHAVADOV,L.-Academy of Sciences USSR, Institute for High Pressure Physics, 142092 Troizk Moscow District, USSR

EGERER, F.-Heavy Industry University Dept. of Mineralogy and Petrography, H-3515 Miskolc-Egyetem-Varo, P.R.Hungary

ELUSZKIEWICZ, J.-Warsaw University, Institute of Geophysics, ul.Pasteura 7, 02-093 Warsaw, P.R.Poland

EULER, H.-Institut für Kristallographie, Theresienstr. 41, 8000 München, FRG

FILIPEK, St.M.-Polish Academy of Sciences, Institute of Physical Chemistry, ul.Kasprzaka 44/52, Warsaw, P.R. Poland

FLASAR, J.- PRAMET National Enterprise, Powder Metallurgy Works, 78753 Sumperk, Czechoslovakia

FÖRSTER, A.-Humboldt-Universität Berlin, Bereich 04, Invalidenstr. 42, 1040 Berlin, GDR

FRANCK, S.-Akademie der Wissenschaften der DDR, Zentralinstitut für Physik der Erde, Telegrafenberg, 1561, Potsdam, GDR

FUKUNAGA.O.- Tokyo Institute of Technology, 2-12-1 Ookayama, Meguro-Ku, Tokyo 152, Japan

GEBERT, H.J.-VEB Rationalisierung, Bereich F/E, Friedrich-Engels-Str. 42, 1500, Potsdam, GDR

GEIGER, CH.-TU Berlin, Institut für Mineralogie u. Kristallographie Ernst-Reuter-Platz, Berlin (West)

GERLACH, U.-Akademie der Wissenschaften der DDR, ZFW Dresden, Helmholtzstr.20, 8027 Dresden, GDR

GLOVER, P.- University of East Anglia, 01.49 NR 47 TJ Norwich Norfolk, Great Britain

GOTTESMANN, B.-Akademie der Wissenschaften der DDR, Zentralinstitut für Physik der Erde, Telegrafenberg, 1561, Potsdam, GDR

GRATZ, R.-Akademie der Wissenschaften der DDR, Zentralinstitut für Physik der Erde, Telegrafenberg, 1561, Potsdam, GDR

GREULING, B.-Akademie der Wissenschaften der DDR, Zentralinstitut für Physik der Erde, Telegrafenberg, 1561, Potsdam, GDR

GREY, K.-Technische Universität Magdeburg, Boleslaw-Bierut Platz 5, 3010, Magdeburg, GDR

HAFNER, ST.-Universität Marburg, Institut für Mineralogie, Lahnberge, 3550, Marburg, FRG

HAHNE, H.-Akademie der Wissenschaften der DDR, Zentralinstitut für Physik der Erde, Telegrafenberg, 1561, Potsdam, GDR

HAN DUC KIM-Hanoi Machinery Research Institute, 25B1-Cat Linh-Dong, Da-Hanoi, P.R. Vietnam

HANIC,F.-Slovak. Academy of Sciences, Institute for Inorganic Chemistry, Dubravska cesta, 84236 Bratislava, Czechoslovakia

HEISE,D.-VEB Rationalisierung, Bereich F/E, Friedrich-Engels-Str. 42, 1500, Potsdam, GDR

HEMMES,H.K.-Vrije Universiteit, Natuurkundig Laboratorium de Boelelaan, 1081 HV Amsterdam, The Netherlands

HESS,H.-Akademie der Wissenschaften der DDR, Zentralinstitut für Physik der Erde, Telegrafenberg, 1561, Potsdam, GDR

HINZE,E.-Institut für Geowissenschaften und Lithosphärenforschung, Senckenbergerstr. 3, 6300 Giessen, FRG

HOCHHEIMER,H.D.-Los Alamos National Laboratory, P-10, M.S.K 764 Los Alamos, N.M.87545, USA

HOHLFELD,CH.-Akademie der Wissenschaften der DDR, Zentralinstitut für Physik der Erde, Telegrafenberg, 1561, Potsdam, GDR

HOLZAPFEL,W.-Universität-Gesamtschule Paderborn, Fachbereich Physik, Warburger Str. 100, 4790, Paderborn. FRG

HURTIG,E.-Akademie der Wissenschaften der DDR, Zentralinstitut für Physik der Erde, Telegrafenberg, 1561, Potsdam, GDR

ITIÉ,J.P.-Université Paris, Laboratoire des Interactions Moleculaires et des Hautes Pressions Avenue Jean-Baptiste Clement, 93430 Villetaneuse, France

JAHODA,A.-VEB Sprengstoffwerk Schönebeck, Schillerstr.40, 3300, Schönebeck, GDR

JEANLOZ,R.- University of California, Dept. of Geology and Geophysics, Berkeley, California 94720, USA

JENNER,G.-Universite Louis Pasteur EHICS, 1 rue Blaise Pascal, 67008, Strasbourg, France

JOLLY,H.-University of East Anglia, Physics Department, 41 Clarence rd., Thorpe, Norwich, Norfolk, Great Britain

JURCZAK,J.-Polish Academy of Sciences, Institute of Organic Chemistry, ul. Kasprazak 44, 01-224 Warsaw, P.R. Poland

KAHLBAUM,T.-Akademie der Wissenschaften der DDR, Zentralinstitut für Physik der Erde, Telegrafenberg, 1561, Potsdam, GDR

KALININ,V.-Academy of Sciences USSR Institute for Physics of the Earth, B. Gruzinskaja 10, Moscow D-242, USSR

KARAMANOW,I.-Bulgarian Academy of Sciences, Research Laboratory of Powder Metallurgy, Sofia, Dsersinsky 122, P.R.Bulgarian

KEELER,N.-Los Alamos National Laboratory, M 6 Groupe, Los Alamos, N.M. 87545, USA

KERN,H.-Mineralogisches Institut der Universität Kiel, Olshausen Str. 40, 2300, Kiel, FRG

KLEMP,D.-VEB Rationalisierung Luckenwalde, Puschkinstr. 54, 1710, Luckenwalde, GDR

KNISCHEWSKI,G.-Akademie der Wissenschaften der DDR, Zentralinstitut für Physik der Erde, Telegrafenberg,1561,Potsdam,GDR

KOPP,J.-Akademie der Wissenschaften der DDR, Zentralinstitut für Physik der Erde, Telegrafenberg, 1561, Potsdam, GDR

KRAFT,A.-Akademie der Wissenschaften der DDR, Zentralinstitut für Physik der Erde, Telegrafenberg, 1561, Potsdam, GDR

KRAVTCHUK,K.-Academy of Sciences USSR, N.S. Kurnakov Institute of General and Inorganic Chemistry, Leninsky prospect 31, 117071 Moscow, USSR

KRIPPSTÄDT,R.-Akademie der Wissenschaften der DDR, Zentralinstitut für Physik der Erde, Telegrafenberg, 1561, Potsdam, GDR

KRISTOFIK,J.-Academy of Sciences of Czechoslovakia, Institute of Physics, Na Slovance 2, 18040,Praha, Czechoslovakia

KROPACEK,V.-Academy of Sciences of Czechoslovakia, Geophysical Institute, Bosni II-1401, 141-31 Praha-Sporilov, Czechoslovakia

KRUKOWSKI,ST.-High Pressure Research Center UNIPRESS, ul. Sokolowska 29/37,01-192 Warsaw, P.R. Poland

KRUGER,ST.-VEB Rationalisierung Bereich F/E, Friedrich-Engels-str. 42, 1500, Potsdam,GDR

KUHNE,U.-Akademie der Wissenschaften der DDR, Zentralinstitut für Physik der Erde, Telegrafenberg, 1561, Potsdam, GDR

KWASNIEWSKI,L.-Silesian Technical University, Rock Mechanics Laboratory, ul. Pstrowskiego 2, 44-100 , Gliwice, P.R. Poland

LAUTERJUNG,J.-Institut für Geowissenschaften und Lithosphärenforschung, Senckenbergstr. 3, 6300, Giessen, FRG

LEBEDEV,T.S.-Academy of Sciences Ukrainian SSR, Institute of Geophysics,Palladin Prospect 32 252680 Kiev-142, USSR

LEGER,J.-M.-C.N.R.S., GEMAP, ER 211 1 place A. Briand, 92190 Meudon, France

LIECKEFETT,R.-VEB Rationalisierung, Bereich F/E, Friedrich-Engels-Str. 42, 1500, Potsdam, GDR

LIPPMANN,H.-ASMW Berlin, FW Druck, Fürstenwalder Damm 388, 1162, Berlin, GDR

LORENZ,B.-Akademie der Wissenschaften der DDR, Zentralinstitut für Physik der Erde, Telegrafenberg, 1561, Potsdam, GDR

LORENZ, H.-Akademie der Wissenschaften der DDR, Zentralinstitut für Physik der Erde, Telegrafenberg, 1561, Potsdam, GDR

LUDLOW, CH.-Queen's University of Belfast, Dept. of Mechanical and Industrial Engineering, Ashby Building,Belfast BT 95 AH Great Britain (Northern Ireland)

LUKASZEWICZ,K.-Polish Academy of Sciences, Institute of Low Temperature and Structure Research P.O. 937, 50-950 Wroclaw, P.R. Poland

MALINOWSKI,M.-Polish Academy of Sciences, Institute of Low Temperature and Structure Research P.O. 937, 50-950 Wroclaw, P.R. Poland

MATING,T.-Akademie der Wissenschaften der DDR, Zentralinstitut für Physik der Erde, Telegrafenberg, 1561, Potsdam, GDR

MATUS,E.-Research Institute of Powder Metallurgy, Revolucni 10c 78753 Sumperk, Czechoslovakia

METZKE,E.-Zentralinstitut für Elektronenphysik, Abt. 541, Hausvogteiplatz 5/7, 1086, Berlin, GDR

MORAWSKI,A.-High Pressure Research Center UNIPRESS, ul. Sokolowska. 29/37, 01-142 Warsaw, P.R.Poland

NEGA,H.- Akademie der Wissenschaften der DDR, Zentralinstitut für Physik der Erde, Telegrafenberg, 1561, Potsdam, GDR

NIEDBALSKA, A.-Institute of Electronic Materials Technology, ul.Wolczynska 133, 01-919 Warsaw, P.R. Poland

NOLL,G.-Klöckner-Wilhelmsburger GmbH, Borsigstr. 24, 2054, Geesthacht, FRG

MÜLLER,H.-J.-Akademie der Wissenschaften der DDR, Zentralinstitut für Physik der Erde, Telegrafenberg, 1561, Potsdam, GDR

OLIJNYK,H.-Universität Baureuth, Bayerisches Geoinstitut, PF 101251, 8580, Bayreuth, FRG

ORGAZALL,I.-Akademie der Wissenschaften der DDR, Zentralinstitut für Physik der Erde, Telegrafenberg, 1561, Potsdam, GDR

PANNASCH,S.-FZ für Umform- und Plastverarbeitungstechnik,Schersingstr.1, 9451, Zwickau, GDR

PEUKERT,K.-VEB Rationalisierung, Bereich F/E, Friedrich-Engels-Str. 42, 1500, Potsdam, GDR

PETKEVITSCH,G.-Academy of Sciences USSR, Institute of Geology and Geochemistry of Comb.Min., Nautschnaja 31, 290047 Lwow,USSR

POLIAN,A.-Universite P. et M.Curie, Tour 13 E4, place Juisseu, F-75252 Paris Cedex 05, France

QI ZENGDU-Academia Sinica, Institute of Ceramics, 865 Changning Road, Shanghai, P.R. China

RAAB,S.-Akademie der Wissenschaften der DDR, Zentralinstitut für Physik der Erde, Telegrafenberg, 1561, Potsdam, GDR

RECHT,H.-Akademie der Wissenschaften der DDR, Zentralinstitut für Physik der Erde, Telegrafenberg, 1561, Potsdam, GDR

RHEDE,D.- Akademie der Wissenschaften der DDR, Zentralinstitut für Physik der Erde, Telegrafenberg, 1561, Potsdam, GDR

RHEDE,K.-VEB Rationalisierung, Bereich F/E, Friedrich-Engels-Str. 42, 1500, Potsdam, GDR

REIN,C.-Akademie der Wissenschaften der DDR, Zentralinstitut für Physik der Erde, Telegrafenberg, 1561, Potsdam, GDR

RICHARD,M.-ASMW Berlin, FG Druck, Fürstenwalder Damm 388, 1162, Berlin, GDR

RICHTER,H.-Akademie der Wissenschaften der DDR, Zentralinstitut für Physik der Erde, Telegrafenberg, 1561, Potsdam, GDR

RIEDEL,M.-Akademie der Wissenschaften der DDR, Zentralinstitut für Physik der Erde, Telegrafenberg, 1561, Potsdam, GDR

SCHMÄDICKE,E.-Akademie der Wissenschaften der DDR, Zentralinstitut für Physik der Erde, Telegrafenberg, 1561, Potsdam, GDR

SCHNELL,E.-Akademie der Wissenschaften der DDR, Zentralinstitut für Physik der Erde, Telegrafenberg, 1561, Potsdam, GDR

SCHOLZ,D.-Akademie der Wissenschaften der DDR, Zentralinstitut für Physik der Erde, Telegrafenberg, 1561, Potsdam, GDR

SCHREYER,W.-Ruhr-Universität Bochum, Vogelrute 15, 4630, Bochum 1, FRG

SCHRÖDER,B.-Akademie der Wissenschaften der DDR, Zentralinstitut für Physik der Erde, Telegrafenberg, 1561, Potsdam, GDR

SEIFERT,F.-Universität Bayreuth, Bayerisches Geoinstitut, PF 101251, 8580 Bayreuth, FRG

SEIPOLD,U.-Akademie der Wissenschaften der DDR, Zentralinstitut für Physik der Erde, Telegrafenberg, 1561, Potsdam, GDR

SHANER, J. -National Laboratory Los Alamos, Los Alamos, N.M. 87545, USA

SHANG BO, -Sichuan Normal University, Institute of Solid State Physics, 610066 Chengdu, P.R. China

SHILOBREEVA, S. -Academy of Sciences USSR, Vernadsky Institute of Geochemistry, ul. Kosygina 19, 117975 Moscow, USSR

SHIMADA, M. -Tohoku University, Dept. of Applied Chemistry, Faculty of Engineering Aoba, Sendai, Hiyagi 980, Japan

SLUTSKY, A. -Academy of Sciences USSR, Vernadsky Institute of Geochemistry, ul. Kosygina 19, 117975 Moscow, USSR

SOBOLEV, N. -Academy of Sciences USSR Institute of Geology and Geophysics, 630090 Novosibirsk 90, USSR

SÖLLNER, R. -Akademie der Wissenschaften der DDR, Zentralinstitut für Physik der Erde, Telegrafenberg, 1561, Potsdam, GDR

SPETZLER, H. -University of Colorado, Dept. of Geological Sciences/ CIRES, Box 449, Boulder, Col 80309, USA

SPLIESS, D. -VEB Rationalisierung, Bereich F/E, Friedrich-Engels-Str. 42, 1500, Potsdam, GDR

STACHEL, D. -Friedrich-Schiller- University, Steiger 3, 6900, Jena, GDR

STILLER, H. -Akademie der Wissenschaften der DDR, Zentralinstitut für Physik der Erde, Telegrafenberg, 1561, Potsdam, GDR

STISHOV, S. -Academy of Sciences USSR Institute of Crystallography, Leninsky pr. 59, 117333 Moscow, USSR

STÖSSER, R. -Humboldt-Universität Berlin, Sektion Chemie, Bunsenstr. 1, 1080 Berlin, GDR

SUBSTYK, M. -Institute of Electronics Materials Technology, Wolczynska ul. 133, 01-910 Warsaw, P.R. Poland

SZYMANSKI, A. -Institute of Electronic Materials Technology, Wolczynka ul. 133, 01-910 Warsaw P.R. Poland

TAUBER, F. -Akademie der Wissenschaften der DDR, Zentralinstitut für Physik der Erde, Telegrafenberg, 1561, Potsdam, GDR

TENZLER, H. -Forschungsinstitut für NE- Metalle, Lessingstr. 21, 9200, Freiberg, GDR

ULLMANN, W. -Akademie der Wissenschaften der DDR, Zentralinstitut für Physik der Erde, Burgweg 11, 6900, Jena, GDR

ULLNER,H.-A.-Akademie der Wissenschaften der DDR, Zentralinstitut für Physik der Erde, Telegrafenberg, 1561, Potsdam, GDR

ULLRICH,B.-Institut für Mineralische Rohstoffe- und Lagerstättenwirtschaft,Oskar-Roeder-Str.3 8036, Dresden, GDR

VELICKÝ,V.-Czechoslovakian Academy of Sciences, Geophysical Institute,Bocni II-1401, 14131 Praha-Sporilov, Czechoslovakia

VOLLSTÄDT,H.-Akademie der Wissenschaften der DDR, Zentralinstitut für Physik der Erde, Telegrafenberg, 1561, Potsdam,. GDR

VRBKA,J.-Technical University of Brno, Faculty of Machinery, Dept. of Mechanics, Tr. Obrancu miru 65, 60200 Brno, Czechoslovakia

WAGNER,CH.-Akademie der Wissenschaften der DDR, Zentralinstitut für Physik der Erde, Telegrafenberg, 1561, Potsdam, GDR

WALZER,U.-Akademie der Wissenschaften der DDR, Zentralinstitut für Physik der Erde, Burgweg 11, 6900, Jena, GDR

WEBER,D.-Akademie der Wissenschaften der DDR, ZFW Dresden, Halbkreisstr.9, 8027, Dresden, GDR

WIRSIG,M.-Akademie der Wissenschaften der DDR, Zentralinstitut für Elektronenphysik, Hausvogteiplatz 5-7,1086 Berlin, GDR

YAKIMOV,K.-Laboratory of Powder Metallorgy and Composition Materials, Dsersinsky 122, Sofia, P.R. Bulgaria

YAKOVLEV,E.-Academy of Sciences USSR, Institute for High Pressure Physics, 142092, Troizk, Moscow district, USSR

ZHAO MIN-GUANG-Sichuan Normal University, Institute of Solid State Physics, 610066, Chengdu, P.R. China

ZIEMANN,M.-Akademie der Wissenschaften der DDR, Zentralinstitut für Physik der Erde, Telegrafenberg, 1561, Potsdam, GDR

ZIMMERMANN,H.D.-Aarhus Universitet, Geologisk Institut, Laboratoriet for Endogen Geolgi, C.F. Mollers Alle, DK-8000 Aarhus C, Denmark

ISBN 3-05-500555-4
ISSN 0138-4600

Bei Fragen zur Produktsicherheit wenden Sie sich bitte an:
If you have any questions regarding product safety,
please contact:

Walter de Gruyter GmbH
Genthiner Straße 13
10785 Berlin
productsafety@degruyterbrill.com